Contents

Introduction to GNVQs in Engineering

GNVQs (General National Vocational Qualifications) are a new type of qualification available for students who want to follow a course linking traditional areas of study with the world of work. Those who qualify will be ideally placed to make the choice between progressing to higher education and applying for employment.

GNVQs are available at three levels:

Foundation   normally studied full time for one year
Intermediate   normally studied full time for one year
Advanced   normally studied full time for two years, and often reffered to as 'vocational A-levels'.

Each GNVQ is divided into various types of units:

Mandatory units   which every one must study
Optional units   from which a student must choose to study
Additional units   offered by the various awarding bodies in order to allow students to top up their skills and knowledge
Key skills units   covering the essential skills related to numeracy, IT and communication.

Units are clearly structured. The principal components are:

elements   which focus on specific aspects of the unit
performance criteria   which tell you what must be able to do
range   which tell you what areas you must be able to apply your knowledge to.

Advanced GNVQ in Engineering

This book is intended primarily for students of the Advanced GNVQ in Engineering. This course is divided as follows:

i) eight mandatory units covering all key aspects of engineering.
ii) various specialist pathways (e.g. mechanical engineering, electrical engineering) which can be studied by selecting particular combinations of the optional units.
iii) additional units chosen from those offered by various awarding bodies. Many students will wish to study at least one additional mathematics unit if they intend to study engineering at a higher level some time in the future.
iv) core skills units in the three areas at level three.

This book covers the BTEC optional units needed to make up the pathway in Electrical and Electronic Engineering. Together with the Stanley Thornes textbook on GNVQ Advanced, Further and Additional Mathematics, the books form a complete package.

Assessment

GNVQ are assessed using two methods:

Units tests   one for each unit, taken be everybody.

Portfolio of evidence   developed individually by each student to demonstrate competence and understanding

Optional units are not assessed by unit tests, and so students are assessed on their optional pathway subjects by means of their portfolio only.

Portfolio of evidence

Each student must develop a portfolio (i.e. organised collection) of evidence that they have achieved the required level of skill and understanding for each of the elements of the units they take. The unit specifications say exactly what type of evidence will be acceptable for assessment, but do not dictate the details of any project to be undertaken, nor exactly how the evidence is to be completed. Written, photographic, video or audio tape records of achievement are all accepted, as are testimonials from work placements.

Regardless of the media employed to deliver evidence, it is absolutely essential that the content is organised and referenced so that it can be easily and effectively assessed. There is no point in assembling masses of evidence if its presentation makes it impossible to assess or understand. An index should be included, and evidence relevant to more than one element cross-referenced, or duplicated if necessary.

Grading

A GNVQ can be awarded as a pass, a merit or a distinction. To achieve a merit or a distinction a student must demonstrate notable ability in a range of skills related to the development of high-quality portfolio evidence. Your lecturer or teacher will be able to give you detailed guidance on how to develop and demonstrate these skills.

Additional titles developed to support GNVQ in Engineering are listed on the back cover of this book, and further details are available from Stanley Thornes Publishers Customer Services Department in 01242 228888.

Key skills: Summary of units at level 3

This summary shows how the activities and assignments provide opportunities to provide evidence of the key skills. When planning a piece of work, think carefully about the key skills you can provide evidence for. For example, use information technology as often as possible when producing written work, charts and diagrams.

Key Skills	Activities	Assignments
Communication		
3.1 Take part in discussion	11.1	6, 7, 20
3.2 Produce written material	1.1, 3.1, 5.1, 6.1, 6.2, 6.3, 7.1, 9.6, 10.4, 11.1, 13.2, 14.2, 14.3, 15.3, 16.4, 16.6, 17.5, 22.2, 22.3, 22.5, 22.6, 22.7, 23.1, 25.3, 25.4	1, 2, 3, 4, 5, 6, 7, 8, 9, 10, 11, 12, 13, 14, 15, 16, 17, 18, 19, 20, 21, 22, 23, 24, 25
3.3 Use images	5.1, 6.2, 7.1, 11.1, 14.2, 14.3, 17.5, 19.1, 19.2, 24.1	5, 6, 7, 8, 9, 10, 11, 12, 13, 16, 17, 19, 23
3.4 Read and respond to written material	6.3, 7.1, 9.6, 11.1, 13.2, 14.3, 16.1, 16.2, 16.3, 16.4, 16.5, 17.1, 22.5, 22.6, 23.4, 23.5	4, 12, 20, 24
Information Technology		
3.1 Prepare information	1.1, 3.1, 5.1, 6.1, 6.2, 6.3, 7.1, 9.6, 11.1, 13.2, 14.2, 14.3, 15.3, 16.4, 16.6, 22.3, 22.5, 22.6, 22.7, 23.1	2, 3, 4, 5, 6, 8, 9, 10, 11, 12, 13, 15, 17, 18, 19, 20, 21, 22, 23, 24, 25
3.2 Process information		2, 3, 4, 5, 6, 8, 9, 10, 11, 12, 13, 15, 17, 18, 19, 20, 21, 22, 23, 24, 25
3.3 Present information	1.1, 3.1, 5.1, 6.1, 6.2, 6.3, 7.1, 9.6, 11.1, 13.2, 14.2, 14.3, 15.3, 16.4, 16.6, 22.3, 22.5, 22.6, 22.7, 23.1	2, 3, 4, 5, 6, 8, 9, 10, 11, 12, 13, 15, 17, 18, 19, 20, 21, 22, 23, 24, 25
3.4 Evaluate the use of information technology	13.2, 24.1	11, 12, 13, 17, 22
Application of Number		
3.1 Collect and record data	3.1, 14.2, 17.1, 17.3	3, 4, 5, 6, 7, 16, 18
3.2 Tackle problems	14.2, 15.2, 17.1, 17.3, 17.8	1, 4, 5, 16
3.3 Interpret and present data	3.1	3, 4, 5, 6, 7, 10, 16, 18

Acknowledgements

I am indebted to Karen Ball and Jacqui Boyce for the final wordprocessing of the manuscript. Special thanks to Jonathan Ball for checking calculations and to Mark Barnwell for contributing some artwork.

Terry Ball

Support material

Learning packages are available to support the practical activities associated with Programmable Logic Controllers, Microelectronics, Analogue and Digital Electronics and Electrical Principles, from Plymouth Open Learning Systems Unit (POLSU).

> For further details contact:
> POLSU
> Plymouth College of Further Education
> Kings Road
> Devonport
> Plymouth
> Devon PL1 5QG

About the authors

Terry Ball B.Sc., M.Sc., C.Eng., MIEE, is a Senior Lecturer in the Computing and Electrical Engineering Department of the Faculty of Technology, College of Further Education, Plymouth. He is the author of another textbook on electronics and has co-authored five learning packages on digital and analogue electronics and programmable logic controllers. He is currently engaged in developing multimedia learning systems for use in the motor industry.

Alan Crooks B.Sc., C.Eng., MIEE, is Deputy Director of the Department of Computing and Electrical Engineering in the faculty of technology, Plymouth College of Further Education. He has co-authored a number of learning packages on programmable logic controllers, analogue electronics and power electronics. He is currently developing learning systems based on multimedia and virtual reality.

PART ONE: ELECTRICAL PRINCIPLES

Chapter 1:   Solving problems in d.c. circuits.
Chapter 2:   Series and parallel a.c. circuits.
Chapter 3:   The response of series circuits to step voltage inputs.

- Over the last century or so, discoveries and developments in the field of electrical engineering (particularly electronics) have had a profound effect on the way we live our lives. In fact, it would be difficult to imagine what our lives would be like if electricity did not exist! Underpinning these developments has been the discovery of basic electrical laws and principles. A sound knowledge of these is essential for anyone working in an electrical engineering environment and the aim of the first part of this book is to describe some of these fundamanetal principles.

- Part 1 investigates and explores the two basic types of electrical circuits called direct current (d.c.) and alternating current (a.c.) circuits. The emphasis is on the understanding and use of some of the more important laws and principles involved.

Chapter 1: Solving problems in d.c. circuits

This chapter covers:
Element 11.1: Solve d.c. circuit problems.

... and is divided into the following sections:
- Basic concepts of an electrical circuit
- Ohm's law
- Series circuits
- Parallel circuits
- Series/parallel circuit combinations
- Electrical energy and power
- Circuit laws and theorems
- The star/delta transformation
- The delta/star transformation.

An electrical circuit is any arrangement of components and sources of electrical energy such as batteries, etc. which allows electrical current to flow in the circuit. If the current flows in one direction only through the components, then the circuit is said to be a **direct current** or d.c. circuit. An understanding of d.c. circuits is a prerequisite for further study in electrical engineering and electronics.

When you complete this chapter you should be able to:

- state theorems and laws used to solve d.c. circuits
- describe circuit reduction techniques
- select appropriate methods of solution for given d.c. circuits
- use appropriate theorems and laws and circuit reduction techniques to solve given d.c. circuits for specified quantities.

Basic concepts of an electrical circuit

A simple d.c. circuit is shown in Figure 1.1(a). If the battery and bulb are working, then the bulb will light and get warm. What is happening is that the energy stored in the battery is being converted into light and heat energy in the filament of the bulb. (If left long enough, the battery would go 'flat' and no more light and heat energy would be generated – think about how often you replace the batteries in your Walkman!).

Closer examination and investigation reveals other important facts about this circuit.

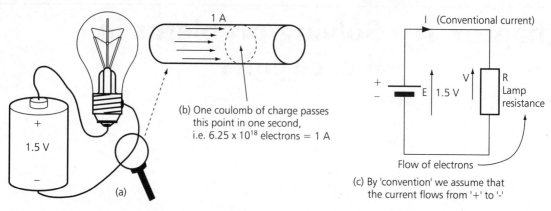

Figure 1.1 A simple d.c. circuit

1 There is a continuous metal path from the '+' terminal of the battery through the lamp to the '−' terminal. In other words, there is a **closed circuit**.

2 If the closed circuit is 'broken' in any way (e.g. by cutting one of the wires or disconnecting one end of the wire from the battery or the bulb) light and heat production ceases. A break in a circuit is called an **open circuit**.

3 If you could 'see' inside the wires and the bulb filament, you would notice high numbers of tiny electric charges called **electrons** moving from the '−' terminal, through the bulb filament and then entering the '+' terminal. Each electron has a tiny electric charge of 1.6×10^{-19} coulombs. Thus one **coulomb** (C) of charge is equivalent to $1/1.6 \times 10^{-19} = 6.25 \times 10^{+18}$ electrons. The rate at which coulombs of charge pass a point in a circuit is called the **electric current** and has the unit of **amperes** (A). When a charge of one coulomb passes a point in a circuit in one second, then 1 A is flowing. This is illustrated in Figure 1.1(b).

4 If different sizes of bulbs are used in the circuit, the current would change even though the battery voltage value stays the same. This is because each bulb has a different **resistance** to the flow of electrons or current through it. The resistance can be thought of as that property of a circuit or component that tries to restrict or control current flowing through it.

Circuit resistance (R)

Figure 1.1(c) is the **circuit diagram** representing the battery, lamp and wire. Here the bulb is defined by its resistance value R. The unit of resistance is the **ohm** (Ω). The wires are shown by solid lines.

Electromotive force (E)

The battery in Figure 1.1 provides the electromotive force (e.m.f.) to produce the current in the circuit. The unit of e.m.f. is the **volt** (V). Most simple batteries produce an e.m.f. of 1.5 V.

Conventional current (I)

Notice carefully that the **current** (I) has been shown on the circuit diagram to flow from the '+' terminal of the battery to the '−' terminal, i.e. *opposite to the flow of electrons in the circuit*. The reason for this is that during the early days of electrical technology, the existence of the electron was not known. The flow of current in a circuit was imagined to be like the flow of water in pipes and an assumption was made that current flowed from 'positive' to 'negative' as in the diagram. This is today called **conventional current flow**. We still use this convention today even though we know that, in fact, electrons are moving in the other direction. The unit of current which is the **ampere** (A), has sub-multiples: **milliamperes** – mA (10^{-3} A) and **microamperes** – μA (10^{-6} A).

Voltage and potential difference (p.d.)

When current (I) flows through a resistance (R), a voltage (V) is produced across it. Thus in Figure 1.1(c) a voltage V is produced across R by the current I. Alternatively, we can say that there is a potential difference (p.d.) or voltage drop of V across R. The 'arrow' used to indicate voltage or p.d. shows the positive end of the voltage across the resistance. Note that E = V.

Ohm's law

One of the most important laws in electrical engineering is Ohm's law. It describes the relationship between values of voltage, current and resistance. It states that in a circuit with stable resistance: 'the current flowing is proportional to the voltage applied' i.e. $V \propto I$.

If the resistance of the circuit is R, then

$$V = I \times R \text{ (volts)}$$

WORKED EXAMPLE 1.1

(a) In the circuit of Figure 1.1 a current of say 0.1 A flows when V is increased to 10 V. Calculate the value of R.
(b) If the applied voltage is increased further to 15 V, what is the new value of current if the resistance remains stable (constant)?
(c) What value of voltage would be needed to produce a current of 0.5 A in the circuit?

Solution

(a) Since V and I are known, we can find R using $V = I \times R$:

$$\text{since} \quad R = \frac{V}{I}$$

$$\text{therefore} \quad R = \frac{10}{0.1} = 100 \ \Omega$$

(b) We now know R and V and need to find I so:

$$I = \frac{V}{R} = \frac{15}{100} = 0.15 \text{ A}$$

(c) Now I = 0.5 A and R = 100 Ω:

$$V = I \times R = 0.5 \times 100 = 50 \text{ V}$$

Progress check 1.1

Measuring current and voltage

Current is measured with an **ammeter** and voltage with a **voltmeter**. For the circuit of Figure 1.1, the ammeter (A) and voltmeter (V) would be connected as in Figure 1.2. Since current flows through the circuit, then the ammeter must be inserted in the current path, as shown. Since the voltage is produced across R by the current I flowing through it, then the voltmeter must be connected across the resistance. We say that:

'the ammeter is placed in **series** and the voltmeter is placed in **parallel**'.

Note in Figure 1.2 that the ammeter placed at the other positions (shown dotted) is in series with R and will measure the same value of current.

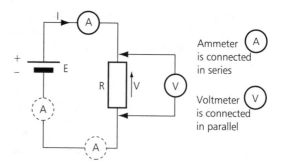

*Figure 1.2
Connecting an
ammeter and
a voltmeter*

Measuring resistance

The resistance of a component is measured with an **ohmmeter**. Unlike the ammeter and voltmeter which are connected in a circuit, the ohmmeter must be used with the component 'out of circuit'. Thus, for example, if it were necessary to measure the resistance of 'R' in Figure 1.2, R would first need to be disconnected from the circuit before the ohmmeter could be used.

Multimeters

In Electrical Engineering, the ability to measure voltage, current and resistance is so crucial that a whole range of instruments is available. One of the most useful is the **multimeter** which allows the measurement of voltage, current and resistance with a single meter. If the measurement is shown by a needle moving over a scale, then it is called an **analogue multimeter**. If the scale uses liquid crystal displays or light-emitting diodes, it is called a **digital multimeter**. Generally, digital multimeters are easier to read and are usually more accurate.

Series circuits

In a series circuit as in Figure 1.2, the current through each part of the circuit is the same. Figure 1.3 shows three resistances connected in series with a battery E.

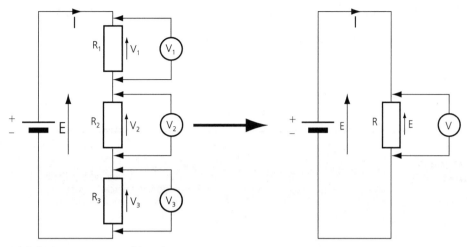

Figure 1.3 Resistances connected in series

As the current I flows through each resistor it produces voltages of V_1, V_2 and V_3. In a series circuit, *the sum of the voltages across each resistance is equal to the applied voltage E.* That is:

$$E = V_1 + V_2 + V_3$$

Each individual resistor voltage is known as a voltage drop (volt drop) or potential difference (p.d.). As shown in Fig. 1.3, the three resistances can be replaced by a single resistance R which would produce exactly the same effect. Using Ohm's law:

$$E = I \times R = (I \times R_1) + (I \times R_2) + (I \times R_3)$$

Dividing through by I gives:

$$R = R_1 + R_2 + R_3 \text{ (ohms)}$$

Thus for a series circuit the **equivalent resistance** is found by adding the individual resistances together. You should remember this equation.

WORKED EXAMPLE 1.2

For the circuit in Figure 1.3, determine the value of the circuit current and the voltage across each resistance if $R_1 = 1\ \Omega$, $R_2 = 3\ \Omega$ and $R_3 = 4\ \Omega$ and E = 12 V.

Solution
The three resistances are connected in series so the equivalent resistance, R, is:

$$R = 1 + 3 + 4 = 8\ \Omega$$

Using Ohm's law $12 = I \times 8$

so I = 1.5 A

7

The voltage therefore across each resistance is (from $V = IR$):

$$V_1 = 1.5 \times 1 = 1.5 \text{ V}$$

$$V_2 = 1.5 \times 3 = 4.5 \text{ V}$$

$$V_3 = 1.5 \times 4 = 6 \text{ V}$$

(Check: $12 = 1.5 + 4.5 + 6$).

Progress check 1.2

If the 4 Ω resistance in Worked Example 1.2 is removed, find the value of the new circuit current and the voltage across each resistance.

Activity 1.1

Electric circuit components which are designed to have particular values of resistance are called **resistors**. Produce a short report detailing how the value of a resistor is indicated on the body of the resistor. Include in your report an explanation of the term **tolerance** as applied to resistors.

Parallel circuits

In a parallel circuit, the *voltage across each component is the same*. Figure 1.4 shows three resistances in parallel. Each resistance has the battery voltage E across it and this would be read on the voltmeter connected across each resistance component. The current through each resistance is different.

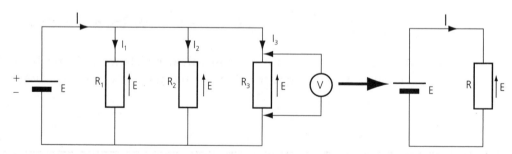

Figure 1.4 Resistances in parallel

The sum of the currents I_1, I_2 and I_3 is equal to the total current I taken from the battery:

$$I = I_1 + I_2 + I_3$$

Since the voltage across each resistance is the same, (i.e. E), using Ohm's law:

$$E = I_1 \times R_1 \quad \text{or} \quad I_1 = \frac{E}{R_1}$$

$$E = I_2 \times R_2 \quad \text{or} \quad I_2 = \frac{E}{R_2}$$

$$E = I_3 \times R_3 \quad \text{or} \quad I_3 = \frac{E}{R_3}$$

For the equivalent resistance R, the same current I = E/R, would be taken from the battery. Therefore:

$$\frac{E}{R} = \frac{E}{R_1} + \frac{E}{R_2} + \frac{E}{R_3}$$

Dividing through by E gives:

$$\frac{1}{R} = \frac{1}{R_1} + \frac{1}{R_2} + \frac{1}{R_3}$$

This is an important equation and should be remembered since it means that the **equivalent resistance** of a parallel circuit can be found from the reciprocals of the individual resistance values.

WORKED EXAMPLE 1.3

For the parallel circuit in Figure 1.4, if R_1 = 1 Ω, R_2 = 100 Ω and R_3 = 1000 Ω and E = 10 V, find the total circuit current and the current through each resistance.

Solution
The first step in finding the overall current is to find the equivalent value of the three resistances:

$$\frac{1}{R} = \frac{1}{1} + \frac{1}{100} + \frac{1}{1000} = 1 + 0.01 + 0.001 = 1.011$$

$$R = \frac{1}{1.011} = 0.989 \ \Omega$$

Notice that the equivalent resistance is *less than the value of the smallest resistance*. This is a general rule. Here it is almost equal to R_1 = 1 Ω because R_2 and R_3 are very much larger than 1 Ω.
 Using Ohm's law, the circuit current I is:

$$I = \frac{10}{0.989} = 10.11 \ A$$

The voltage across each resistance is the same, so by Ohm's law:

$$I_1 = \frac{10}{1} = 10 \ A$$

$$I_2 = \frac{10}{100} = 0.1 \ A$$

$$I_3 = \frac{100}{1000} = 0.01 \text{ A}$$

(Check: $I = 10.11 = 10 + 0.1 + 0.01 = 10.11$ A)

Note, as expected, that *in a parallel circuit, most of the current flows through the smallest resistance value*. Current will always take the path of least resistance.

Progress check 1.3

Firstly, two 3 Ω resistances and secondly, three 3 Ω resistances, are connected in parallel. Calculate the equivalent resistance in each case. Can you spot a general rule for *equal resistances in parallel*?

Series-parallel circuit combinations

Many circuits in electrical/electronic engineering consist in practice of combinations of series and parallel circuits. In calculating circuit values, simplification is achieved by replacing the series and parallel groups with their equivalent resistances. Study the following worked example carefully.

WORKED EXAMPLE 1.4

For the circuit in Figure 1.5, calculate:
(a) total circuit current,
(b) the voltage across the parallel combination, the voltage across R_3 and R_4,
(c) the currents through R_1 and R_2.

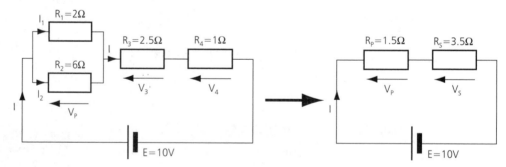

Figure 1.5 Worked Example 1.4

Solution

(a) The first step is to replace the parallel part of the circuit with its equivalent resistance R_P.

$$\frac{1}{R_P} = \frac{1}{R_1} + \frac{1}{R_2}$$

i.e. $R_P = \dfrac{R_1 \times R_2}{R_1 + R_2}$

Notice that the equivalent value equals product/sum. You should remember this.

$$R_P = \dfrac{2 \times 6}{2 + 6} = 1.5\ \Omega$$

R_3 and R_4 in series is equivalent to a resistance:

$$R_S = R_3 + R_4 = 2.5 + 1 = 3.5\ \Omega$$

The total resistance R_T of the circuit R_P in series with R_S is:

$$R_T = 1.5 + 3.5 = 5\ \Omega$$

By Ohm's law, the circuit current I is:

$$I = \dfrac{10}{R_T} = \dfrac{10}{5} = 2\ A$$

(b) The voltage across the parallel circuit is:

$$V_P = I \times R_P = 2 \times 1.5 = 3\ V$$

Voltage across R_3 and R_4 is:

$$V_3 = 2 \times 2.5 = 5\ V$$

$$V_4 = 2 \times 1 = 2\ V$$

(Check: $E = V_P + V_3 + V_4 = 3 + 5 + 2 = 10\ V$)

(c) Both R_1 and R_2 have the same voltage (V_P) across them.

Therefore,

$$I_1 = \dfrac{V_P}{2} = \dfrac{3}{2} = 1.5\ A$$

$$I_2 = \dfrac{V_P}{6} = \dfrac{3}{6} = 0.5\ A$$

(Check: $I = I_1 + I_2 = 1.5 + 0.5 = 2\ A$)

Electrical energy and power

When current flows through a resistance R, energy is consumed or dissipated by the resistance and it gets warm. If the current flows for a time t and produces a voltage drop V across R, then the energy U consumed is:

$$U = V \times I \times t\ \text{joules (J)}$$

Since $V = I \times R$ this equation can also be written as:

$$U = \frac{V^2}{R} \times t \quad \text{or} \quad U = I^2 R t$$

Power (P) is the *rate of dissipating energy*, so

$$P = \frac{\text{energy}}{\text{time}} = \frac{U}{t}$$

Dividing each of the above equations for U by t will give an equation for P, i.e.

$$P = V I \quad \text{(watts)}$$

or

$$P = \frac{V^2}{R} \quad \text{(watts)}$$

or

$$P = I^2 R \quad \text{(watts)}$$

Progress check 1.4

Calculate the power dissipated in the resistances R_1 and R_2 in Worked Example 1.4.

Circuit laws and theorems

In order to solve problems in electrical circuits, a number of laws and theorems have been discovered. Some of these will now be investigated.

Kirchoff's laws

The current law

'At a junction in any circuit the current flowing towards the junction is equal to the current flowing away from that junction.'

Thus in Figure 1.6(a) $I_1 + I_4 = I_2 + I_3$. Although we did not use the name of Kirchoff, we have already used this law when we found the equation for

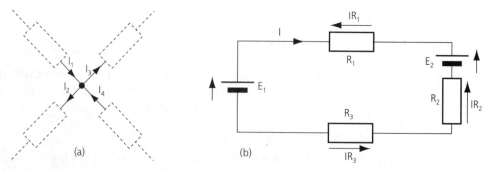

(a)   (b)

Figure 1.6 Kirchoff's laws

resistors in parallel Figure 1.4). The current into the junction of the three resistances (I) was equal to the currents leaving the junction and flowing through the resistances (I_1, I_2 and I_3).

The voltage law

'In any closed voltage loop of a circuit the algebraic sum of the voltage drops (current × resistance) is equal to the resultant e.m.f. within that loop.'

Study Figure 1.6(b) carefully. The voltage arrowheads point to the most positive voltage point so on the batteries E_1 and E_2 they are as shown. We have assumed here that the current I flows clockwise as shown. Therefore, the 'voltage arrows' on each resistance will be as indicated. The value of each voltage is current × resistance. By convention, *if current flows away from the positive terminal of a source, that source is considered to be positive.* Therefore, in Figure 1.6(b) E_1 is positive and E_2 is negative. Thus:

$$E_1 - E_2 = IR_1 + IR_2 + IR_3$$

or we could write:

$$E_1 - E_2 - IR_1 - IR_2 - IR_3 = 0$$

A good way to remember is to say 'in the closed loop clockwise arrows are positive and anti-clockwise are negative'. The sum of the clockwise arrows is equal to the sum of the anti-clockwise arrows.

Have you realised that Kirchoff's voltage law has been used already? Remember the three resistances in series – the sum of the voltages across them was equal to the applied e.m.f. E. We used this fact to find the equation for resistances in series.

WORKED EXAMPLE 1.5

Use Kirchoff's laws to find the current in each branch of the circuit in Figure 1.7.

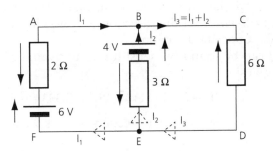

Figure 1.7 Worked Example 1.5

Solution

1 Label the currents in all the branches. Which direction is chosen is irrelevant. Notice how at point B we have applied Kirchoff's current law, i.e. $I_1 + I_2 = I_3$. (Of course, this applies at point E as well.)

2 Mark in the voltage drop arrows, making sure their direction is correct, i.e. the voltage arrowhead opposes the current through that resistance. The voltage arrows on the batteries have the arrowhead on the positive terminal.

3 We now need two equations involving I_1 and I_2, which we can then solve. There are three loops in the circuit. Consider loop ABEF; by adding volt drops and taking 'clockwise arrows as positive' gives:

$$+ 6 - 2I_1 - 4 + 3I_2 = 0$$

i.e.

$$2 = 2I_1 - 3I_2 \qquad (1)$$

For the outer loop ACDF the sum of the voltage drops gives:

$$+ 6 - 2I_1 - 6(I_1 + I_2) = 0$$
$$6 - 2I_1 - 6I_1 - 6I_2 = 0$$
$$6 = 8I_1 + 6I_2 \qquad (2)$$

We now have two equations (1) and (2) having two unknowns (I_1 and I_2) which can now be solved. Multiplying (1) by +4 gives:

$$8 = 8I_1 - 12I_2$$

i.e.

$$8 + 12I_2 = 8I_1 \qquad (3)$$

Rearranging (2) gives:

$$6 - 6I_2 = 8I_1 \qquad (4)$$

Equating (3) and (4) gives:

$$8 + 12I_2 = 6 - 6I_2$$
$$8 - 6 = -12I_2 - 6I_2$$
$$I_2 = -2/18 = -0.1111 \text{ A}$$

Note, we haven't made a mistake! The negative sign just means that the original choice for the direction of I_2 was incorrect. In the real circuit it is flowing in the opposite direction.

I_1 can be found by substituting the value of I_2 into equation (1) or (2). Using (1) gives:

$$2 = 2I_1 - 3(-0.1111) = 2I_1 + 0.333$$
$$I_1 = +0.8334 \text{ A (direction is as chosen)}$$

4 The remaining branch current can now be calculated. The current I_3 through the 6 Ω resistance is $I_1 + I_2$.

$$I_3 = I_1 + I_2 = 0.8334 - 0.1111$$
$$= +0.7223 \text{ A (direction as chosen)}$$

5 *Checking*: The circuit has three loops. The equation for the third loop can be used to check that the solution is correct. Considering loop BCDE, summing the voltage drops for 'clockwise arrows positive':

$$+4 - 6I_3 - 3I_2 = 0$$

$$4 - 6(0.7223) - 3(-0.1111) = 0$$

$$4 - 4.3338 + 0.3333 = 0$$

$$-0.0005 = 0$$

(The slight error is just because of 'rounding up' the calculated values.)

Progress check 1.5

The 6 Ω resistance in Worked Example 1.5 is replaced with a 10 Ω resistance. Calculate the values of the new currents in the circuit and the power dissipated in the 10 Ω resistance.

The Superposition theorem

'In a circuit containing more than one e.m.f., the current in each branch is the algebraic sum of the currents in that branch which would be produced by each e.m.f. acting separately, the other e.m.f.'s being replaced by any internal resistance they might have.'

WORKED EXAMPLE 1.6

Determine the currents in the branches of the circuit in Figure 1.7 using the Superposition Theorem.

Solution

1 As shown in Figure 1.8(a) we temporarily remove one of the e.m.f.'s (the 4 V) and find the currents I, I_1 and I_2 in each branch. The 3 Ω and 6 Ω resistances are in parallel and are equivalent to 2 Ω. This is in series with the other 2 Ω resistance. Therefore:

$$I = \frac{6}{2+2} = 1.5 \text{ A}$$

The voltage across the 3 Ω and 6 Ω in parallel (at points BC/ED) is 6 V less the voltage drop across the 2 Ω resistance = 6 − 2 × I = 3 V. Therefore:

$$I_1 \times 3 = 3 \qquad I_1 = +1 \text{ A}$$

$$I_2 \times 6 = 3 \qquad I_2 = +0.5 \text{ A}$$

2 The next step is to temporarily remove the 6 V battery and find the currents I_3, I_4 and I_5 as in the circuit of Fig. 1.8(b). As shown, the 2 Ω and 6 Ω are in parallel and equivalent to a 1.5 Ω resistance in series with the 3 Ω. Thus:

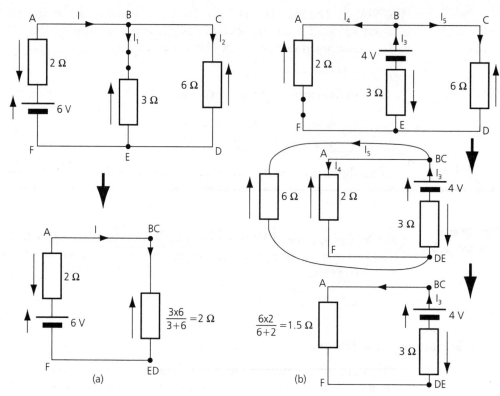

Figure 1.8 Worked Example 1.6

$$I_3 = \frac{4}{3 + 1.5} = +0.8889 \text{ A}$$

The voltage across the parallel resistances is $4 - 3 \times I_3 = 1.3333$ V. Therefore:

$$I_4 = \frac{1.3333}{2} = +0.6667 \text{ A}$$

and

$$I_5 = \frac{1.3333}{6} = +0.2222 \text{ A}$$

3 Finally, for the real circuit we need the algebraic sum of the currents. The current from the 6 V battery is:

$$I - I_4 = 1.5 - 0.6667 = +0.8333 \text{ A}$$

The current from the 4 V battery is:

$$I_3 - I_1 = 0.8889 - 1 = -0.1111 \text{ A}$$

The current down through the 6 Ω resistance is:

$$I_2 + I_5 = 0.5 + 0.2222 \text{ A} = +0.7222 \text{ A}$$

Apart from the slight rounding errors the currents are identical to those found using Kirchoff's laws in Worked Example 1.5.

Thévenin's theorem

Thévenin (a Frenchman) developed one of the most important circuit theorems. Although quite a mouthful to state, it is straightforward to apply in practice. It may be stated as follows:

'The current through a resistance R connected between points X and Y of an active circuit (i.e. a circuit having one or more sources of e.m.f.) can be found by dividing the voltage measured between X and Y with R *disconnected* by a resistance of value R + r where r is the resistance measured between points X and Y with R disconnected and all the sources of e.m.f. replaced by their own internal resistances.'

Thévenin's theorem is illustrated in Figure 1.9. Study the diagram carefully.

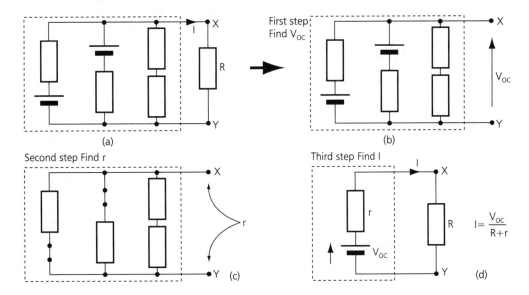

Figure 1.9 Illustrating Thévenin's theorem

The resistance R is part of a complicated circuit and we need to find the current through it. The first step is to remove R and find (or measure) the voltage between X and Y. This is called the **open circuit voltage** V_{OC}. Next the e.m.f.'s are replaced by their internal resistances (just remove the battery symbols and replace with a wire). Find (or measure) the resistance between X and Y. The current through R is found by simply applying Ohm's law to the equivalent circuit in Figure 1.9(d). i.e.:

$$V_{OC} = I \times (R + r)$$

Therefore

$$I = \frac{V_{OC}}{R + r}$$

It is important that you realise that in effect you have 'replaced' the complicated circuit shown in the 'dotted box' with two simple components in series – V_{OC} and r.

Therefore an alternative (and probably better) way of describing Thévenin's Theorem would be:

'An active circuit can be replaced between any two terminals X and Y with a voltage source V_{OC} in series with a resistance r. V_{OC} is the open circuit voltage between X and Y and r is the resistance measured at the open circuit between X and Y with all sources of e.m.f. replaced by their internal resistances.'

We will now use Thévenin's theorem to find the current through the 6 Ω resistance in Worked Example 1.5.

WORKED EXAMPLE 1.7

Use Thévenin's theorem to find the current through the 6 Ω resistance in Figure 1.7.

Solution

1 Firstly, remove the 6 Ω resistance and find V_{OC}. See Figure 1.10(a).

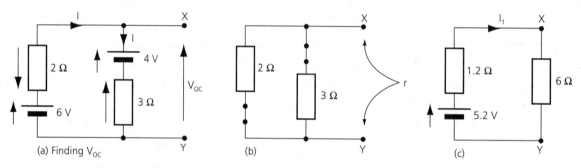

Figure 1.10 Worked Example 1.7

V_{OC} is the voltage across the 4 V e.m.f. in series with the 3 Ω resistance or the voltage across the 6 V e.m.f. in series with the 2 Ω resistance – they are the same since the branches are in parallel. Let the current flowing in the loop be I. Make sure you understand that there is NO current flowing out 'towards' X because it is an open circuit.

Applying Kirchoff's voltage law to the loop gives:

$$+6 - 2I - 4 - 3I = 0$$

$$I = 0.4 \text{ A}$$

Now $V_{OC} = 6 - 2I = 6 - 2 \times 0.4 = 5.2 \text{ V}$

(Check: V_{OC} is also given by $V_{OC} = 4 + 3I$. Remember the arrow on the 3 Ω resistance is in the same direction as the 4 V e.m.f. Thus, $V_{OC} = 4 + 3 \times 0.4 = 5.2$ V.)

2 The circuit for finding r is shown in Figure 1.10(b). When the e.m.f.'s are removed, the 2 Ω and 3 Ω resistances are in parallel. Therefore:

$$r = \frac{2 \times 3}{2 + 3} = \frac{6}{5} = 1.2 \, \Omega$$

Using Figure 1.10(c), the current through the 6 Ω, I_1, is

$$I_1 = \frac{5.2}{6 + 1.2} = \frac{5.2}{7.2} = 0.7222 \, A$$

As before, I_1 flows down from X to Y and (apart from 'rounding errors') is exactly the same as was found by just using Kirchoff's law or by the Superposition theorem.

Progress check 1.6

Determine the current through, and the power dissipated in, the 1 Ω resistance in Figure 1.11 using Thévenin's theorem.

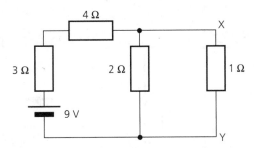

Figure 1.11 Progress check 1.6

Norton's theorem

When using Thévenin's theorem we have effectively replaced whatever circuit exists between X and Y with a **constant voltage source** V_{OC}, in series with an **internal resistance** r. This provides the electrical energy source to drive current through whatever else is connected between X and Y. An alternative method is to represent the electrical energy source by a **constant current source** in parallel with a resistance. This is illustrated in Figure 1.12.

Norton's theorem describes how the values of I_{SC} and r can be calculated. It states:

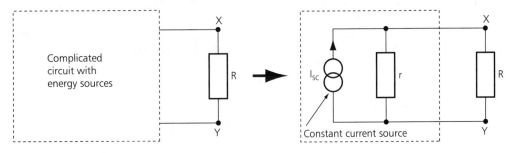

Figure 1.12 Circuit 'replacement' with a current source

'An active circuit can be replaced between any two terminals X and Y by a constant current source in parallel with a resistance. The value of the current source I_{SC} is the **short circuit current** that would flow between X and Y. The value of the resistance r is that measured at the open circuit between X and Y with all the sources of e.m.f. replaced by their internal resistances.'

Figure 1.13 shows how the current through a resistance R between X and Y can be found using Norton's theorem.

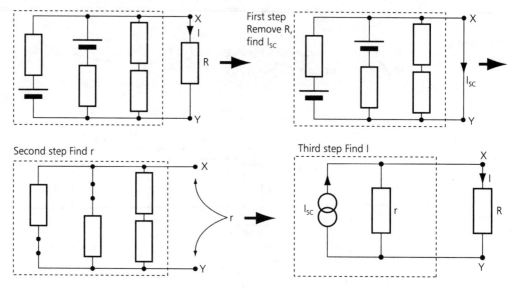

Figure 1.13 *Illustrating Norton's theorem*

Make sure you understand how the value of I is obtained in the third step. The voltage across r and R in parallel is:

$$V = I_{SC} \times \left(\frac{r \times R}{r + R} \right)$$

But this voltage is the voltage across R alone (I x R), therefore:

$$I \times R = I_{SC} \times \left(\frac{r \times R}{r + R} \right)$$

Cancelling R on each side gives:

$$I = I_{SC} \times \left(\frac{r}{r + R} \right)$$

Note also that the definition for r is exactly as for Thévenin's theorem.

WORKED EXAMPLE 1.8

Repeat Progress Check 1.6 using Norton's theorem.

Solution

1 The first step is to remove the 1 Ω resistance and replace it with a short circuit (see Figure 1.14(b)).

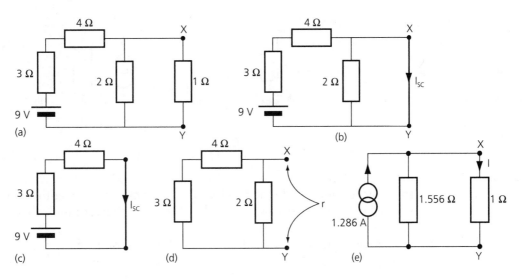

Figure 1.14 Worked Example 1.8

The short circuit shorts the 2 Ω resistance and takes it out of the circuit completely. (See Figure 1.14(c).)

Therefore: $I_{SC} = \dfrac{9}{3+4} = 1.286$ A

2 r (Figure 1.14(d)) is 2 Ω in parallel with the series combination of 3 Ω and 4 Ω.

Using Figure 1.14(e), the current I through the 1 Ω resistance is:

$$I = I_{SC} \times \left(\frac{r}{r+R}\right) = 1.286 \times \left(\frac{1.556}{1.556+1}\right) = 0.782 \text{ A}$$

This is the same value found using Thévenin's theorem in Progress Check 1.6. The power dissipated is the same, of course.

The link between Thévenin's and Norton's theorems

Thévenin's and Norton's theorems allow an active circuit to be 'replaced' between two terminals with simpler equivalent circuits. Since they can replace the *same* circuit, there must be a link between them. If we establish this link, then what it means in practice is that having found one equivalent circuit it is possible to find the other. Consider Figure 1.15. Notice right away that the r's are exactly the

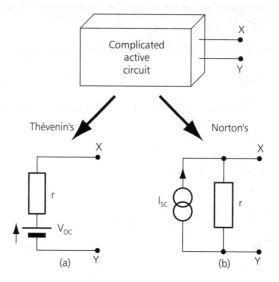

Figure 1.15 The link between Thévenin's and Norton's theorems

same since they are defined in exactly the same way for both theorems. We now need the link between V_{OC} and I_{SC}. Suppose the terminals X and Y of Figure 1.15(a) are short circuited.

The short circuit current that would flow is equal to

$$\frac{V_{OC}}{r}$$

If Figure 1.15(b) is short circuited, then the current that flows is I_{SC}. If the two circuits in Figure 1.15(a) and (b) are to be equivalent, then these short circuit currents must be equal, i.e.

$$I_{SC} = \frac{V_{OC}}{r}$$

or

$$V_{OC} = I_{SC} \times r \ \text{(volts)}$$

Thus, in practice, if we know one equivalent circuit we can easily find the other.

The star/delta transformation

Consider the three resistors in Figure 1.16(a). They are connected in what is known as (for obvious reasons) a **star connection**. It is possible to replace the star connection with the equivalent circuit called the **delta connection** in Figure 1.16(b). The use of this transformation sometimes aids the simplification of more complicated circuits. If the two circuits are to be equivalent, then the resistance measured between any two terminals in Figure 1.16(a) must be the same as that between the same pair of terminals in Figure 1.16(b).

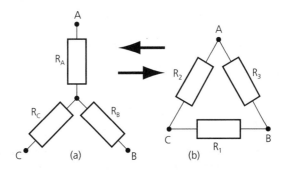

Figure 1.16 Star/delta and delta/star transformations

It is found that:

$$R_1 = R_B + R_C + \frac{R_B \times R_C}{R_A}$$

$$R_2 = R_A + R_C + \frac{R_A \times R_C}{R_B}$$

$$R_3 = R_A + R_B + \frac{R_A \times R_B}{R_C}$$

From the equations you can see that:

'the equivalent delta resistance between two terminals is the sum of the two star resistances connected to those terminals plus the product of the same two star resistances divided by the third star resistance.'

The delta/star transformation

It is possible, of course, to go from a delta connection to a star connection in Figure 1.16.

The equations then are:

$$R_A = \frac{R_2 \times R_3}{R_1 + R_2 + R_3}$$

$$R_B = \frac{R_1 \times R_3}{R_1 + R_2 + R_3}$$

$$R_C = \frac{R_1 \times R_2}{R_1 + R_2 + R_3}$$

You can see that:

> 'the equivalent star resistance connected to a given terminal is equal to the product of the two delta resistances connected to the same terminal divided by the sum of the delta resistances.'

WORKED EXAMPLE 1.9

Use the (a) star/delta transformation and (b) delta/star transformation to calculate the equivalent resistance between the terminals A and B of the circuit of Figure 1.17(a).

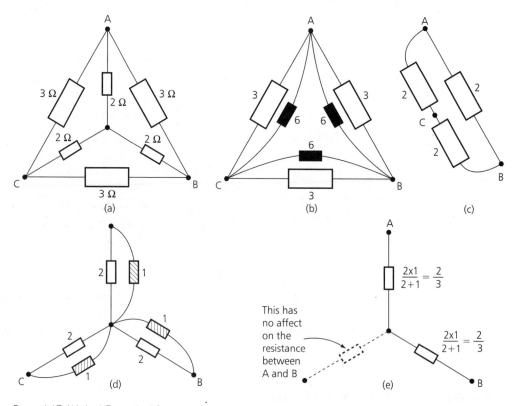

Figure 1.17 Worked Example 1.9

Solution

(a) As shown in Figure 1.17(b), the $2\,\Omega$ star can be changed into a delta. Each delta resistance is:

$$2 + 2 + \frac{2 \times 2}{2} = 6\,\Omega$$

We now have two delta circuits superimposed. The equivalent delta resistance in each branch is $3\,\Omega$ in parallel with $6\,\Omega$ i.e.:

$$\frac{3 \times 6}{3 + 6} = \frac{18}{9} = 2\,\Omega$$

The resistance between A and B is 2 Ω in parallel with 2 Ω in series with 2 Ω as shown in Figure 1.17(c) i.e.:

$$\frac{2 \times 4}{2 + 4} = \frac{8}{6} = 1.333 \ \Omega$$

(b) As shown in Figure 1.17(d), the 3 Ω delta can be changed into a star. Each resistance value is:

$$\frac{3 \times 3}{3 + 3 + 3} = \frac{9}{9} = 1 \ \Omega$$

We now have two stars superimposed. The equivalent resistance in each branch is 2 Ω in parallel with 1 Ω. i.e.:

$$\frac{2 \times 1}{2 + 1} = \frac{2}{3} = 0.667 \ \Omega$$

The resistance between A and B (as shown in Figure 1.17(e)) is two resistances in series i.e.:

$$\frac{2}{3} + \frac{2}{3} = \frac{4}{3} = 1.333 \ \Omega$$

This is the same value obtained above. Notice in Figure 1.17(e) that the delta resistance connected to C has no effect on the resistance between A and B.

Do this 1.1

Checking out some theorems and laws

Equipment and components required:

- digital multimeter
- Two power supplies (0–15 V)
- Resistors 20 Ω, 40 Ω, 60 Ω (5%).

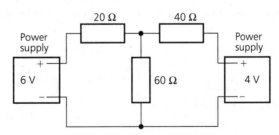

Figure 1.18 Do This 1.1

Procedure:

1 Construct the circuit in Figure 1.18. Accurately set the power supply voltages with the digital meter.
2 Design a procedure to verify Kirchoff's laws.
3 Calculate the current through the 60 Ω resistor using: the Superposition theorem, and Thévenin's theorem.

4 Apply each theorem practically and take the necessary measurements as accurately as possible.

5 Compare the values obtained practically and theoretically.

Assignment 1

This assignment provides evidence for:
Element 11.1: Solve d.c. curcuit problems
and the following key skills:
Communication 3.2
Application of Number 3.2

The evidence indicators for this element require you to provide solutions for five circuit problems. Three separate circuits are shown in Figure 1.19. For the circuit in Figure 1.19(a), determine the current taken from the supply, the voltage across the 2 Ω parallel circuit and the power dissipated in the 1 Ω resistance.

From Figure 1.19(b), determine the current through the 10 Ω resistance using:

(a) Kirchoff's laws only,

(c) Norton's theorem,

(b) Thévenin's theorem,

(d) the Superposition theorem.

For Figure 1.19(c), determine the current taken from the 6 V supply using the star/delta transformation and the formulae for resistance combinations.

Report

In your report include, for each solution, a statement of the Theorems and Laws you are using and a detailed description of how you arrive at your final results.

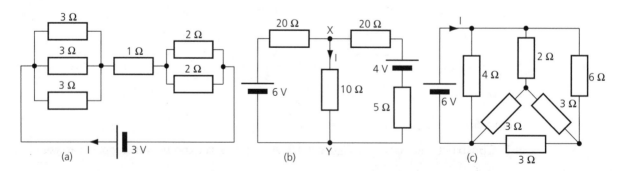

Figure 1.19 Circuits for Element Assignment

Chapter 2: Series and parallel a.c. circuits

This chapter covers:
Element 11.2: Investigate the response of series and parallel circuits to sinusoidal inputs.

... and is divided into the following sections:
- Alternating voltage and current
- Sinusoidal waveform equations and phasors
- Inductors and capacitors
- Resonance in a series RLC circuit
- Power dissipation in a.c. circuits
- Parallel a.c. circuits.

In Element 11.1 the electrical energy sources connected to the circuits were batteries which produced current flow in one direction only in the circuit. When this current flowed through the circuit resistances, a voltage of fixed direction was generated across each resistance. Thus in d.c. circuits, the currents and voltages are unidirectional. Alternatively, we could describe them as being of **constant polarity**. In practice electrical energy is produced at power stations by electrical machines called generators and distributed throughout the country via the National Grid of transmission lines. It turns out that it is easier to generate and distribute electricity that is alternating in nature and not direct. An alternating voltage or current changes its direction or polarity systematically in a predetermined and controlled manner. In this element we will investigate alternating voltage and current and the response of series and parallel circuits to them.

When you complete this element you should be able to:

- define sinusoidal alternating quantities
- represent sinusoidal quantities
- determine electrical quantities for steady-state series circuits and parallel circuits with sinusoidal inputs
- describe conditions for resonance in circuits with sinusoidal inputs.

Alternating voltage and current

An alternating voltage or current is one that changes its direction (polarity) periodically. If a graph of voltage or current is plotted against time then the resulting

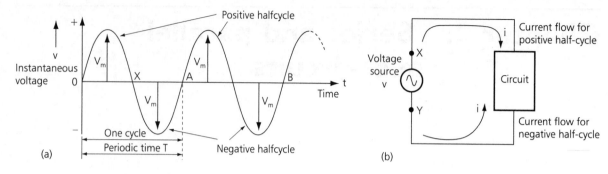

Figure 2.1 A sinusoidal voltage waveform

graph is called a **waveform**. By far the most important waveform in electrical engineering is one whose shape is that of a sinewave. It is called a **sinusoidal waveform**. The standard mains voltage supply produces a waveform which is sinusoidal.

A sinusoidal voltage waveform is drawn in Figure 2.1(a). Study the waveform carefully and note the following points:

● The voltage waveform repeats itself after one **cycle**, i.e. the first cycle is from 0 to A, the next from A to B and so on.
● A complete cycle is divided into two half cycles. The first is called the **positive half cycle** (0 to X) and the next is called the **negative half cycle** (X to A).
● The value of the voltage at any instant is called its instantaneous value and is represented by a lower case letter – v in the case of voltage.
● The time taken to complete one cycle is called the **periodic time**, T (or **period**). T is measured in seconds.
● The number of cycles that occur in one second is called the **frequency,** f. Frequency is measured in **hertz** (Hz) and is related to T by the equation:

$$f = \frac{1}{T} \text{ (cycles per second) (hertz)}$$

● The largest value reached during a half cycle is called the **amplitude, peak value** or **maximum value**. Capital letters are used – V_M in this case.
● The **peak-to-peak value** is the difference between the maximum and minimum values reached in a cycle. For the sinusoidal voltage the peak-to-peak value = $2 V_M$.

Figure 2.1(b) shows what happens when an alternating voltage is connected to a circuit so that current flows. Note the symbol used for the voltage source v. For the positive half cycle point X is positive with respect to Y and the current i flows 'down' into the circuit. For the negative half cycle Y is positive with respect to X and i flows 'up' into the circuit. Thus the alternating voltage produces an alternating current i in the circuit – first flowing in one direction and then in the opposite direction for as long as the voltage is connected.

Progress check 2.1

The U.K. mains frequency is 50 Hz. What is:

(a) the period
(b) the time required to complete a half cycle
(c) the time taken for voltage to reach its first peak value.

Complex waveforms

Although the sinusoidal waveform is the most common, many other waveform shapes are encountered in electrical engineering. Some non-sinusoidal wave-forms are drawn in Fig. 2.2. Although usually described by their **waveform shape** if possible (e.g. sawtooth, etc.) they are all examples of complex waveforms.

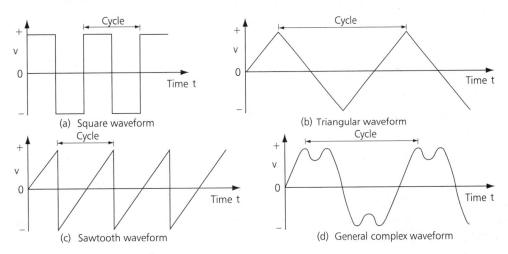

Figure 2.2 Some complex waveforms

Each of these complex waveforms is alternating because it changes its polarity regularly, however, they are definitely *not* sinusoidal in shape.

The average value of a sinusoidal waveform

A sinewave is symmetrical, i.e. has exactly the same shape above and below the zero line. Consequently, over one complete cycle (or a whole number of cycles) the average value is zero. Thus in electrical engineering an average value is taken over one half of a cycle. Usually, the positive half cycle is chosen for convenience. All the waveforms in Figure 2.2 are symmetrical and the average value for these is also defined over a half cycle only.

Several methods can be used to calculate the average value. The method shown in Figure 2.3 is called the **mid-ordinate method** and will give an approximate value for the average value. In the diagram a sinusoid of current has been drawn. The positive half cycle is divided into a number of parts and the mid-ordinate value is measured. The average value, I_{AV} is:

$$I_{AV} = \frac{i_1 + i_2 + i_3 + i_4 + \ldots + i_n}{n}$$

where n = number of mid-ordinates.

In Figure 2.3 n = 6. Obviously, to get good accuracy n should be as large as possible.

Advanced mathematics shows that the average value of a sinusoidal waveform is given by:

> Average value = 0.637 × Maximum value

The average value is often called the **mean value** of the waveform.

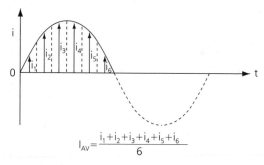

$$I_{AV} = \frac{i_1 + i_2 + i_3 + i_4 + i_5 + i_6}{6}$$

Figure 2.3 *Using the mid-ordinate method to find* I_{AV}

WORKED EXAMPLE 2.1

A sinusoidal current waveform of peak value 1 A has its positive half cycle divided into six 30° sections. If the mid-ordinate values are:

15° = 0.259 A   75° = 0.966 A   135° = 0.707 A

45° = 0.707 A   105° = 0.966 A   165° = 0.259 A

determine the approximate average value of the waveform. Compare the value with the value expected from the equation.

Solution

$$I_{AV} = \frac{0.259 + 0.707 + 0.966 + 0.966 + 0.707 + 0.259}{6} = \frac{3.864}{6} = 0.644 \text{ A}$$

The exact value is:

$$I_{AV} = 0.637 \times 1 = 0.637 \text{ A}$$

If the half sinewave were divided into smaller sections so that more mid-ordinates could be used then we would be closer to the exact value. In practice always use the equation for sinusoidal waveforms.

The effective value of a sinusoidal waveform

If a sinusoidal current flows through a resistance, power is dissipated and the resistance would get warm. The effective value of the sinusoidal current is defined as being equal to that direct current (d.c.) which would produce exactly the same heating effect as the sinusoidal (a.c.) waveform. The effective value in practice is usually called the **root-mean-square value** (r.m.s.) of the waveform. It is calculated from the (current)^2 waveform over a *complete cycle* and can be estimated from the mid-ordinate method if required.

Advanced mathematics however shows that the equation for the r.m.s. value of a sinusoid is:

$$\text{r.m.s. value} = 0.707 \times \text{maximum value}$$

By convention, whenever the value of an alternating quantity is given it is the r.m.s. value that is quoted. For example, the mains voltage in the UK is 230 volts r.m.s. The maximum value that the voltage reaches is therefore, 230/0.707 = 325.32 V.

Progress check 2.2

The square voltage waveform in Figure 2.2(a) has a peak value of 1 V. Calculate its average value using the mid-ordinate method.

Sinusoidal waveform equations and phasors

Consider the diagram in Figure 2.4(a).

Here a line OA is rotating in an anticlockwise direction at a constant angular velocity of ω (pronounced omega) radians/second. After time t_1, OA has turned through an angle of ωt_1. The line AB, perpendicular to the horizontal is such that:

$$\frac{AB}{OA} = \sin \omega t_1$$

(Remember: 'sine' = opposite divided by hypotenuse.) At some later time t_2, OA has rotated to the position shown and $A'B' = OA \sin \omega t_2$. In general for any time t, the vertical component = OA sin ωt, i.e.

$$AB = OA \sin \omega t_1.$$

If all the vertical components (AB, A'B', etc.) are drawn on a graph against ωt (in radians) then a sinusoid is the resulting waveform with peak value = OA. The rotating line OA is called a **phasor** and any quantity such as voltage or current that varies sinusoidally can be represented by a rotating phasor. Thus for voltage, the vertical component would be the instantaneous voltage v and OA would be V_M. The equation for the voltage sinusoid would be:

$$v = V_M \sin \omega t$$

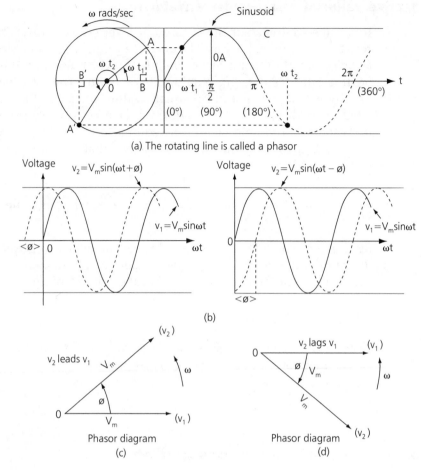

(a) The rotating line is called a phasor

(b)

Phasor diagram
(c)

Phasor diagram
(d)

Figure 2.4 How a rotating phasor represents a sinusoid

ω and the frequency f are related by the equation $\omega = 2\pi f$

or

$$f = \frac{\omega}{2\pi}$$

Thus, for example the above equation for v could be written as:

$$v = V_M \sin 2\pi ft$$

For a sinusoid of current the equation would be:

$$i = I_m \sin 2\pi f\,t$$

Phase angle

The sinuoid in Figure 2.4(a) is a true sinewave because it starts at 0°. However, a sinusoid of voltage or current may not start at 0° in all situations. Consider the two voltage waveforms in Figure 2.4(b). v_2 starts before v_1 by ϕ rads and so reaches its first peak value earlier. We say that 'v_2 **leads** v_1 **by the phase angle** ϕ'. Therefore, the equation for v_1 and v_2 are:

$$v_1 = V_M \sin \omega t$$

and $v_2 = V_M \sin (\omega t + \phi)$

Note very carefully how this is represented in the **phasor diagram** in Figure 2.4(c). v_1 is drawn as a horizontal line of length V_M. v_2 has the same length but leads v_1 by the phase angle ϕ rads.

Figure 2.4(b) shows the situation when v_2 starts after v_1. We say that 'v_2 lags v_1 by the phase angle ϕ'. The equations are now:

$$v_1 = V_M \sin \omega t$$

and $v_2 = V_M \sin (\omega t - \phi)$

v_2 lagging v_1 is represented on the phasor diagram in Figure 2.4(d). Always remember that really the lines on a phasor diagram are 'rotating anticlockwise. In summary, voltage and current sinusoids are represented by the general equations:

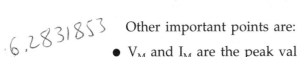

$$v = V_M \sin (\omega t \pm \phi) \quad \text{(volts)}$$

$$i = I_M \sin (\omega t \pm \phi) \quad \text{(amps)}$$

6.2831853

Other important points are:

- V_M and I_M are the peak values.
- frequency $f = \omega/2\pi$
- ϕ is the phase angle measured with respect to waveforms of i and v
- periodic time $T = 1/f = 2\pi/\omega$
- $0.707 \, I_M$ and $0.707 \, V_M$ are the r.m.s. values of the waveforms.

WORKED EXAMPLE 2.2

A sinusoidal alternating current is given by the equation i = 10 sin 942 t. Calculate:

(a) the r.m.s. value
(b) the frequency
(c) the period
(d) the instantaneous value of current 1 ms after the start of the waveform
(e) the time taken for the instantaneous value of current to reach +6 A for the first and second time.

Solution

(a) The peak value of the current, $I_M = 10$ A. Therefore:

$$I_{RMS} = 0.707 \times 10 = 7.07 \text{ A}$$

(b) $\omega = 942$, i.e. $2\pi f = 942$. Therefore:

$$f = \frac{942}{2\pi} = 150 \text{ Hz}$$

(c) $T = \dfrac{1}{f} = \dfrac{1}{150} = 6.67$ ms

(d) When t = 1 ms, i = 10 sin $(942 \times 1 \times 10^{-3})$ = 10 sin (0.942). It is important to remember that the '0.942' is in radians. You can *either* make sure the 'mode' on your calculator is set to radians so you can take the sine directly *or* change the '0.942 rads' into degrees as follows:

π rads   = 180°

1 rad   = 180°/π

0.942 rads = $(0.942 \times 180/\pi)°$ = 53.97°

Therefore,

i = 10 sin (53.97°) = 8.09 A

(e) Suppose the time taken to reach +6A is t_1 secs. Therefore,

6   = 10 sin $(942t_1)$

0.6   = sin $(942t_1)$

$942t_1$ = arc sin(0.6) = 0.644

t_1 = 0.683 ms

Remember (0.6) is in radians.

The current will next reach +6 A one cycle later i.e. at a time t_2 = (0.683 + period) = (0.683 + 6.67) = 7.353 ms.

WORKED EXAMPLE 2.3

An alternating sinusoidal voltage has a peak-to-peak value of 100 V and a periodic time of T = 2 ms. At t = 0 the instantaneous voltage v = 35.36 V. Write an equation for v in the form v = V_M sin($\omega t \pm \phi$). Express ϕ in radians and degrees.

Solution

The peak-to-peak value = $2V_M$ = 100 V; V_M = 50 V

The periodic time $T = \dfrac{2\pi}{\omega}$ so $\omega = \dfrac{2\pi}{T} = \dfrac{2\pi}{2 \times 10^{-3}} = 1000\pi$

Because v is positive at t = 0 then the phase angle ϕ is positive. Therefore, the equation for v is:

v = 50 sin $(1000\pi t + \phi)$ V

We now need to calculate ϕ. At t = 0, v = 35.36 V so:

35.36 = 50 sin $(0 + \phi)$ = 50 sin ϕ

$$\sin \phi = 35.36/50 = 0.7072$$

$$\phi = \text{arc } \sin (0.7072) = 0.7855 \text{ radians}$$

(In degrees this would be $\left(0.7855 \times \dfrac{180}{\pi}\right) = 45°$)

The final equation for v is:

$$v = 50 \sin(1000\pi t + 0.7855) \text{ V}$$

Note: The phase could be more conveniently expressed as $\pi/4$

Progress check 2.3

A sinusoidal voltage has an r.m.s. value of 28.28 V and a period of I ms. At t = 0 the instantaneous voltage v = −20 V. Find the equation for v. Express ϕ in radians and degrees.

Addition of phasors

57.29578

When d.c. voltages are connected together in series the final voltage is just the sum of the individual voltages. For example, if two 1.5 V batteries are connected in series then the final voltage is 1.5 + 1.5 = 3 V. For alternating quantities it's a little more complicated because the phase angle between them must be taken into account. The resultant of two or more phasors can be found by either drawing an accurate phasor diagram to scale or by calculation from a sketch of the phasor diagram. You should study the next two worked examples carefully.

WORKED EXAMPLE 2.4

Two sinusoidal voltages have the equations $v_1 = 40 \sin 100\pi t$ and $v_2 = 50 \sin (100\pi t - \pi/4)$. Determine by drawing a phasor diagram to scale and by calculation the equation for the resultant, $v_r = v_1 + v_2$

Solution

v_1 and v_2 have the same angular frequency but v_2 lags v_1 by $\pi/4 = 45°$. A sketch of the phasor diagram is shown in Figure 2.5(a). To draw the phasor diagram to scale we choose say 10 V to be equivalent to 1 cm as in Figure 2.5(b). V_{M1} and V_{M2} are then drawn to scale. The resultant is the diagonal of the parallelogram. By measurement, $V_R = 8.4$ cms so $V_R = 84$ V. The phase angle ϕ is about 26°, and V_r lags V_1. In radians ϕ is:

$$\phi_{rads} = \frac{\pi}{180} \times 26 = 0.45 \text{ rads}$$

The equation for v_r is:

$$v_r = 84 \sin (100\pi t - 0.45) \text{ V}$$

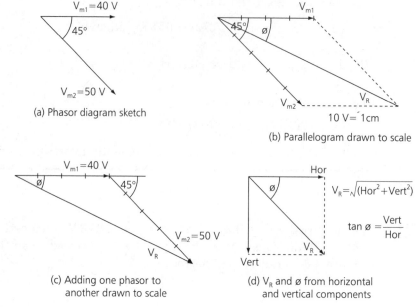

(a) Phasor diagram sketch

(b) Parallelogram drawn to scale

(c) Adding one phasor to
another drawn to scale

(d) V_R and ø from horizontal
and vertical components

Figure 2.5 Worked example 2.4

A slightly quicker method is shown in Figure 2.5(c). It saves you having to draw an accurate parallelogram. The phasor for v_2 is drawn on the end of the phasor for v_1 at the correct angle. The resultant V_R is the line that completes the triangle.

Obviously, an exact solution is obtained if V_R and ϕ are found by calculation. Probably the easiest method is to realise that:

- horizontal component of V_R = net horizontal component of v_1 and v_2
- vertical component of V_R = net vertical component of v_1 and v_2

As shown in Figure 2.5(d) once the net horizontal component (Hor) and the net vertical component (Vert) are found V_R and ϕ can be calculated. From Figure 2.5(c), the net horizontal component is:

$$\text{Hor} = V_{M1} + V_{M2} \cos 45° = 40 + 50(0.7071) = 75.355 \text{ V}$$

When finding vertical components the convention 'downwards components are negative' is used. Since V_1 does not have a vertical component,

$$\text{Vert} = V_{M2} \sin 45° = -50(0.7071) = -35.355 \text{ V}$$

Therefore,

$$V_R = \sqrt{[(75.355)^2 + (-35.355)^2]} = 83.24 \text{ V}$$

The phase angle ϕ is given by:

$$\tan \phi = \frac{-35.355}{75.355} = -25.13° = -0.439 \text{ rads}$$

You can see why the above convention is adopted. The phase angle comes out *negative* showing that v_r is *lagging* v_1. So by calculation:

$$v_r = 83.24 \sin (100\pi t - 0.439)$$

Notice that we were fairly close to the correct value by drawing the phasor diagram to scale. Obviously, the larger the scale is the more accurate we could be. In practice, a sketch of the phasor diagram followed by calculation is the preferred method since it will give an exact result.

Progress check 2.4

Determine the resultant produced by the following voltages:

- 40 sin ωt + 50 sin ωt
- 40 sin ωt + 50 sin (ωt − π/2)

WORKED EXAMPLE 2.5

Two current sinusoidal waveforms are given by

$$i_1 = 30 \sin \omega t \quad \text{and} \quad i_2 = 50 \sin (\omega t - \pi/6).$$

Determine an expression for the resultant waveform which is the *difference* between i_1 and i_2, i.e. $i_r = i_1 - i_2$

Solution

We can find $(i_1 - i_2)$ by *adding* i_1 to $(-i_2)$. The phasor for $-i_2$ is equal and opposite to that for i_2 as shown in Figure 2.6(a). From the phasor diagram you can see that i_r will lead i_1 by an angle greater than 90°.

The horizontal and vertical components are:

$$\text{Hor} = I_{M1} - I_{M2} \cos 30° = 30 - 50 \times (0.8660) = -13.301$$

(The negative sign shows that the net horizontal component is to the *left*).

$$\text{Vert} = 50 \sin 30° = 25 \text{ A}$$

Therefore,

$$I_R = \sqrt{[(-13.301)^2 + (25)^2]} = 28.318 \text{ A} \quad \text{say } 28.32 \text{ A}$$

We have got to be careful with the phase angle. As shown in Figure 2.6(b) the vertical and horizontal components allow us to find the angle θ *not* the phase angle ϕ.

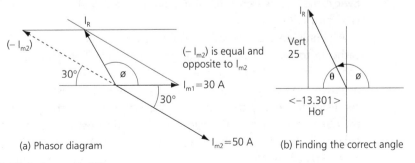

(a) Phasor diagram

(b) Finding the correct angle

Figure 2.6 Worked example 2.5

From the diagram,

$$\tan \theta = 25/13.301$$

$$\theta = 61.98° \quad \text{say } 62°$$

The phase angle $\phi = 180 - \theta = 118° = 2.06$ rads

The equation for i_r is, $i_r = 28.32 \sin(\omega t + 2.06)$ A

Using r.m.s. values and phasor diagrams

For an alternating sinusoidal waveform it is the r.m.s. value that is important rather than the peak value since it is the r.m.s. value that will be measured on ammeters and voltmeters. Consequently, phasor diagrams are often drawn with the phasors scaled to the r.m.s. value rather than the peak value. This, of course, has no effect on the *shape* of the diagram or the phase angle. It just means that the resultant will be an r.m.s. value and *not* a peak value. In this book we will always use peak values for drawing the diagrams. The r.m.s. value of the resultant is found, as usual, by multiplying the peak value by 0.707.

Do this 2.1

Basic use of an oscilloscope

Equipment and components required:

- oscilloscope (double beam)
- signal generator
- resistors 22R and 1R0.

Basic operation

One of the most versatile pieces of measurement and display equipment available is the **cathode ray oscilloscope** (CRO). The instrument allows the visual presentation and measurement of voltage, phase difference and frequency (current can also be measured and displayed by measuring the voltage produced by the current across a known resistance – see later). At the heart of the CRO is a **cathode ray tube** (CRT) upon the face of which the waveform is displayed. Even if you have never seen an oscilloscope you have certainly seen a CRT. Everytime you watch television you are seeing electrical signals displayed on the face of a CRT!

A simplified diagram of a CRT is drawn in Figure 2.7(a). The heater causes the cathode to emit negatively charged electrons. The grid has a negative voltage applied to it whilst the anodes have a positive voltage applied. Since the grid is negative with respect to the cathode only the fastest electrons can overcome the repulsive force and pass through the small hole in the grid face. Once past the grid they are accelerated to a high speed by the positive voltage on the anodes and eventually strike the tube face. The tube face is coated with phosphor which 'glows' when struck by electrons. The number hitting the face determines the brightness

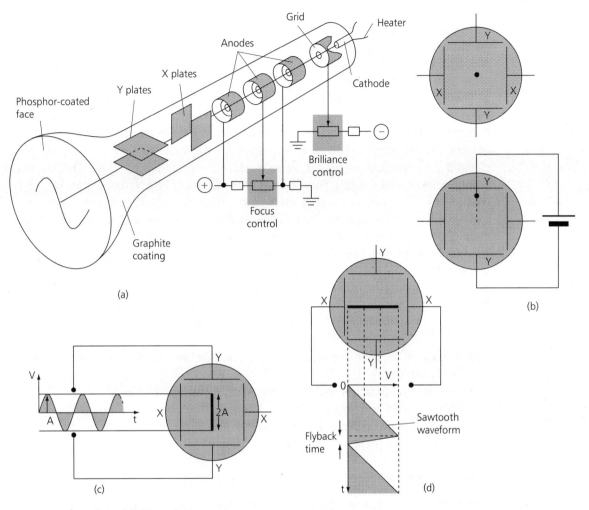

Figure 2.7 The cathode ray tube

and hence it is possible to control the **brilliance** of the glow by varying the negative voltage on the grid. After striking the face the electrons find their way back to the cathode via the graphite coating on the inside of the tube body. On leaving the grid, the beam tends to 'spread out' because of the mutual repulsion of the electrons. The anodes focus the beam of electrons as well as providing the necessary acceleration.

The beam passes between two sets of plates called the X and Y plates. When a voltage is applied to either set of plates an electric field is produced and the beam is deflected. Figure 2.7(b) shows what happens when a d.c. voltage is applied between the Y plates. As expected the beam moves towards the positive terminal with the distance moved depending on the size of the voltage. Figure 2.7(c) shows what happens when a sinusoidal voltage is connected to the Y plates with the X plates left disconnected. If the frequency is low (less than about 10 Hz) you will see the spot moving up and down. If the frequency is high then the spot will move so rapidly that a continuous line will be seen. The length of the line will be the peak-to-peak value of the voltage. Although this is a useful measurement the shape of the waveform is not fully displayed on the screen. What is necessary is

39

to apply a voltage to the X plates that moves the beam at a linear rate across the screen whilst the signal being examined simultaneously deflects the beam vertically. As shown in Figure 2.7(d) a **sawtooth voltage waveform** is used for this purpose. In a CRO this sawtooth is produced by a circuit called the **timebase**. Now examine your oscilloscope.

The tube face has a grid on it marked in cms. The timebase section controls the frequency of the internally generated sawtooth waveform. There will be a variable control (usually a red knob) and a stepped switch giving various time/cm options (μs/cm, ms/cm, s/cm). There are two inputs to the Y plates called channel 1 and channel 2 to which the voltages to be displayed are connected. (The CRT produces a double beam not a single beam). Each channel will have a variable control (red knob) and a stepped switch marked in volts/cm.

Depending on the complexity of your oscilloscope there will be various other controls. Locate the one called 'trigmode' and set this to INT (internal). This connects the internal sawtooth generator to the X plates to 'drag' the beam across the screen.

Procedure:

1 Initial setup:

● Switch on and set the intensity, focus, X-shift and Y-shift controls to about mid-position.
● Set the timebase control to somewhere in the ms/cm range.
● Wait a while and you should see a line on the tube face. If not, adjust the X and Y shift controls until you have a line in the centre of the face. Adjust intensity and focus so that the line is 'sharp'.

2 Displaying a waveform

● Connect a frequency generator to channel 1 input. Set the frequency to 250 Hz and the output amplitude control about half-way. Set the volts/cm switch on channel 1 to maximum.
● Switch on the frequency generator and adjust the volts/cm and timebase controls to display a few cycles of the sinusoid on the face. (If necessary use the trig level control to get a stationary waveform).

3 Making measurements

When making measurements you must set the variable controls on the timebase and channel 1 to the CAL (calibrated) position. Suppose, for example, that the sinusoid has a peak-to-peak deflection of 6 cms and a complete cycle of the sinusoid takes 4 cms. If the oscilloscope controls are for example 1 V/cm and 1 ms/cm then:

$$\text{peak-to-peak value} = 6\,\text{cm} \times 1\,\text{V/cm} = 6\,\text{V}$$

Therefore,

$$\text{peak value} = 3\,\text{V}$$

$$\text{period } T = 4\,\text{cm} \times 1\,\text{ms/cm} = 4\,\text{ms}$$

The frequency f is:

$$f = \frac{1}{T} = \frac{1}{4 \times 10^{-3}} = \frac{1000}{4} = 250 \text{ Hz}$$

You should experiment with different frequencies and output levels from the frequency generator. If it will also give squarewaves, display these as well.

4 Displaying current waveforms

An oscilloscope will only display voltage waveforms. Very often, however, we would need to examine the waveform of the alternating current in a circuit. The simplest way is to measure the voltage produced across a 1 Ω resistor placed in the circuit. By Ohm's law the voltage measured will be equal to the current because v = i × 1.

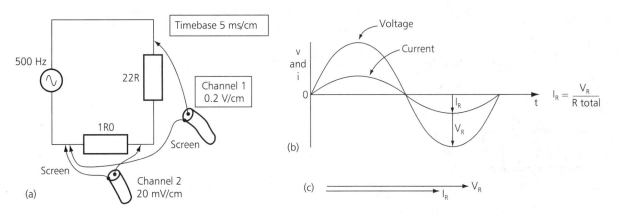

Figure 2.8 Voltage and current waveform

- Set up the circuit in Figure 2.8. Make sure the 'screen braid' on the leads are connected correctly.
- Set the timebase and 'volts/cm' controls to those shown and adjust the output of the frequency generator so that the voltage measured on channel 1 is about 0.8 V peak-to-peak. Channel 2 shows the waveform of the current through the circuit. Channel 1 waveform is the voltage across the two resistors in series.

Progress check 2.5

What can you say about the phase difference between the voltage and current waveforms in Figure 2.8?

Voltage and current in a circuit just having resistance

A circuit that just contains resistance is called a purely resistive circuit. If you answered the Progress Check correctly then you would have worked out that the voltage and current are **in phase** with one another. Consider the circuit in Figure 2.8(a). A voltage of $v_R = V_R \sin \omega t$ is connected across the total resistance R. V_R is the peak value of the voltage. By Ohm's law the current flowing:

$$i = \frac{V_R}{R} = \frac{V_R}{R} \sin \omega t = I_R \sin \omega t$$

The waveform and phasor diagrams are drawn in Figure 2.8(b) and (c).

Inductors and capacitors

Two other components are important in electrical circuits. Inductors have the property of **inductance** and capacitors have the property of **capacitance**. We need to investigate each further before we can extend our understanding of alternating currents and voltages.

Inductors

An inductor is a coil of wire wound on a former as shown in Figure 2.9(a). The former can be iron or ferrite or air. Because of its construction an inductor is often called a **coil**. The property of inductance that the coil possesses is caused by a **magnetic effect** which occurs *when the current through the coil changes.*

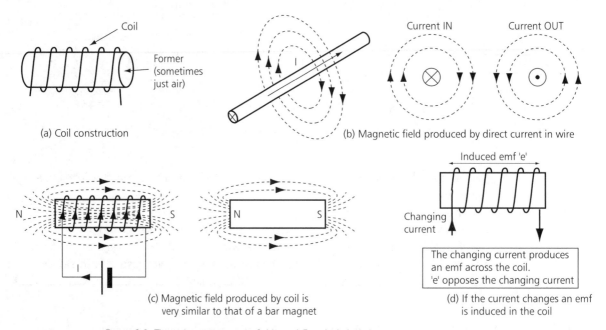

(a) Coil construction

(b) Magnetic field produced by direct current in wire

(c) Magnetic field produced by coil is very similar to that of a bar magnet

(d) If the current changes an emf is induced in the coil

Figure 2.9 The inductor, magnetic fields and Faraday's laws

The reason is as follows. When current flows through a wire a **magnetic field** is produced around the wire. Figure 2.9(b) shows the field produced when a direct current I flows through a wire. The magnetic field lines are in the shape of circles around the wire. The direction of the lines is given by gripping the wire with your *right hand with the thumb in the direction of current*. Your fingers will curl around the wire in the direction of the magnetic field. If the wire is wound into a coil then the magnetic field produced is very like that of a normal bar magnet as shown in Figure 2.9(c). To find the direction of the equivalent North Pole of the coil's field the right hand is used. Now the coil is 'gripped' so that the fingers are in the direction of the current flowing in the turns of the coil. The thumb then points in the direction of the North Pole. Study Figure 2.9(c) carefully and note that the coil is lying in a magnetic field which is produced by the d.c. current I flowing through it.

The important question to answer is: 'What happens if the current changes so that the coil is in a magnetic field which is changing?' The answer was discovered by Michael Faraday. His research was wide ranging and very important. It led to **Faraday's laws of electromagnetic induction**. Applied to our coil they can be stated as follows:

- If the magnetic field linking the coil changes, because of the changing current an e.m.f. 'e' is produced (or induced) in that coil. The coil is said to have the property of self-inductance. Self-inductance is given the symbol L. The unit of L is the **henry** (H).
- The size of the induced e.m.f. 'e' is proportional to the rate of change of the current through the coil. Mathematically we find that e = L × rate of change of current.
- A rule for finding the direction of the induced e.m.f. was given by Lenz. The direction of the induced e.m.f. is such that it tries to produce a current which *opposes the change in magnetic field that produced the e.m.f.*

These important laws are shown diagrammatically in Figure 2.9(d). We can now see what happens if a coil is used in a circuit through which alternating current is flowing. *Since an alternating current is always changing then an e.m.f. will always be induced in a coil through which that current is flowing.*

Voltage and current in a circuit just having inductance

Suppose a coil of inductance L has a voltage of $v_L = V_L \sin \omega t$ across it as shown in Figure 2.10(a). (Even though a real coil is made of wire which has resistance we will assume for the moment that the resistance is zero). A coil with no resistance is called an **ideal inductor**.

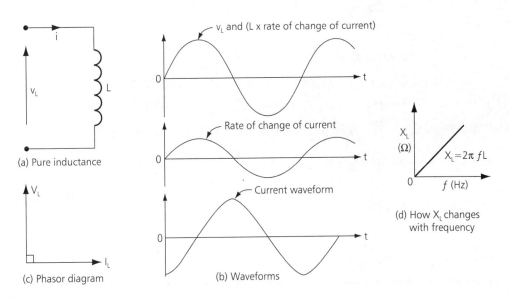

(a) Pure inductance

(c) Phasor diagram

v_L and (L x rate of change of current)

Rate of change of current

Current waveform

(b) Waveforms

$X_L = 2\pi fL$

(d) How X_L changes with frequency

Figure 2.10 Voltage and current in a circuit just having inductance

Since voltage across L = L × rate of change of current, the waveform for the *rate of change of current* will have exactly the same shape as that for v_L (See Figure 2.10(b)).

The waveform for the current can be found by the mathematical process of integration. (The technique is beyond the scope of this book). Study the waveform for the current carefully. It is sinusoidal but *lags the voltage by 90°*. The equation for the current i could be written as:

$$i = I_L \sin \left(\omega t - \frac{\pi}{2} \right)$$

The phasor diagram is drawn in Figure 2.10(c).

The opposition to the flow of current in a purely inductive circuit is called the **inductive reactance** X_L. X_L has the units of ohms.

$$X_L = \frac{V_L}{I_L} = \omega L = 2\pi fL \text{ (ohms)}$$

So

$$I_L = \frac{V_L}{2\pi fL} \quad \text{and} \quad V_L = I \times 2\pi fL$$

Notice that X_L is directly proportional to the frequency f. Thus *if the frequency is increased* while the voltage is kept constant, *the current will decrease in size* and X_L will get larger. A graph of X_L against frequency is drawn in Figure 2.10(d).

WORKED EXAMPLE 2.6

A voltage of v = 10 sin ωt is applied to a pure inductance of L = 5 mH. Determine equations for the current flowing at a frequency of (a) 100 Hz (b) 1 kHz.

Solution

(a) At f = 100 Hz, $X_L = 2\pi \times 100 \times 5 \times 10^{-3} = \pi \ \Omega$

The peak value of current, I_L, is, $I_L = 10/\pi = 3.183$ A

The equation for i is, i = 3.183 sin $(200\pi t - \pi/2)$

(b) At f = 1 kHz, $X_L = 2\pi \times 1000 \times 5 \ 10^{-3} = 10\pi \ \Omega$

The peak value of current, I_L, is, $I_L = 10/10\pi = 0.3183$ A

The equation for i is, i = 0.3183 sin $(2000\pi t - \pi/2)$

Notice that the peak value of current falls. The current always lags the voltage by 90°.

Resistance and inductance in series

A real coil or inductor of inductance L will always have some resistance because of the resistance of the wire. Thus the inductor will be equivalent to a resistance R in series with an inductance L as shown in Figure 2.11(a).

In any series circuit the current (peak value = I) is common to each component. When drawing the phasor diagram (Figure 2.11(b)), I is taken as the **reference phasor** and is drawn horizontally. Thus V_R is in phase with I and V_L leads I by 90°.

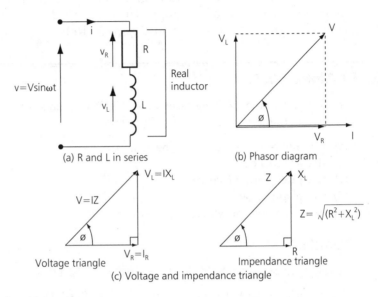

(a) R and L in series

(b) Phasor diagram

(c) Voltage and impendance triangle

Figure 2.11 R and L in series

Kirchoff's voltage law is used in much the same way as in d.c. circuits *except that the phase angles must be taken into account*. Thus the supply voltage V is the **phasor** sum of V_R and V_L.

Using Pythogoras's Theorem

$$V^2 = V_R{}^2 + V_L{}^2$$

Because

$$V_R = IR \quad \text{and} \quad V_L = IX_L$$

$$V^2 = I^2R^2 \text{ and } I^2X_L{}^2$$

$$V^2 = I^2(R^2 + X_L{}^2)$$

$$V = I \times \sqrt{(R^2 + X_L{}^2)}$$

In a.c. circuits (alternating current circuits) the ratio of the applied voltage V, to the current I is called the **impedance of the circuit**, Z. Thus:

$$Z = \sqrt{(R^2 + X_L{}^2)} = (R^2 + X_L{}^2)^{1/2} \text{ (ohms)}$$

$$V = I \times Z$$

The **phase angle** ϕ of the circuit is given by,

$$\tan \phi = \frac{V_L}{V_R} = \frac{IX_L}{IR} = \frac{X_L}{R}$$

Although the phasor diagram gives all the information required, two other diagrams are often drawn from it. They are called the **voltage triangle** and **impedance triangle** and are shown in Figure 2.11(c).

WORKED EXAMPLE 2.7

A 50 Hz alternating voltage peak value 100 V is connected across an inductor of inductance 0.1 H and resistance 12 Ω. Determine:

(a) the inductive reactance of the coil
(b) the impedance
(c) the peak value of current
(d) the phase angle
(e) equations for the voltage and current

Solution

(a) $X_L = 2\pi fL = 2\pi \times 50 \times 0.1 = 10\pi = 31.42 \ \Omega$

(b) $Z = (R^2 + X_L^2)^{1/2} = (12^2 + 31.42^2)^{1/2} = 33.63 \ \Omega$

(c) Peak value of current $I = V/Z = 100/33.63 = 2.94$ A

(d) Tan $\phi = X_L/R = 2.618$, $\phi = 69.1°$

(e) Suppose we write the voltage as $v = 100 \sin 100\pi t$. The current i lags v by angle ϕ so:

$$i = 2.97 \sin(100\pi t - 1.21).$$

(The 1.21 is 69.1° converted to radians).

Capacitance and capacitors

If two metal plates are separated by an insulator such as plastic or air then a capacitor is formed. The capacitor has the ability to store electric charge. A simple capacitor is drawn in Figure 2.12(a). The two plates have an area A and are separated by a distance d. The insulating material in between (air in the diagram) is called the **dielectric**.

Figure 2.12(c) shows what happens when a d.c. voltage V is connected across C. Electrons gradually leave the top plate and move around to the bottom plate when the switch is closed. Thus at any instant, the bottom plate is acquiring a gradually increasing negative charge of –q and the top plate has an equal and

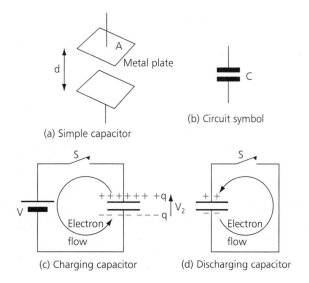

(a) Simple capacitor

(b) Circuit symbol

(c) Charging capacitor

(d) Discharging capacitor

Figure 2.12 A capacitor charging and discharging

opposite charge of $+q$. The rate of flow of charge is the current in the circuit. Because of the imbalance of charge there is a gradually increasing voltage v_c across the capacitor. The ratio of charge to voltage at any instant is the capacitance C of the capacitor.

$$C = \frac{q}{v_c}$$

Eventually, there will be no more movement of electrons and so no current and we then say that the capacitor is **fully charged**. The voltage across C at this point is equal to the battery voltage V. At this instant $C = Q/V$ where Q is the final charge on the plates.

The unit of capacitance is the **farad**, F. Components designed to have particular values of capacitance are called **capacitors**. For example, the capacitor in Figure 2.12(a) is called a parallel-plate capacitor. For this type, C depends on the area of the plates, the distance between then, and the dielectric material.

$$C = \frac{\varepsilon A}{d}$$

where ε = the permittivity of the dielectric.

Figure 2.12(d) shows what happens when a charged capacitor is **discharged**. When the switch is closed, the electrons return from the bottom plate to the top plate. While the electrons are returning there is a current flowing in the circuit. When there is no more flow of electrons, the current is zero and we say that the capacitor is **fully discharged**. (The charge and discharge of a capacitor will be investigated in more detail in Element 11.3).

If an alternating voltage is connected across a capacitor, alternating current flows in the circuit as the capacitor is continually charged and discharged. The equation $C = q/v_c$ can be written slightly differently, i.e.

$$q = C \times v_c$$

Therefore, since C is a constant because it depends on the dimensions and insulating material used, rate of change of q = C × rate of change of v_c.

But rate of change q is the circuit current and so:

current = C × rate of change of capacitor voltage.

Voltage and current in a circuit with just capacitance

Suppose a capacitor of capacitance C has an alternating voltage v across it as shown in Figure 2.13(a). C will continually charge and discharge so let the current i in the circuit be i = I_C sin ωt.

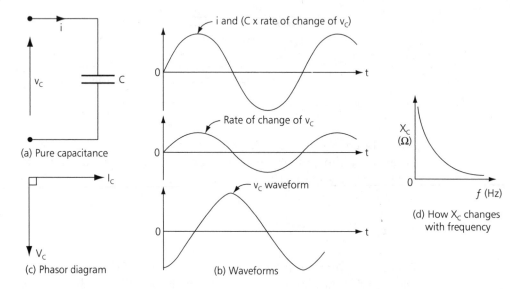

(a) Pure capacitance

(c) Phasor diagram

(b) Waveforms

(d) How X_C changes with frequency

Figure 2.13 Voltage and current in a circuit just having capacitance

Since, current = C × rate of change of capacitor voltage, the *waveform for rate of change of voltage* will have exactly the same shape as that for i (see Figure 2.13(b)). The waveform for the capacitor voltage is found by the mathematical technique of integration. Study the waveform for the voltage carefully. It is sinusoidal but *lags the current by 90°*. We can write the equation for voltage as:

$$v_c = V_C \sin\left(\omega t - \frac{\pi}{2}\right)$$

The phasor diagram is drawn in Figure 2.13(c).

The opposition to the flow of alternating current in a purely capacitive circuit is called the **capacitive reactance** X_C. X_C has the units of ohms.

$$X_C = \frac{V_C}{I_C} = \frac{1}{\omega C} = \frac{1}{2\pi f C} \text{ (ohms)}$$

so

$$I_C = 2\pi fC \times V_C \quad \text{and} \quad V_C = \frac{I_C}{2\pi fC}$$

Notice that X_C changes with frequency as shown in Figure 2.13(d). X_C gets *smaller as the frequency increases*.

WORKED EXAMPLE 2.8

A capacitor of 10 μF has an alternating voltage of peak value 10 V connected across it. Calculate the peak value of current at:

(a) f = 100 Hz.
(b) f = 10 kHz.

Write suitable equations for v_c and i in each case.

Solution

(a) At f = 100 Hz,

$$X_C = \frac{1}{2\pi \times 100 \times 10 \times 10^{-6}} = 159.155 \ \Omega$$

The peak value of current, I_C, is

$$I_C = \frac{V_C}{X_C} = \frac{10}{159.155} = 62.8 \ \text{mA}$$

Equations for i and v_c are:

$$i = 62.8 \times 10^{-3} \sin 200 \ \pi t$$

$$v_c = 10 \sin \left(200\pi t - \frac{\pi}{2}\right)$$

(b) At f = 10 kHz,

$$X_C = \frac{1}{2\pi \times 10 \times 10^3 \times 10 \times 10^{-6}} = \frac{10}{2\pi} = 1.592 \ \Omega$$

$$I_C = \frac{V_C}{X_C} = \frac{10}{1.592} = 6.28 \ \text{A}$$

Equations for i and v_c are:

$$i = 6.28 \sin 20\pi \times 10^3 t$$

$$v_c = 10 \sin \left(20\pi \times 10^3 t - \frac{\pi}{2}\right)$$

Notice that the peak value of current increases. The voltage always lags the current by 90°.

Resistance and capacitance in series

Any real capacitor will always have some resistance R. Thus the capacitor will be equivalent to a resistance R in series with a capacitance C as shown in Figure 2.14(a).

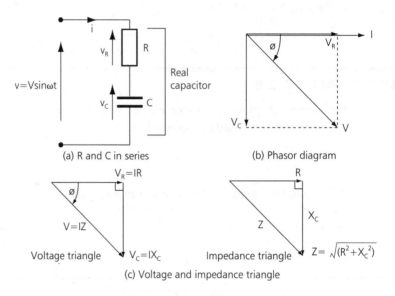

(a) R and C in series   (b) Phasor diagram

(c) Voltage and impedance triangle

Figure 2.14 R and C in series

The current is common to each component so the phasor diagram can be drawn as in Figure 2.14(b). V_R is in phase with I and V_C lags I by 90°. The supply voltage V is the phasor sum of V_R and V_C. V will lag I by angle ϕ.

By Pythagoras's theorem,

$$V^2 = V_R^2 + V_C^2$$

Because $V_R = IR$ and $V_C = IX_C$

$$V^2 = I^2(R^2 + X_C^2)$$
$$V = I \times \sqrt{(R^2 + X_C^2)}$$

The impedance Z and phase angle ϕ are:

$$Z = \sqrt{(R^2 + X_C^2)} = (R^2 + X_C^2)^{1/2} \ \text{(ohms)}$$

$$\tan \phi = \frac{V_C}{V_R} = \frac{IX_C}{IR} = \frac{X_C}{R}$$

If necessary, the **voltage and impedance triangles** can be drawn as shown in Figure 2.14(c).

WORKED EXAMPLE 2.9

A 22 μF capacitor of negligibly small resistance is connected in series with a resistance of 50 Ω. The circuit is connected across an alternating voltage of peak value 100 V at 100 Hz. Determine:

(a) the impedance
(b) peak value of current
(c) phase angle between the supply voltage and circuit current

Solution

(a) $X_c = \dfrac{1}{2\pi fC} = \dfrac{1}{2\pi \times 100 \times 22 \times 10^{-6}} = \dfrac{10^4}{44\pi} = 72.343 \; \Omega$

Impedance $Z = \sqrt{(50^2 + 72.343^2)} = 87.94 \; \Omega$

(b) Since $V = I \times Z, I = \dfrac{V}{Z} = \dfrac{100}{87.94} = 1.137 \; A$

(c) The phase angle is given by $\tan \phi = \dfrac{X_C}{R} = \dfrac{72.343}{50} = 1.447$

Therefore, $\phi = 55.35°$.

The supply voltage lags the circuit current by $55.35°$.

R, L and C in series

The phasor diagram for the circuit in Figure 2.15(a) will depend upon the relative values of X_L and X_C.

The phasor diagram is drawn as usual with I as the reference. V_R is in phase with I, V_L leads I by 90° and V_C lags I by 90°.

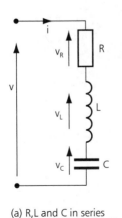

(a) R,L and C in series

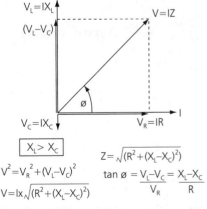

(b) $X_L > X_C$ – circuit is inductive

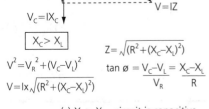

(c) $X_C > X_L$ – circuit is capacitive

Figure 2.15 R, L and C in series

In Figure 2.15(b), $V_L > V_C$ since $X_L > X_C$ (remember $V_L = IX_L$ and $V_C = IX_C$). The supply voltage V is the phasor sum of $(V_L - V_C)$ and V_R. The supply voltage V leads I by ϕ so the circuit is inductive, i.e. the circuit behaves as an inductance. If $X_C > X_L$ then V_C is greater than V_L so the phasor diagram in Figure 2.15(c) results. V will lag by ϕ and so the circuit is capacitive.

WORKED EXAMPLE 2.10

A coil of resistance $10\,\Omega$ and inductance $150\,mH$ is connected in series with a capacitor of $100\,\mu F$ and negligible resistance. The circuit is connected across a $100\,V$ supply at $50\,Hz$. Determine:

(a) X_L and X_C
(b) the impedance
(c) the current
(d) voltages across R, L and C
(e) the phase angle between the current and supply voltage.

Solution

(a) When solving problems of this type always calculate X_L and X_C first to decide which is the largest. This verifies which phasor diagram will result – either Figure 2.15(b) or (c).

$$X_L = 2\pi fL = 2\pi \times 50 \times 150 \times 10^{-3} = 47.124\,\Omega$$

$$X_C = \frac{1}{2\pi fC} = \frac{1}{2\pi \times 50 \times 100 \times 10^{-6}} = 31.83\,\Omega$$

Since $X_L > X_C$ the circuit will be inductive and V will lead I.

(b) $Z = \sqrt{(R^2 + (X_L - X_C)^2)} = \sqrt{(10^2 + (47.124 - 31.83)^2)} = 18.27\,\Omega$

(c) The peak value of supply voltage $= 100\,V$.

$$\text{Current, } I = \frac{V}{Z} = \frac{100}{18.27} = 5.473\,A \text{ (peak value)}$$

(d) $V_R = IR = 54.73\,V$

$V_L = IX_L = 257.77\,V$

$V_C = IX_C = 174.1\,V$

(e) $\phi = \arctan\left(\frac{X_L - X_C}{R}\right) = 56.82°$

Resonance in a series R, L, C circuit

Since X_L is directly proportional to f, but X_C is inversely proportional to f a frequency exists where $X_L = X_C$. On the phasor diagram in Figure 2.16(b), $V_L = V_C$ when this happens and so the supply voltage V is in phase with I. The circuit behaves electrically as if it were just a resistance R.

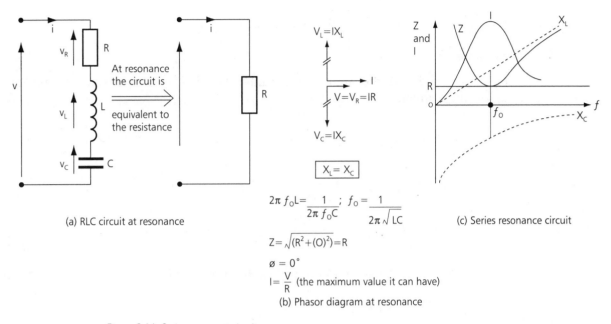

(a) RLC circuit at resonance

$$2\pi f_o L = \frac{1}{2\pi f_o C} \;;\; f_o = \frac{1}{2\pi\sqrt{LC}}$$

$$Z = \sqrt{(R^2+(O)^2)} = R$$

$$\emptyset = 0°$$

$$I = \frac{V}{R} \text{ (the maximum value it can have)}$$

(b) Phasor diagram at resonance

(c) Series resonance circuit

Figure 2.16 Series resonant circuit

The frequency at which this occurs is called the **resonant frequency** f_o and the circuit is described as being in **resonance**. Study Figure 2.16 carefully and note the following relationships at resonance:

- Since $V_L = V_C$, $X_L = X_C$ so $2\pi f_o L = 1/2\pi f_o C$

$$f_o^2 = \frac{1}{(2\pi)^2 LC}$$

$$f_o = \frac{1}{2\pi\sqrt{(LC)}} \text{ Hz}$$

- $Z = R$ and this is the minimum value the impedance can be.
- $I = V/R$ and this is the maximum value that the current can have.
- The circuit is equivalent to a single resistance of value R.
- Since I is a maximum, V_L and V_C can be very large.

Typical graphs of how I and Z change with frequency are shown in Figure 2.16(c). X_C is drawn below the axis to show that V_C and V_L are 180° out of phase with each other. Notice how I peaks and Z reaches its minimum value of R.

WORKED EXAMPLE 2.11

A variable capacitor C of negligible resistance is connected in series with a 500 mH inductance and a 10 Ω resistance. The series circuit is connected across a 100 V, 50 Hz supply and C is varied until the circuit is resonant. Determine:

(a) value of C to give resonance
(b) the circuit current at resonance
(c) the voltage across C and the 500 mH inductance.

Solution

(a) C is varied until the circuit is resonant at f_o = 50 Hz. First make C the subject of the frequency formula:

$$f_o = \frac{1}{2\pi\sqrt{(LC)}}$$

$$4\pi^2 f_o{}^2 = \frac{1}{LC}$$

$$C = \frac{1}{4\pi^2 f_o{}^2 L} = \frac{1}{4\pi^2 \times 50^2 \times 500 \times 10^{-3}} = 20.3 \ \mu F$$

(b) At resonance, Z has its minimum value of R = 10 Ω

$$I = \frac{100}{10} = 10 \ A$$

(c) The voltage across C, $V_C = I\,X_C$ so,

$$V_C = 10 \times \frac{1}{2\pi \times 50 \times 20.3 \times 10^{-6}} = 1568 \ V$$

Since $V_L = V_C$ at resonance then V_L = 1568 V

Notice how much larger V_L and V_C are compared to the supply voltage V. Touching capacitors or inductors in a series resonant circuit can be a risky business!!

Progress check 2.6

A series resonant circuit consists of a 0.1 μF capacitor, a 15 Ω resistance, and an inductor of 25 mH and negligible resistance. If the circuit is resonant with a supply of 19 V peak determine:

● the resonant frequency
● the current
● the voltage across C and L at resonance.

Q-factor of a resonant series circuit

The Worked Example 2.11 and the above Progress Check show that if R is small compared to X_L and X_C, V_L and V_C can be very much larger than the supply V. Thus there is a *voltage magnification at resonance*. A measure of the magnification is given by the Q-factor of the circuit. (Q means quality). Q is defined as:

$$Q = \frac{\text{voltage across L (or C) at resonance}}{\text{supply voltage}}$$

We can obtain three equations for Q.

$$Q = \frac{V_L}{V} = \frac{IX_L}{IR} = \frac{X_L}{R} = \frac{2\pi f_o L}{R} \quad \text{(no units)}$$

Also,

$$Q = \frac{V_C}{V} = \frac{IX_C}{IR} = \frac{X_C}{R} = \frac{1}{2\pi f_o CR} \quad \text{(no units)}$$

Alternatively, if $f_o = \dfrac{1}{2\pi\sqrt{(LC)}}$ is substituted in either equation we find

$$Q = \frac{1}{R}\sqrt{\left(\frac{L}{C}\right)} \quad \text{(no units)}$$

This final equation is useful because it gives Q directly from the component values – you do not need to know what the resonant frequency is.

Progress check 2.7	What is the Q of the circuit used in the previous Progress Check?

Power dissipation in a.c. circuits

A real circuit always has some resistance and power will be dissipated in the resistance. In d.c. circuits, the power dissipated can be calculated using the formulae given in Chapter 1. These are once again:

$$P = V \times I$$

or $\quad P = I^2 \times R$

or $\quad P = V^2/R$

In a.c. circuits, the phase angle between the voltage and current must be taken into account. The actual power dissipated in an a.c. circuit, (which is often called the **true power**) will, of course, still be dissipated in the resistive part of the circuit and can be found using either of the following formulae:

$$\text{True power} = P = I^2_{RMS}R \text{ (watts)}$$

or

$$\text{True power} = P = V_{RMS}I_{RMS} \cos \phi \text{ (watts)}$$

Notice how the phase angle is used in the second formula. The factor cos ϕ is called the *power factor of the a.c. circuit*.

Two other power definitions are also used in a.c. circuits. The product, $V_{RMS} \times I_{RMS}$ is known as the **apparent power** and has the unit of 'volt-amps' or VA. i.e.

$$\text{Apparent power} = V_{RMS} \times I_{RMS} \text{ (volt-amps)}$$

The **reactive power** is defined as:

$$\text{Reactive power} = V_{RMS} \times I_{RMS} \sin \phi \text{ (volt-amps reactive)}$$

The unit 'VA_r' is used, pronounced 'VAR', for 'volt-amp reactive'.

WORKED EXAMPLE 2.12

(a) Calculate the power dissipated in the circuit of Worked Example 2.10.
(b) How much power is dissipated in a pure inductance and a pure capacitance?

Solution

(a) The peak value of current I = 5.473 A. Therefore,

$I_{RMS} = 0.707 \times 5.473 = 3.869$ A

$P = I^2_{RMS} \times R = (3.869)^2 \times 10 = 149.69$ W

or

$V_{RMS} = 0.707 \times 100 = 70.7$ V

$P = V_{RMS} I_{RMS} \cos \phi = 70.7 \times 3.869 \times \cos 56.82° = 149.69$ W

(b) In a pure L or C, R = 0 and $\phi = 90°$

So, $P = I^2_{RMS} \times 0 = 0$ W or $P = V_{RMS} I_{RMS} \cos 90° = 0$ W since cos 90° = 0.

Progress check 2.8

Calculate the power dissipated in the resonant circuit of Worked Example 2.11.

Parallel a.c. circuits

In parallel circuits it is the *voltage which is common to each branch of the circuit.* Therefore, the *voltage is taken as the reference phasor.* Parallel R and L and R and C are shown in Figure 2.17.

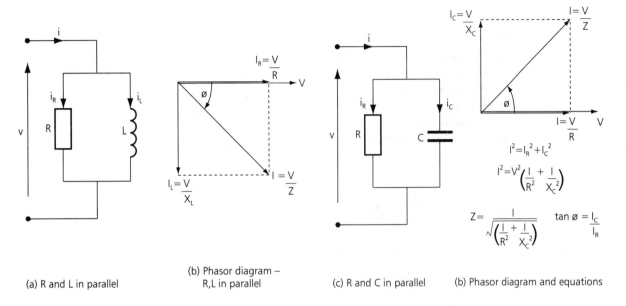

(a) R and L in parallel (b) Phasor diagram – (c) R and C in parallel (b) Phasor diagram and equations
R,L in parallel

Figure 2.17 R and L and R and C in parallel

Consider Figure 2.17(a) and (b) which show R and L in parallel. V is the reference phasor. I_R is in phase with V. I_L lags V by 90°. The supply current I is the phasor sum of I_R and I_L.

By Pythogoras's theorem,

$$I^2 = I_R^2 + I_L^2$$

But $I_R = V/R$ and $I_L = V/X_L$ so

$$I^2 = V^2 \left(\frac{1}{R^2} + \frac{1}{X_L^2} \right)$$

Taking the square root and rearranging gives

$$V = I \times \frac{1}{\sqrt{\left(\frac{1}{R^2} + \frac{1}{X_L^2} \right)}} \text{ (volts)}$$

The phase angle ϕ is given by

$$\tan \phi = \frac{I_L}{I_R} = \frac{V}{X_L} \cdot \frac{R}{V} = \frac{R}{X_L}$$

57

Figure 2.17(c) and (d) shows how the R, C parallel circuit is solved. Now I_C leads V by 90°. The equations for the impedance Z and phase angle ϕ are given in Figure 2.17(d).

WORKED EXAMPLE 2.13

A pure inductance of 500 mH is connected in parallel with a 50 Ω resistance and placed across a 100 V, 50 Hz supply. Determine:

(a) the current in each branch
(b) the supply current
(c) the impedance of the circuit
(d) the phase angle
(e) the power factor
(f) the power dissipated.

Solution

(a) $I_R = \dfrac{V}{R} = \dfrac{100}{50} = 2 \text{ A}$

$X_L = 2\pi fL = 2\pi \times 50 \times 500 \times 10^{-3} = 50\pi = 157.08 \ \Omega$

$I_L = \dfrac{V}{X_L} = 0.637 \text{ A}$

(b) $I = \sqrt{(I^2_R + I^2_L)} = \sqrt{[(2)^2 + (0.637)^2]} = 2.099 \text{ A}$

(c) $Z = \dfrac{V}{I} = \dfrac{100}{2.099} = 47.642 \ \Omega$

(d) $\tan \phi = \dfrac{I_L}{I_R} = \dfrac{0.637}{2}, \ \phi = 17.667°$

(e) Power factor $= \cos \phi = 0.953$

(f) Power is dissipated in the resistance of 50 Ω.

$P = (0.707 \times 2)^2 \times 50 = 99.97 \text{ W}$

R, L and C in parallel

Since real capacitors have a very much smaller resistance than real inductors then the circuit in Figure 2.18(a) will represent a practical inductor and capacitor in parallel. Notice that the inductive branch is now a series R, L circuit in its own right. The current flowing in this branch is labelled as i_{LR}.

Figure 2.18(b) shows the phasor diagram for each branch considered separately. The phasor diagram for the complete circuit can be found by superimposing each diagram with V as the reference phasor – see Figure 2.18(c). Study the diagram

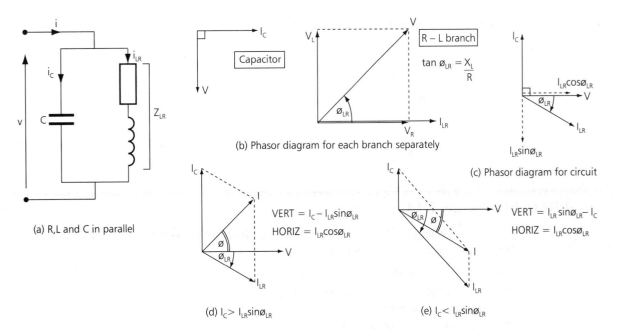

(a) R,L and C in parallel

(b) Phasor diagram for each branch separately

(c) Phasor diagram for circuit

(d) $I_C > I_{LR} \sin \phi_{LR}$

(e) $I_C < I_{LR} \sin \phi_{LR}$

Figure 2.18 R, L and C in parallel

carefully. I_{LR} has a horizontal component of $I_{LR} \cos \phi_{LR}$ and a vertical component of $I_{LR} \sin \phi_{LR}$. The supply current I is the phasor sum of I_C and I_{LR}. There are three possible conditions for the circuit depending on the relative sizes of I_C and $I_{LR} \sin \phi_{LR}$:

● If $I_C > I_{LR} \sin \phi_{LR}$ Figure 2.18(d) is the result. I leads V so the circuit is overall **capacitive**.
● If $I_C < I_{LR} \sin \phi_{LR}$, Figure 2.18(e) is the result. I lags V so the circuit is overall **inductive**.
● If $I_C = I_{LR} \sin \phi_{LR}$ the circuit is resonant and I and V are **in phase**. This will be investigated later.

Of course current I could be found by constructing a phasor diagram to scale as we did earlier in this element. It is more accurate however, to find I mathematically. The best way is to find the horizontal and vertical components of I_C and I_{LR} and noting that:

$$I = \sqrt{[(\text{Hor})^2 + (\text{Vert})^2]}$$

From the phasor diagrams in Figure 2.18(d) and (e):

$$I = \sqrt{[(I_{LR} \cos \phi_{LR})^2 + (I_C \sim I_{LR} \sin \phi_{LR})^2]}$$

where the sign '~' means 'take the difference'.
The other important equations are:

$$\text{Impedance of circuit } Z = \frac{V}{I}$$

$$\tan \phi = \frac{I_C \sim I_{LR} \sin \phi_{LR}}{I_{LR} \cos \phi_{LR}}$$

59

For the R L circuits on its own,

$$I_{LR} = \frac{V}{Z_{LR}}$$

$$Z_{LR} = \sqrt{(R^2 + X_L^2)}$$

$$\tan \phi_{LR} = \frac{X_L}{R}$$

WORKED EXAMPLE 2.14

A coil of inductance 150 mH and resistance 20 Ω is connected in parallel with a 20 μF capacitor across a 100 V, 50 Hz supply. Determine:

(a) the current in the coil and its phase angle
(b) the capacitor current and phase angle
(c) the supply current
(d) the phase angle of the circuit
(e) the impedance of the circuit
(f) the power dissipated.

Solution

(a) For the inductance $X_L = 2\pi fL = 2\pi \times 50 \times 150 \times 10^{-3} = 47.124 \, \Omega$

Impedance $Z_{LR} = \sqrt{[(20)^2 + (47.124)^2]} = 51.192 \, \Omega$

Inductor current $I_{LR} = \frac{V}{Z_{LR}} = 1.953$ A

$\tan \phi_{LR} = \frac{X_L}{R}$, $\phi = 67.003°$ lagging supply voltage

(b) Capacitance $X_C = \frac{1}{2\pi fC} = \frac{1}{2\pi \times 50 \times 20 \times 10^{-6}} = \frac{500}{\pi} = 159.155 \, \Omega$

Capacitor current $I_C = \frac{V}{X_C} = \frac{100}{159.155} = 0.628$ A

Capacitor current leads V by 90°.

(c) $I = \sqrt{[(I_{LR} \cos \phi_{LR})^2 + (I_C \sim I_{LR} \sin \phi_{LR}{}^2)]}$

$I_{LR} \cos \phi_{LR} = 1.953 \times \cos 67.003° = 0.763$ A

$I_{LR} \sin \phi_{LR} = 1.798$ A

Since $I_{LR} \sin \phi_{LR} > I_C$ the phasor diagram in Figure 2.18(e) is the result. So I will lag V.

$$I = \sqrt{[(0.763)^2 + (1.798 - 0.628)^2]} = \sqrt{[(0.763)^2 + (1.17)^2]} = 1.397 \text{ A}$$

(d) Phase angle is given by $\tan \phi = \dfrac{I_C \sim I_{LR} \sin \phi_{LR}}{I_{LR} \cos \phi_{LR}} = \dfrac{1.17}{0.763}$

Therefore $\phi = 56.89°$ (with I lagging V)

(e) Impedance $Z = \dfrac{V}{I} = \dfrac{100}{1.397} = 71.582\ \Omega$

(f) Power will be dissipated in the resistance in the inductive branch.

$$P = I^2_{LR(RMS)} \times R = (0.707 \times 1.953)2 \times 20 = 38.13\ \text{W}$$

Alternatively,

$$P = V_{RMS} I_{RMS} \cos \phi$$

$$= (0.707 \times 100) \times (0.707 \times 1.397) \cos 56.89°$$

$$= 38.14\ \text{W}$$

(The slight error is due to 'rounding')

Resonance in a parallel circuit

Resonance occurs at a frequency which makes the supply current I and supply voltage V, in phase. Thus the circuit behaves as a resistance. The phasor diagram at resonance is drawn in Figure 2.19. (The phasor diagram for the inductive branch is also drawn again).

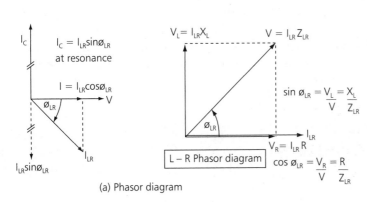

(a) Phasor diagram

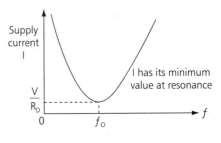

(b) Variation of supply current with frequency

Figure 2.19 Phasor diagram for parallel circuit at resonance

At resonance:

$$I_C = I_{LR} \sin \phi_{LR}$$

Therefore,

$$\frac{V}{X_C} = \frac{V}{Z_{LR}} \times \frac{X_L}{Z_{LR}}$$

Rearranging gives:

$$Z_{LR}^2 = X_C \times X_L = \frac{1}{2\pi f_o C} \times 2\pi f_o L = \frac{L}{C}$$

Now

$$Z_{LR} = \sqrt{[(R)^2 + (X_L)^2]}$$

so

$$\{\sqrt{[(R)^2 + (X_L)^2]}\}^2 = R^2 + X_L{}^2 = \frac{L}{C}$$

$$(2\pi f_o L)^2 = \frac{L}{C} - R^2$$

$$f_o = \frac{1}{2\pi L} \sqrt{\left(\frac{L}{C} - R^2\right)}$$

Taking the 'L' term inside the square root gives:

$$f_o = \frac{1}{2\pi} \sqrt{\left(\frac{1}{LC} - \frac{R^2}{L^2}\right)} \quad \text{(hertz)}$$

Notice that the equation is slightly more complicated than that for series resonance. (If R = 0, however, they are exactly the same).

Value of supply current at resonance

If you compare the phasor diagram in Figure 2.18 and the phasor diagram for resonance, you can see that *the supply current I has its smallest value at resonance*. This means that the impedance of the circuit is just a resistance value because V and I are in phase, and must have its maximum value at resonance. From the phasor diagram, in Figure 2.19:

$$I = I_{LR} \cos \phi_{LR} = \frac{V}{Z_{LR}} \times \frac{R}{Z_{LR}} = \frac{V \times R}{Z_{LR}^2}$$

However, from above $Z_{LR}{}^2 = L/C$ so

$$I = \frac{V\,RC}{L}$$

Dynamic resistance

The impedance of the circuit of resonance is V/I and is called the dynamic resistance, R_D.

$$R_D = \frac{V}{I} = \frac{L}{RC} \quad \text{(ohms)}$$

The variation of supply current with frequency is shown in Figure 2.19(b). I reaches its minimum value at f_o. Just because the supply current is a minimum does not mean that the current circulating in the parallel circuit is also a minimum. In fact this circulating current can be very much larger than I. The reason is that when C is discharging and losing its energy, L is storing that energy and vice versa. This interchange of energy occurs at the resonant frequency f_o. The small supply current is drawn to allow for the power loss in the circuit due to the resistance R.

Q-factor of a parallel resonant circuit

The Q-factor is a measure of the size of the circulating current compared to the supply current.

$$Q = \frac{\text{circulating current}}{\text{supply current}}$$

The circulating current is I_C which is also equal to $I_{LR} \sin \phi_{LR}$, so:

$$Q = \frac{I_{LR} \sin \phi_{LR}}{I} \quad \text{or} \quad Q = \frac{I_C}{I}$$

From the phasor diagram, $I = I_{LR} \cos \phi_{LR}$ so using the first equation:

$$Q = \frac{I_{LR} \sin \phi_{LR}}{I_{LR} \cos \phi_{LR}} = \tan \phi_{LR}$$

But

$$\tan \phi_{LR} = \frac{I_{LR} X_L}{I_{LR} R}$$

$$Q = \frac{X_L}{R} = \frac{2\pi f_o L}{R} \quad \text{(no units)}$$

This is the same formula as for the series circuit. Remember however that in the series circuit, Q measures the voltage magnification. In the parallel circuit it is a measure of the **current magnification**.

WORKED EXAMPLE 2.15

A parallel circuit with R = 5 Ω, L = 585 μH and C = 120 pF is connected across a 100 V variable frequency supply. The supply frequency is varied until the circuit is resonant. Determine:

(a) the resonant frequency
(b) dynamic resistance
(c) supply current
(d) the Q-factor
(e) the circulating current.

Solution

(a) $f_o = \dfrac{1}{2\pi} \sqrt{\left(\dfrac{1}{LC} - \dfrac{R^2}{L^2} \right)}$

$= \dfrac{1}{2\pi} \sqrt{\left(\dfrac{1}{585 \times 10^{-6} \times 120 \times 10^{-12}} - \dfrac{5^2}{(585 \times 10^{-6})^2} \right)}$

$= \dfrac{\sqrt{1.425 \times 10^{13} - 7.305}}{2\pi}$

Notice that the second term in the bracket is so much smaller than the first that we can ignore it. Effectively we can use the formula for series resonance.

Therefore, $f_o = 600.8$ kHz

(b) $R_D = \dfrac{L}{CR} = \dfrac{585 \times 10^{-6}}{120 \times 10^{-12} \times 5} = \dfrac{585 \times 10^{6}}{120 \times 5} = 975 \text{ k}\Omega$

(c) $I = \dfrac{V}{R_D} = \dfrac{100}{975 \times 10^{3}} = 102.56 \text{ } \mu A$

(d) $Q = \dfrac{2\pi f_o L}{R} = 2\pi \times 600.8 \times 10^{3} \times 585 \times 10^{-6} = 441.67$

(e) The circulating current is the capacitor current I_C, and $Q = \dfrac{I_C}{I}$

Therefore, $I_C = Q \times I = 441.67 \times 102.56 \times 10^{-6} = 45.3 \text{ mA}$

Notice in this example how high R_D is and hence how small the supply current is at resonance. The Q of the circuit is high so the circulating current is also high compared to the supply current.

Progress check 2.9

For a parallel resonant circuit, R = 15 Ω, L = 25 mH and C = 0.1 μF. If the circuit is placed across a 100 V supply and brought to resonance determine:

● the approximate resonant frequency
● dynamic resistance
● supply current
● Q-factor
● circulating current.

Assignment 2

This assignment provides evidence for:
Element 11.2: Investigate the response of series and parallel curcuits to sinusoidal inputs
and the following key skills:
Communication 3.2
Information Technology 3.1, 3.2, 3.3

The evidence indicators for this element require you to:

1   provide a description of sinusoidal quantities, and
2   analyse and measure a series and a parellel circuit.

The circuits given below are suitable series and parallel circuits.

Equipment and components required:

- Signal generator
- Dual channel oscilloscope
- Digital multimeter (if possible)

- Resistors 1K0, 100R0, 1R0
- Inductor 100 mH
- Capacitor 0.1 μF.

Series resonance Construct a series circuit with R = 1 kΩ, L = 100 mH and C = 0.1 μF. Connect across generator and set output to say, 3 V peak-to-peak. *Maintain this level throughout the measurement.* Vary the frequency from 500 Hz to 3 kHz in 100 Hz steps. At each frequency, measure the supply current and calculate the circuit impedance Z = V/I. Plot graph of supply current and impedance versus frequency. Estimate the resonant frequency of the circuit. Sketch waveform diagrams of supply voltage and current at 1 kHz, 3 kHz and at resonance. Decide whether the circuit is inductive or capacitive at 1 kHz and 3 kHz.

Parallel resonance Construct a parallel circuit with R = 100 Ω, L = 100 mH and C = 0.1 μF. Perform the same measurements as for the series circuit and produce waveform sketches at 1 kHz, 3 kHz and at resonance.

Calculations for each circuit Calculate f_o. Sketch phasor diagrams at 1 kHz, 3 kHz. Decide whether each circuit is inductive or capacitive at these frequencies.

Report

Include in your report a description of all your measurements and accurately drawn graphs. Compare your measured results with those calculated.

Hints and suggestions
You can measure the supply current by including an ammeter or by 'scoping' the voltage across a 1 Ω resistor placed in series with each circuit.

Chapter 3: The response of series circuits to step voltage inputs

This chapter covers:
Element 11.3: Investigate the response of series circuits to a step voltage input.

. . . and is divided into the following sections:
- Charging a capacitor
- Discharging a capacitor
- Growth of current in an RL circuit
- Decay of current in an RL circuit.

When a d.c. voltage source, such as a battery, is connected to a circuit just containing resistance, the current immediately reaches its final value and remains at that value for as long as the battery is connected. When the battery is disconnected the current immediately drops to zero. With circuits containing capacitors and/or inductors this is not the case. When the battery is connected the current will gradually rise to its final value. The current will gradually fall to zero when the battery is removed. The period of time that elapses while changes occur in the circuit is called the **transient period**. When the transient period is complete the circuit is said to be in its **steady-state condition** .

In this chapter the transient behaviour of circuits containing capacitors and inductors will be described.

When you complete this chapter you should be able to:

- define the time constant of series circuits
- describe the transient response of series circuits to a step voltage
- state equations to describe d.c. transient response of series circuits
- describe the effect of the circuit time constant on the d.c. transient response of series circuits
- calculate circuit quantities for the transient state.

Charging a capacitor

The circuit diagram of a capacitor C connected in series with a resistor R is drawn in Figure 3.1(a). We will assume that C is initially uncharged and investigate what

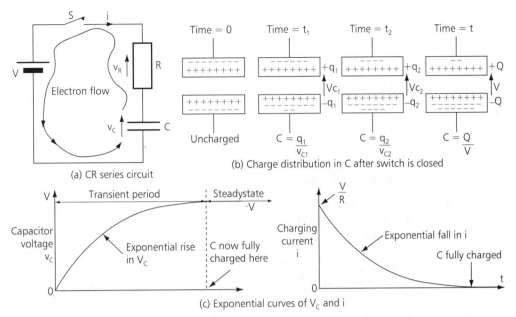

(a) CR series circuit

(b) Charge distribution in C after switch is closed

(c) Exponential curves of V_c and i

Figure 3.1 Charging a capacitor

happens when the switch S is closed at time t = 0 so that the d.c. voltage V is connected across the series R-C circuit.

The charging process was briefly described in Element 11.2. A more in-depth study will now be given. C consists of two metal conductors separated by an insulating material. The atomic structure of metals is such that when the atoms combine to form the solid material the outermost electrons become free to move. The electrically positive cores of the atoms are fixed however and cannot move. The metal conductors are electrically neutral because the number of positive cores is equal to the number of negative electrons. This is the situation shown at t = 0, just before the switch is closed in Figure 3.1(b). As soon as S is closed, the free electrons in the top plate will feel the attractive force of the positive terminal of the battery and begin to move through the circuit. The electrons move from the top plate, flow through the circuit, and collect on the bottom plate. At time t_1 the plates have a charge of $\pm q_1$ and there is a voltage v_{c1} across the capacitor. This initial **inrush of charge** when S is closed will give the maximum circuit current because the electrons will feel the full attractive force from the positive terminal of the battery. As time goes on electrons continue to leave the top plate but the net attractive force on them is less because of the positive charge on the top plate which is 'trying to hold them back'. Thus the rate of flow of charge (i.e. the current) gradually decreases and the rate of increase of capacitor voltage also decreases. Eventually at some time t, no more electrons leave the top plate and the current is zero. We say that *'the capacitor is fully charged'*. The final voltage across C will be the battery voltage V and its plates will be charged to $\pm Q$.

At any instant, the ratio of the capacitor charge to voltage is a constant. This constant value is the capacitance of C. The sequence of events is shown pictorially in Figure 3.1(b). We can also analyse the charging process mathematically, because Kirchoff's voltage law must be obeyed when the switch is closed, i.e. $V = v_c + v_R$. Since $v_R = i \times R$, where i = instantaneous current then:

$$V = v_c + iR$$

Let us examine this equation at various times:

1 Just at switch on, $t = 0$ and $v_c = 0$ so $V = i_o R$. The initial current, $i_o = V/R$ is the maximum current that flows in the circuit.

2 At $t = t_1$, $V = v_{c1} + i_1 R$.

3 At $t = t_2$, $V = v_{c2} + i_2 R$. Since v_{c2} is bigger than v_{c1} then i_2 must be less than i_1. As the capacitor voltage grows the current falls because the rate of flow of charge is decreasing.

4 When the capacitor is fully charged $i = 0$ and so $V = V_c$.

Graphs of v_c and i versus time are shown in Figure 3.1(c). Each curve is called an **exponential curve**. Notice:

- The maximum current is at $t = 0$.
- The rate of rise of v_c and rate of fall of i gradually decreases with time in the transient period.

Transient equation

It is possible to find equations which describe the transient curves. They are:

$$v_c = V(1 - e^{-t/CR}) \quad \text{(volts)}$$

$$i = \frac{V}{R} e^{-t/CR} \quad \text{(amps)}$$

You should remember these equations and be able to use them to solve problems. Notice the factor $C \times R$ which appears in each equation. This is called the **time constant** of the circuit and has the units of seconds. We'll see later how the time constant determines how long it takes C to charge. From these equations two other important ones can be derived:

$$\text{Initial rate of rise of } v_c = \frac{V}{CR} \quad \text{(V/sec)}$$

$$\text{Initial rate of fall of } i = \frac{V}{CR^2} \quad \text{(A/sec)}$$

WORKED EXAMPLE 3.1

A capacitor of $C = 0.25\ \mu F$ is connected in series with a resistor of $R = 10\ k\Omega$ and switched across a 100 V d.c. supply. Determine:

(a) the circuit time constant
(b) the capacitor voltage after a time equal to one time constant
(c) the voltage after 12.5 ms
(d) the initial charging current
(e) the current after 5 ms
(f) the voltage across R and C after 5 ms.

Solution

(a) Time constant = $C \times R = 0.25 \times 10^{-6} \times 10 \times 10^3 = 2.5$ ms

(b) $v_c = V(1 - e^{-t/CR}) = 100(1 - e^{-2.5 \times 10^{-3}/2.5 \times 10^{-3}}) = 100(1 - e^{-1})$

$v_c = 100(1 - 1/e) = 100(1 - 0.368) = 63.2$ V

Now 63.2 V is 63.2% of the final voltage. This gives us a definition of time constant.

> *The time constant of a CR circuit is the time it takes the capacitor voltage to reach 63.2% of its final value.*

(c) $v_c = 100(1 - e^{-12.5 \times 10^{-3}/2.5 \times 10^{-3}}) = 100(1 - e^{-5}) = 100(1 - 1/e^5) = 99.33$ V

This is an important result because for all practical purposes C is fully charged since it has reached over 99% of its final value of 100 V. Thus the rule for capacitor charging is

> *A capacitor can be considered to be fully charged after a time of 5 time constants.*

(d) $i\,(t = 0) = \dfrac{V}{R} e^{-0/CR} = \dfrac{V}{R} \cdot \dfrac{1}{e^0} = \dfrac{V}{R} = \dfrac{100}{10 \times 10^3} = 10$ mA

(e) $i = \dfrac{V}{R} e^{-5 \times 10^{-3}/2.5 \times 10^{-3}} = \dfrac{V}{R} e^{-2} = \dfrac{100 \times e^{-2}}{100 \times 10^3} = 1.353$ mA

(f) Always $v_R = i \times R$ so $v_R = 1.353 \times 10^{-3} = 13.53$ V

We could calculate v_c by using the exponential equation but because we have v_R it is easy to determine v_c because Kirchoff's law always holds:

$100 = v_R + v_c$

$v_c = 100 - 13.53 = 86.47$ V

Time taken to charge a capacitor

The rule stated in Worked Example 3.1 Part (c) is very important and can be proved quite easily. We need to determine the time required for v_c to reach 99% of V. Let the time taken be t_1, so:

$0.99\,V = V(1 - e^{-t_1/CR})$

$e^{-t_1/CR} = 1 - 0.99 = 0.01$

Therefore

$e^{+t_1/CR} = \dfrac{1}{0.01}$

$\dfrac{t_1}{CR} = 4.61$

$t_1 = 4.61 \times CR$

We normally round 4.61 up to 5 so a capacitor can be considered to be fully charged after 5 time constants.

**Progress
check 3.1**

A 10 μF capacitor is connected in series with a 40 kΩ resistor and placed across a 100 V d.c. supply. Determine:

● the initial charging current
● initial rate of rise of capacitor voltage
● initial rate of fall of current
● the time taken for C to reach 40% of its final voltage

Discharging a capacitor

Suppose a capacitor C, charged to a voltage V, is switched in series with a resistor R as shown in Figure 3.2.

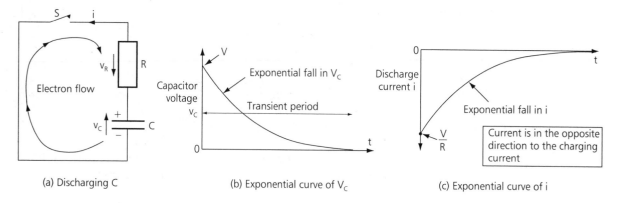

(a) Discharging C   (b) Exponential curve of V_C (c) Exponential curve of i

Figure 3.2 Discharging a capacitor

When S is closed, the process that occurs is very similar to that which took place when C was charged except that free electrons will return from the bottom plate to the top plate. The electrons return because they sense the attractive force of the positive charge in the top plate. As time goes on, the rate of flow of the electrons (i.e. the current) decreases because the attractive force gets less. The voltage across C gradually falls until, when all the electrons are transferred, the voltage across C is zero.

The current i which flows is in the *opposite* direction to the charging current. Kirchoff's law still applies and v_c is now always *equal and opposite* to $v_R = i \times R$.

1 At t = 0, v_c = V = $i_0 \times$ R. The initial current i_0 is the maximum value of discharge current. The voltage across R is V.
2 When fully discharged, v_c = 0 = 0 × R i.e. i = 0.

As shown in Figure 3.2(b) and (c), both v_c and i fall exponentially. Notice that i is drawn negative just to indicate that it is in the *opposite* direction to the charging current.

The equations for the transient curves are:

$$v_c = V e^{-t/CR} \text{ (volts)}$$

$$i = \frac{V}{R} e^{-t/CR} \text{ (amps)}$$

Initial rate of fall of $v_c = \dfrac{V}{CR}$ (V/sec)

Initial rate of fall of $i = \dfrac{V}{CR^2}$ (A/sec)

Time taken to discharge a capacitor

How long will it take to discharge a charged capacitor? We suspect that you have probably guessed the answer already. C is considered to be fully discharged when its voltage has fallen to 1% of the initial voltage. Suppose the time taken is t_1 seconds. Using the equation for v_c above gives:

$$0.01 \times V = V e^{-t_1/CR}$$

$$e^{t_1/CR} = \frac{1}{0.01} = 100$$

so $\qquad t_1 = 4.61 \times CR$

This rounds to 5 so a capacitor can be considered to be fully discharged after 5 time constants.

You should remember that capacitors can retain their charge for a very long time after the supply has been removed. It can therefore be dangerous to work with circuits containing capacitors even though the supply may be switched off. Sometimes the circuit is designed so that when the supply is switched off a suitable resistor is automatically switched across the capacitor to safely discharge it. When in doubt, note the capacitor value and connect a high value resistor across. Wait for at least five time constants so that C is discharged. NEVER short the terminals together with a pierce of wire! The current can be so large that it would melt the wire.

Summary of main findings for C-R circuits

- A capacitor takes 5 time constants to charge and discharge.
- In a d.c. circuit, once charged no more current will flow in that circuit. The capacitor prevents or blocks d.c. current flow. For this reason capacitors are often called **d.c. blockers or blocking capacitors**.
- If the d.c. voltage changes, current will flow while C adjusts its charge. Then no more current flows.
- As you discovered in Chapter 2, if a sinusoidal alternating voltage is connected across C, it continually goes through a charge/discharge cycle and current

always flows in the circuit. The reactance $X_C = 1/2\pi fC$ is the impedance to this current flow. Capacitors are often called **a.c. couplers** because they allow alternating current to flow in the circuit.

Electrical energy stored in a capacitor

When charged to a voltage V, electrical energy U is stored in C. The quantity of energy is given by:

$$U = \tfrac{1}{2} CV^2 \text{ joules}$$

WORKED EXAMPLE 3.2

A capacitor of C = 12 μF is charged to 10 V through a resistor of R = 1 kΩ. The capacitor is then quickly removed and connected across a resistor of R = 1 MΩ. Determine:

(a) the approximate time taken for C to charge
(b) the time taken for the voltage to fall to 5 V from the beginning of discharge
(c) the current flowing at a time of 6 s from the beginning of discharge
(d) the approximate time for C to discharge
(e) the energy given to the circuit as a result of discharging.

Solution

(a) Charging time constant $= C \times R = 12 \times 10^{-6} \times 1 \times 10^3 = 12$ ms. It will take approximately $5 \times 12 = 60$ ms to fully charge.

(b) The discharge time constant $= C \times R = 12 \times 10^{-6} \times 1 \times 10^6 = 12$ s. Suppose the time taken to fall to 5 V is t_1. Using the discharge equation $v_C = Ve^{-t/CR}$ gives:

$$5 = 10\, e^{-t_1/12} \quad \text{so} \quad e^{-t_1/12} = 0.5$$

Therefore

$$e^{+t_1/12} = \frac{1}{0.5} = 2 \quad \text{so} \quad \frac{t_1}{12} = 0.6931 \quad \text{and} \quad t_1 = 8.32 \text{ s}$$

(c) $i = \dfrac{V}{R} e^{-t/CR} = \dfrac{10}{1 \times 10^6} e^{-6/12} = 10\, e^{-0.5} \times 10^{-6} = 6.07 \text{ μA}$

(d) C will take 5 discharge time constants to discharge, so time taken is

$$5 \times 12 = 60 \text{ s}$$

(e) When charged the energy stored $U = \tfrac{1}{2} CV^2$

$$U = \tfrac{1}{2} \times 12 \times 10^{-6} \times 5^2 = 150 \text{ μjoules}$$

This amount of energy is given to the circuit when C discharges.

Progress check 3.2

A 10 μF capacitor is charged to 100 V and then connected across a 50 kΩ resistor. Determine:

● the initial discharge current
● the capacitor voltage after a time equal to one time constant
● the current two seconds after the beginning of discharge

Do this 3.1

Plotting current and voltage in a circuit with capacitance

This practical activity will form part of the assignment at the end of this element. The aim is to plot graphs of v_C and i for a capacitor being charged from a d.c. supply. If you can work with two colleagues it'll make the measurement easier.

Equipment and components required:

● power supply unit (0–25 V) with switch
● ammeter, voltmeter or two digital multimeters (DMMs)
● resistor: 4K7; electrolytic capacitor: 4700 μF (25 V working)
● watch or clock.

Procedure:

1 Set up the circuit in Figure 3.3. Set the power supply to, say, 10 V.

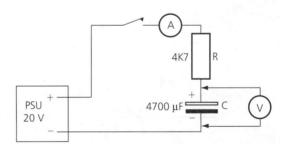

Figure 3.3 Do this 3.1 circuit diagram

The capacitor will be electrolytic so make sure you connect it to the correct polarity.

2 The time constant of the circuit is $4700 \times 10^{-6} \times 4700 = 22.09$ s. Let's say 22 s. You will have plenty of time to watch the charging process. Switch on and at 10 s intervals note the current i and capacitor voltage v_C. (Obviously working with two colleagues will help here). Continue for say 150 s.

3 Plot accurate graphs of i and v_C against time.

Do this 3.2

The effect of a CR circuit on a square wave

If a squarewave is connected to a CR circuit then the shape of the output will be affected by the time constant of the circuit. You can see the effect with an oscilloscope.

Equipment and components required:

- Dual beam oscilloscope
- Signal generator with squarewave option
- Resistor: IK0.
- Capacitors: 10 nF, 100 nF, 1 μF.

Procedure:

1 Set up the circuit in Figure 3.4(a). Set the squarewave generator frequency to f = 1 kHz and a convenient peak-to-peak value. The period T = 1/f = 1 ms, so the pulse width t_p is 0.5 ms.
2 For each value of C sketch the waveform of v_C. The first one is done for you for C = 10 nF. In this case CR = 0.01 ms. Since C will take 5 time constants to charge and discharge, it has plenty of time to do this during the lifetime of the pulse t_p = 0.5 ms. The voltage across C is almost a squarewave.

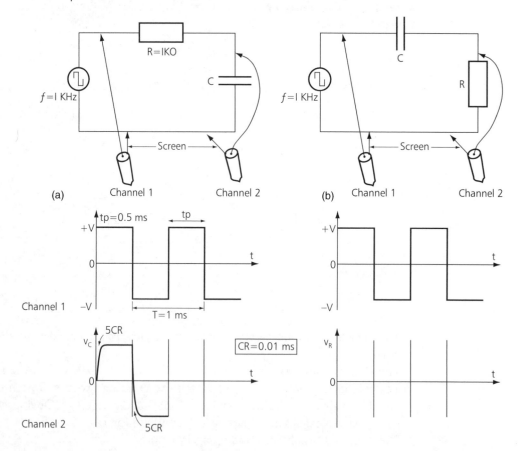

Figure 3.4 Effect of CR circuit on a squarewave

3 Change the circuit to that in Figure 3.4(b) so that now you are measuring the voltage across R. Sketch v_R for the three values of time constant.

Activity 3.1

Produce a short report detailing the findings in Do This 3.2. Give your reasoning to explain the shape of v_C and v_R for each time constant value.

Growth of current in an RL circuit

In Element 11.2 Faraday and Lenz's laws were described when we investigated the inductance of a coil. When the current through a coil changes an e.m.f. is induced which opposes the change in current. You should expect therefore that if a voltage is suddenly connected across a circuit containing resistance and inductance (RL), the current will not immediately reach its final value but will take a certain time to establish itself.

Consider the R-L circuit in Figure 3.5(a). In a real circuit R *would be the sum of the circuit resistance and the inductor resistance.*

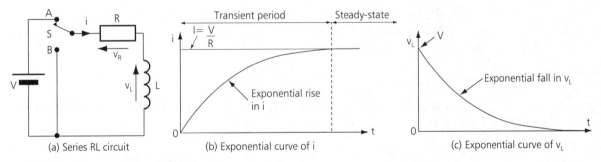

(a) Series RL circuit   (b) Exponential curve of i   (c) Exponential curve of v_L

Figure 3.5 Connecting a d.c. voltage across an RL circuit

Suppose the switch S is closed to terminal A at t = 0. Kirchoff's voltage law will always be obeyed so, $V = v_R + v_L$, where v_L is the induced e.m.f. across L and is given by:

$v_L = L \times$ rate of change of current.

Mathematically we show the rate of change of current as $\dfrac{di}{dt}$ so

$$v_L = L \times \frac{di}{dt}$$

Thus Kirchoff's voltage law can be written as:

$$V = (i \times R) + \left(L \times \frac{di}{dt}\right)$$

Let us examine the equation from the time that the switch is just closed at $t = 0$.

1 At switch-on $t = 0$, the current $i = 0$, but its rate of change is *not* zero because the current begins to rise. The e.m.f. induced in the coil is equal and opposite to V, i.e.

$$V = 0 + v_L = v_L$$

2 A little time later (t_1), a current i_1 is flowing so:

$$V = i_1 \times R + L \times \frac{di_1}{dt}$$

The rate of change of current

$$\left(\frac{di_1}{dt}\right)$$

must now be less than it was at $t = 0$ because the supply V does not change.

3 At $t = t_2$, the current increases to i_2 and so:

$$V = i_2 \times R + L \times \frac{di_2}{dt}$$

The rate of change of i_2 must be less than it was at t_1.

4 Eventually, the current reaches its final steady-state value and the transient period is complete. There is no voltage across L because the current does not vary and

$$V = I \times R$$

The growth of current and the decay of v_L is shown in Figure 3.5(b). The curves are exponential and are given by the equations:

$$i = \frac{V}{R}(1 - e^{-tR/L}) \text{ (amps)}$$

$$v_L = V e^{-tR/L} \text{ (volts)}$$

They are very similar to the equations for v_C and i when a capacitor is charged. The following points should be remembered:

● The ratio L/R is called the **time constant** of the RL circuit.
● The time constant is the time taken for the current to reach 63.2% of its final value when establishing itself.
● It will take about 5 time constants for i to reach its final value.
● Initial rate of rise of i = V/L (A/s)
● Initial rate of fall of v_L = VR/L (V/s)

Decay of current in an RL circuit

If in Figure 3.5(a) with current i flowing, the switch is moved to position B, the current begins to decay to zero. *It does not fall immediately to zero however because of the effects of Faraday and Lenz's laws.* As i begins to decrease, an e.m.f. is induced in L (= L × di/dt) which tries to keep i flowing. By Kirchoff's voltage law, v_L will be equal and opposite to $v_R = i \times R$ as the current decays. The decay of the current and $v_L = v_R$ is exponential and is given by the equations:

$$i = \frac{V}{R} e^{-tR/L} \text{ (amps)}$$

$$v_L = v_R = V e^{-tR/L} \text{ (volts)}$$

The decay of current is exponential. It takes about 5 time constants.

Electrical energy stored in an inductor

When current flows in an R-L circuit energy is stored in the inductance L. The quantity of energy is given by:

$$U = \tfrac{1}{2} LI^2 \text{ (joules)}$$

WORKED EXAMPLE 3.3

An inductor of inductance 2 H and resistance 10 Ω is connected to a 20 V d.c. supply. Determine:

(a) the time constant
(b) the initial rate of rise of i
(c) the current after 30 ms
(d) the current after a time equal to 5 time constants
(e) the final energy stored.

Solution

(a) Time constant $= \dfrac{L}{R} = \dfrac{2}{10} = 0.2$ s

(b) Initial rate of rise of i $= \dfrac{V}{L} = \dfrac{20}{2} = 10$ A/s

(c) $i = \dfrac{V}{R}(1 - e^{-tR/L}) = \dfrac{20}{12}(1 - e^{-0.3/0.2}) = 1.554$ A

(d) 5 time constants = 5 × 0.2 = 1 s

$$i = \frac{20}{10}(1 - e^{-1/0.2}) = 2(1 - e^{-5}) = 1.987 \text{ A}$$

For all practical purposes this is equal to the final current of 20/10 = 2 A

(e) $U = \frac{1}{2}LI^2 = \frac{1}{2} \times 2 \times 2^2 = 4$ joules

WORKED EXAMPLE 3.4

A single voltage pulse of peak value 10 V and duration 10 ms is connected to an inductor of L = 100 mH and R = 100 Ω. Determine the time taken for the current in the circuit to:

(a) rise to 50 mA
(b) fall to 50 mA

Solution

(a) Time constant $= \dfrac{L}{R} = \dfrac{0.1}{100} = 1$ ms.

Suppose the time taken to reach 50 mA is t_1. Then:

$$i = \frac{V}{R}(1 - e^{-tR/L})$$

$$50 \times 10^{-3} = \frac{10}{100}(1 - e^{-t_1/1\times10^{-3}})$$

$$500 \times 10^{-3} = 1 - e^{-t_1/1\times10^{-3}}$$

$$e^{+t_1/1\times10^{-3}} = 2$$

$$t_1 = 0.693 \text{ ms}$$

The current will continue to increase until, after about 5 time constants = 5 ms, it reaches its final value of 100 mA.

(b) When the pulse ends, the current will begin to fall to zero. Let t_2 be the time taken from the *end of the pulse* for i to fall to 50 mA. Then:

$$i = \frac{V}{R}e^{-tR/L}$$

$$50 \times 10^{-3} = \frac{10}{100}e^{-t_2/1\times10^{-3}}$$

$$0.5 = e^{-t_2/1\times10^{-3}}$$

$$t_2 = 0.693 \text{ ms} \quad \text{(from end of pulse)}$$

Because the pulse is 10 ms wide the current is 50 mA after 0.693 ms and 10 + 0.693 = 10.693 ms. A waveform sketch is drawn in Figure 3.6.

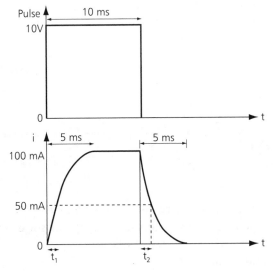

Figure 3.6 Worked example 3.4

Activity 3.2

Design and report on a procedure similar to that in Do This 3.2 to show the effect of a squarewave on an RL circuit. Choose suitable inductors and resistors and remember to take the inductor resistance into account when calculating the total resistance.

Summary of main findings for RL circuits

- It takes 5 time constants for the current to grow to its final value and decay to zero.
- Once the current has grown to its final value it is the resistance of the circuit which limits the current.
- If the current changes in any way, an e.m.f. is induced in L which tries to prevent the changes. For this reason inductors are often called **chokes** because they try to prevent or 'choke-off' changes in currents.
- As you discovered in Chapter 2, if a sinusoidal alternating voltage is connected across L, an e.m.f. is always induced in L. The reactance $X_L = 2\pi fL$ is the 'hinderance' to a.c. current flow.

Progress check 3.3

An inductor of R = 4 Ω and L = 2 H is switched across a 20 V d.c. supply. Calculate:

- the time constant
- final value of current
- the final energy stored
- the value of i, one second after connection.

Assignment 3

This assignment provides evidence for:
Element 11.3: Investigate the response of series circuits to a step voltage input
and the following key skills:
Communication 3.2
Information Technology 3.1, 3.2, 3.3
Application of Number 3.1, 3.3

The evidence indicators for this element require you to investigate the response of a CR and LR circuit to a step voltage input.

You could:

(a) Extend the procedure in Do This 3.1 to include the discharge of the capacitor as well. Choose a smaller or larger time constant and repeat the measurements.
(b) Select suitable L and R values and design and perform a procedure to show the growth and decay of current in the circuit. Choose a smaller or larger time constant and repeat the measurements.

Report

Provide a procedure report which:

- describes the measurement procedures
- contains graphs illustrating the transient response of the circuits
- describes the effect of the circuit time constant on the responses
- contains calculations of the circuit quantities for the transient state using the exponential equations.

PART TWO: ELECTRICAL POWER TECHNOLOGY

Chapter 4:   **Three-phase systems**
Chapter 5:   **Investigating transformers**
Chapter 6:   **Investigating d.c. machines**
Chapter 7:   **Three-phase induction motors**

- The distribution of electrical energy is fundamental to our everyday lives; without it our homes would have very little lighting or heating. The electrical distribution system makes use of three-phase circuits and transformers; it is therefore important to have a basic understanding of the principles involved. Industry would not be able to run without electrical energy with the different types of motors used to drive various processes. It is therefore important to understand the characteristics and applications of different types of d.c. and a.c. motors.
- Part 2 builds on the work of Part 1 and introduces three-phase a.c. circuits investigating their characteristics and applications. In addition to the basic types of three-phase circuits, components such as transformers and devices such as d.c. machines and induction motors are also investigated.

Chapter 4:   Three-phase systems

This chapter covers:
Element 12.1:   Investigate three-phase systems.

... and is divided into the following sections:
- Features of three-phase systems
- Balanced three-phase systems
- Electrical quantities
- Measurement of power in three-phase systems.

In heavy industrial applications the current levels required cannot be supported by a single phase supply system and therefore a multi-phase system, in particular, a three-phase system is used instead. To widen your understanding of the applications of electrical principles an investigation of three-phase supplies is essential.

When you complete this element you should be able to:

- describe relationships between electrical quantities in balanced three-phase systems
- describe the features of three-phase supplies using phasor diagrams
- determine circuit quantities in three-phase loads.

Features of three-phase systems

The generation and transmission of electrical power is much more efficient using a multi-phase or poly-phase system employing combinations of three sinusoidal voltages. It is also found that three-phase motors start and run much better than single-phase motors.

Three-phase voltage generation

The first thing to consider is how a three-phase supply is generated and to do this requires an understanding of how an a.c. generator operates. An a.c generator consists of a rotating coil (rotor) and a stationary magnet (stator) as shown in Figure 4.1(a).

As the coil rotates in the magnetic field an e.m.f. is induced in the wires which in turn causes an a.c. voltage to appear across the terminals of the coil. For three-phase generation three separate coils are required each being positioned 120°

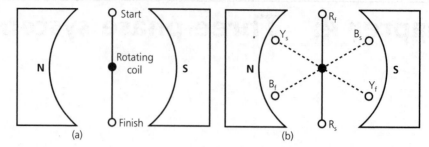

Figure 4.1 Simple a.c. generator; s = start of winding, f = finish of winding

apart as shown in Figure 4.1(b). In practice a.c generators are constructed with three fixed coils and a rotating magnet, this overcomes the problem of making electrical connections to the coils.

Waveform displacement

As the magnet rotates an a.c. voltage appears across the terminals of each winding in turn and as can be seen in Figure 4.2(a) each voltage waveform is displaced from the next by 120°.

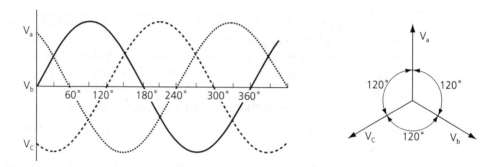

Figure 4.2 Generated waveforms and phasor diagram

Phase sequence

Assuming that the generator rotor always rotates in the same direction and that each winding voltage is represented by Va, Vb and Vc the relationship between these voltages can be shown on a phasor diagram where each phasor represents the magnitude and phase of each winding voltage as shown in Figure 4.2(b).

In a practical three-phase system the phase voltages Va, Vb and Vc are placed in the order in which they attain their peak amplitudes known as the **phase sequence**. To distinguish between phases the following **phase colours** are used: red, yellow and blue. From now onwards we will refer to these voltages in relation to their colours, that is Vr, Vy and Vb.

The whole point of generating three-phase voltages is to supply power to an **electrical load** but how are the loads to be connected to the individual phase windings? Each winding could be connected to its own load, however this would require six wires to connect between the loads and the generator as shown in Figure 4.3(a).

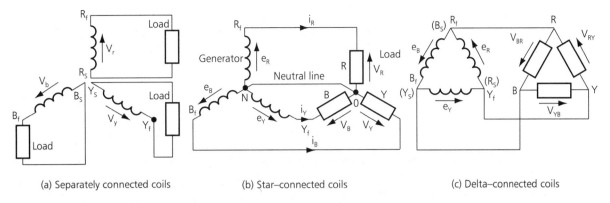

(a) Separately connected coils   (b) Star–connected coils   (c) Delta–connected coils

Figure 4.3 Six-wire, four-wire and three-wire connection methods

This method would be uneconomic in practice and so a more efficient method is required.

Balanced three-phase systems

If we assume that each load connected to each phase is the same then the system is said to be balanced. In a balanced system the e.m.f's, currents and phase angles would be the same. As long as the system is balanced the sum of the phase e.m.f's will be zero and the sum of the phase currents will also be zero. This implies that we do not need a separate return connection between each load and the generator.

Two common methods are used in practice to connect balanced loads to a three-phase generator, called **star** and **delta**.

Star connection method

In the star connection the three generator windings have one end commoned together to form a star point or neutral, as shown in Figure 4.3(b).

The voltage measured between any pair of lines is equal to the phasor difference between two of the phase voltages and is called the **line voltage**. The value of this voltage is given by the expression:

$$V_L = \sqrt{3} \times V_P \text{ (volts)}$$

where (V_L) is the line voltage and (V_P) is the phase voltage.

In the star connection *the line current is equal to the phase current* and the phase relationships between line and phase voltages can be represented by a triangle as shown in Figure 4.4(a)

From this triangle it can be seen that *the line voltages lead the phase voltages* by 30°. It is interesting to note that if the loads are balanced then no current flows into the star point or neutral wire.

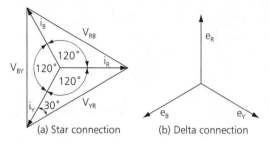

Figure 4.4 Three-phase phasor diagram

WORKED EXAMPLE 4.1

(a) A three-phase star-connected system has a phase voltage (Vp) of 240 V. Calculate the line voltage (V_L)

(b) The system is connected to three identical loads of 30 Ω. Calculate the phase and line currents.

Solution

(a) To calculate the line voltage we use the equation $V_L = \sqrt{3} \times V_P$

Therefore $V_L = \sqrt{3} \times 240 = 415$ V

(b) The phase current (I_P) is calculated using Ohm's law ($I = V_P/R_L$) where V_P is the phase voltage and R_L is the load resistance.

$I_P = 240/30 = 8$ A

Progress check 4.1

Three identical loads each of resistance 30R are connected in STAR to a three-phase 415 V supply. Find:

● phase voltage
● phase current
● line current.

In a practical situation loads are very rarely perfectly balanced and in many cases may be a mixture of domestic and industrial loads. Domestic requirements can normally be met using a single phase with a voltage of 230 V r.m.s. whereas industrial loads are usually supplied at $\sqrt{3} \times 230$ V.

Domestic supplies

A large number of domestic loads can be supplied from a three-phase source by connecting single-phase loads to each of the three phases. In a typical street the first house would be connected to the first phase, the second house to the second phase, the third house to the third phase and then the fourth house would be connected to the first phase and so on. However, as the loads will never be exactly the same an additional conductor is required to allow for the imbalance in the load currents. A so-called four wire system is therefore used where the star point becomes the **neutral conductor** and is connected to each household (Figure 4.3(b)).

Activity 4.1

Using the circuit arrangement and values specified in the last progress check, sketch a complete phasor diagram showing the relationships between line and phase currents and voltages.

Delta connection

In the Delta connection the ends of the generator windings are connected end to end as shown in Figure 4.3(c).

With this form of connection there is no neutral point and under balanced load conditions the line voltage is the same as the phase voltage, that is

$$V_L = V_P$$

The opposite is true of the phase and line currents, the line currents are now equal to the phasor difference between the phase currents and therefore the line current:

$$I_L = \sqrt{3} \times I_P \text{ (amps)}$$

The phase relationships between phase and line currents are shown in the phasor diagram in Figure 4.4(b).

WORKED EXAMPLE 4.2

In a balanced delta-connected three-phase system with a line voltage of 240 V, the phase current is found to be 3.3 A. Calculate the value of the line current.

Solution
To calculate the value of the line current we use the equation $I_L = \sqrt{3} \times I_P$

therefore $I_L = \sqrt{3} \times 3.3 = 5.72$ A

Progress check 4.2

Three identical loads each of 60 Ω are delta connected to a 415 V, three-phase supply. Find:

- phase voltage
- phase current
- line current.

Electrical quantities

Having considered the relationships between line current and phase current and phase voltage and line voltage we now need to turn our attention to the determination of *power in three phase circuits*.

If we assume that the systems are balanced then the load currents will be equal and the phase power will be one third of the total power. To take account of the fact that the load may contain a reactive element, as covered in chapter 2, we must apply the idea of power factor as follows.

Star connection power

In a star-connected system phase power (Pp) is given by the expression: $P_P = V_P I_L \cos(\phi)$ where $\cos(\phi)$ is the power factor. The total power (P_T) is three times the phase power. Therefore total power is given by

$$P_T = 3 V_P I_L \cos(\phi)$$

As $V_L = \sqrt{3}\, V_P$ total power is in fact equal to:

$$P_T = \sqrt{3}\, V_L I_L \cos(\phi)\ \text{W}$$

WORKED EXAMPLE 4.3

Measurements taken on a balanced three-phase star-connected system indicates a line voltage of 415 V, a line current of 20 A and a power factor of 0.8. Calculate the total power consumed.

Solution
To calculate the total power we use the equation ($P_T = \sqrt{3}\, V_L I_L \cos(\phi)$)

Thus $P_T = \sqrt{3}\, 415 \times 20 \times 0.8 = 11.5\ \text{kW}$

Delta connection power

In the delta connection the phase voltage (Vp) is equal to the line voltage (V_L) and therefore the phase power is given by:

$$P_P = V_L I_L \cos(\phi)$$

The total power will be three times the value of the phase power and therefore:

$$P_T = 3 V_L I_P\ \text{(W)}$$

The load current is equal to the $\sqrt{3}$ multiplied by the phase current therefore:

$$P_A = \sqrt{3}\, V_L I_L \cos(\phi)\ \text{(W)}$$

This expression is exactly the same as the star power expression proving that the same amount of power can be delivered using either star or delta connections.

Progress check 4.3

Measurements were taken on a three-phase 415 V system as follows:

- phase current = 25 A
- line current = 43 A
- power factor cos ϕ = 0.6.

From this calculate the total power dissipated.

Measurement of power in three-phase systems

Power can be measured in three-phase systems using the following methods.

One-wattmeter method

This method can be used in both star and delta-connected circuits but only where the loads are balanced. Figure 4.5(a) shows the connections required to make the power measurement; it is important to note the way in which the windings of the power meter is connected.

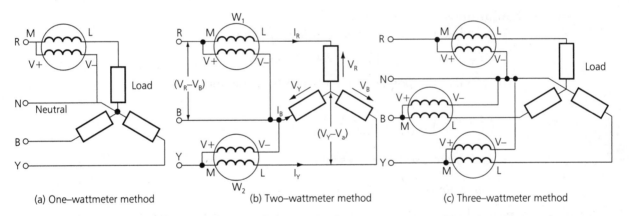

(a) One–wattmeter method (b) Two–wattmeter method (c) Three–wattmeter method

Figure 4.5 Three-phase power measurement

From the wattmeter reading obtained the total power is calculated by simply multiplying the wattmeter reading by three, thus:

Total power = 3 × wattmeter reading (watts)

Two-wattmeter method

This method can be used for both star and delta-connected circuits and for balanced or unbalanced loads. Figure 4.5(b) shows the wattmeter connections required for a star connected system.

The total power is calculated from the sum of the two readings obtained, thus:

$$\text{Total power} = (P_1 + P_2) \text{ (watts)}$$

The phase angle ϕ, power factor $\cos \phi$ and hence the overall power factor can be calculated from:

$$\tan \phi = \sqrt{3}\ \frac{(P_1 - P_2)}{(P_1 + P_2)}$$

WORKED EXAMPLE 4.4

Measurements where taken on a three-phase unbalanced system using the two-wattmeter method. The readings obtained were: $P_1 = 1200$ W; $P_2 = 750$ W. Calculate the total power and the power factor.

Solution
To calculate the total power, use the equation:

$$P_T = P_1 + P_2 = 1200 + 750 = 1950 \text{ W}$$

To calculate the power factor we use the equation:

$$\tan \phi = \sqrt{3}\ \frac{(P_1 - P_2)}{(P_1 + P_2)} = 0.399$$

To find ϕ we use the expression:

$$\phi = \tan \phi^{-1} = 21.75°$$

Power factor is then found from: $\cos \phi = \cos (21.75) = 0.885$

Three-wattmeter method

This method is used to measure power in four-wire three-phase systems using either balanced or unbalanced loads. The connections required for this method are shown in Figure 4.5(c).

The total power is obtained from this method by summing the wattmeter readings hence

$$\text{Total power} = (P_1 + P_2 + P_3) \text{ (watts)}$$

Progress check 4.4

The two-wattmeter method was used to find the power dissipated in a three-phase system. The readings taken from the two wattmeters were: $P_1 = 300$ W; $P_2 = 200$ W.

From these readings calculate:

- total power
- power factor.

Assignment 4

This assignment provides evidence for:
Element 12.1: Investigate three-phase systems
and the following key skills:
Communication 3.2, 3.4
Information Technology 3.1, 3.2, 3.3
Application of Number 3.1, 3.2, 3.3

For this assignment you are required to find out what configuration of three-phase supply comes into your school, college or company and then to estimate the load per phase, the phase and line current and the phase and line voltages. From these try to estimate the total power consumed.

Report

Your report is to include details of electrical circuits, if available, a detailed phasor diagram for the system and details of your calculations.

The evidence indicators for this element require you to:

- describe the relationships between electrical quantities in balanced three-phase systems.
- describe the features of three-phase supplies using phasor diagrams of the voltages and currents in each circuit.
- determine electrical quantities in balanced three-phase loads with calculations supporting the measurements taken.
- determine electrical quantities in unbalanced three-phase loads with calculations supporting the measurements taken.

Key skills in Application of Numbers are likely to be used in this element. If circuit simulation or analysis software is available, Information Technology may also be covered.

Chapter 5:   Investigating transformers

The transformer is probably one of the most widely used electrical components and finds applications in virtually every conceivable type of power supply from low voltage, low power to high voltage, high power. It takes on many different physical forms from the miniature construction found in electronic systems to the physically large forms found in power distribution networks. It is therefore important to understand the characteristics of the transformer and its associated applications.

When you complete this element you should be able to:

- describe the operation of ideal transformers using appropriate laws and principles
- describe the construction of transformers
- describe the applications of transformers
- determine the characteristics of transformers.

Transformer principles

To understand the characteristics of the transformer some basic principles need to be understood and therefore we start by considering the transformer as an 'ideal' component.

The ideal transformer

An ideal transformer consists of a ferromagnetic core with two coils wound around each arm of the core as shown in Figure 5.1(a).

When an alternating voltage is applied across the terminals of one of the coils (called the **primary**), an alternating current flows through the coil and this in turn generates an alternating magnetic flux ϕ in the core. Since the primary coil has this changing flux through it, then by Faraday's law (chapter 2), an

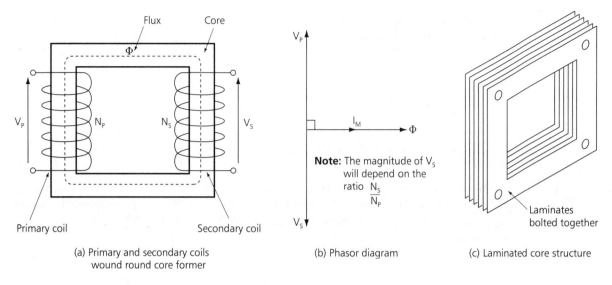

Figure 5.1 *Ideal transformer*

induced emf equal and opposite to the applied voltage is induced in it. Furthermore, since the other coil (called the **secondary**) also has the same changing flux through it, an emf is induced in it as well. The actual size of the voltage depends on the number of turns used in both coils. Mathematically, the relationship between the voltages at the coil terminals and the coil windings is given by:

$$\frac{Vp}{Vs} = \frac{Np}{Ns}$$

Therefore

$$Vs = Vp\,\frac{Ns}{Np} \quad \text{(volts)}$$

where Vs = secondary voltage, Vp = primary voltage, Np = number of primary turns and Ns = number of secondary turns.

WORKED EXAMPLE 5.1

A transformer has 1000 primary turns and 100 secondary turns, the primary voltage is 240 V. Calculate the magnitude of the secondary voltage.

Solution
Using the equation Vs = VpNs/Np:

$$Vs = 240 \times 100/1000 = 24\ V$$

Progress check 5.1

A transformer is required to have a 240 V primary and a 12 V secondary. What would the ratio between primary and secondary windings need to be?

It is useful at this point to introduce some basic definitions relating to the transformer as follows:

- *Turns ratio*   This is the ratio of secondary turns to primary turns (Ns/Np).
- *Voltage ratio*   This is the ratio of secondary voltage to primary voltage (Vs/Vp).
- *Current ratio*   This is the ratio of secondary current to primary current (Is/Ip).

The relationship between primary, secondary turns and primary and secondary voltages has already been stated; however, there is also a relationship between turns and primary and secondary currents. This relationship is given by the following expression:

$$\text{Is} = \text{Ip}\,\frac{\text{Np}}{\text{Ns}}\ \text{(amps)}$$

WORKED EXAMPLE 5.2

A transformer has a primary to secondary turns ratio of 10:1. If the primary current is 2.5 A what is the value of the secondary current?

Solution
Using the equation Is = IpNp/Ns:

Is = 2.5 × 10 = 25 A

If the ratio of primary to secondary windings is changed the voltage generated at the secondary terminals can be made to be greater than or less than the primary voltage, therefore indicating that transformers can be designed to be both **step up** or **step down**.

The inverse is true for the secondary current thus indicating that transformers can again be designed to supply very high secondary currents with very small primary currents.

Progress check 5.2

If the primary current in a transformer is measured to be 1.5 A and the primary to secondary turns ratio is 15:1, what is the value of the secondary current?

The relationship between core flux and primary and secondary voltages can be shown with the aid of a phasor diagram as indicated in Figure 5.1(b).

With no load connected to the secondary a so-called magnetising current I_M flows in the primary generating magnetic flux Q in the core. The voltage induced leads the flux by 90° and the secondary voltage lags behind the primary voltage by 180° as shown in Figure 5.1(b). The flux Φ is the reference phasor since it is common to both coils.

Transformer power losses

So far the transformer has been considered to be ideal and by ideal we mean 100% efficient. This implies that there are no **power losses** in an ideal transformer. This is not true of course for a real transformer, which will have an efficiency approaching 95%. These losses need to be accounted for and generally occur in either the windings or the core.

Winding power losses (copper losses)
Both the primary and secondary windings will have resistance of Rp and Rs Ω respectively. When current flows in both primary and secondary windings, power will be lost in their respective resistances as follows:

$$\text{Copper losses Pc} = \text{Ip}^2\text{Rp} + \text{Is}^2\text{Rs}$$

Core power losses (iron losses)
These losses relate to the energy required to change the direction of flow of magnetic flux in the transformer core. In general by carefully selecting the core material these losses can be minimised.

WORKED EXAMPLE 5.3

A transformer has a primary current of 10 A and a primary resistance of 1.5 Ω, the secondary current is 2.5 A and the secondary resistance is 6 Ω. Calculate the copper losses.

Solution
$$\text{Pc} = (10^2 \times 1.5) + (2.5^2 \times 6) = 187.5 \text{ W}$$

Hence the copper losses in this transformer are 187.5 W

Eddy current losses
Eddy currents are generated within the core and the general structure of the transformer due to the alternating flux within the core. This type of loss can be significantly reduced by using a laminated core structure as shown in Figure 5.1(c).

Total no-load power losses
Having considered the losses in a practical transformer we can now revisit the no-load phasor diagram and incorporate the effect of losses as shown in Figure 5.2(a).

The primary magnetising current I_M is now added to the phasor diagram with the core losses being represented by an extra current component I_C, giving the magnitude of the no-load current I_O as :

$$I_O = \sqrt{I_C^2 + I_M^2} \text{ (amps)}$$

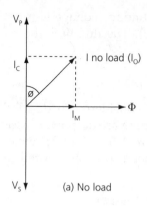
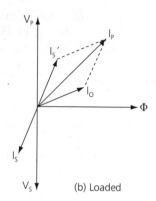

Figure 5.2 Practical transformer phasor diagrams

The no-load power factor is given by the expression:

$$\cos \phi = \frac{I_C}{I_O}$$

Progress check 5.3

Explain the different type of losses encountered in a transformer.
Explain how each of these losses can be minimised.

Transformer on load

The effect of connecting a load to the secondary of the transformer is shown by the phasor diagram in Figure 5.2(b).

In this case the effect of the secondary load current is to reflect back to the primary a current demand equal to the secondary current multiplied by the secondary to primary turns ratio. The total primary current is then given by the phasor sum of this current and the no-load primary current.

Efficiency

Having considered the losses within a transformer the overall efficiency of a transformer can now be defined as:

$$\text{Efficiency} = \frac{\text{Output power}}{\text{Output power} + \text{losses}} \times 100\%$$

where losses are: (copper losses + iron losses). Typically most transformers have an efficiency of around 95%.

Voltage regulation

One other important characteristic is that of voltage regulation and relates to the changes in secondary voltage with changing load. This is defined as:

$$\text{Voltage regulation} = \frac{V_O - V_L}{V_O} \times 100 \ (\%)$$

where V_O is the no-load secondary voltage and V_L is the full-load secondary voltage. For a well designed transformer this figure is usually less than 5%.

Progress check 5.4

Sketch the phasor diagram representing a transformer on no-load and explain how it changes when a load is connected to the transformer. Define the following terms:

● voltage regulation
● efficiency

State typical values for both.

Transformer types

There are a number of different types of transformer each having a specific range of applications.

Single-phase double-wound

This is probably the most common type of transformer, the construction of which is shown in Figure 5.3(a). It is available in a wide range of sizes and is widely used in power supplies for electrical/electronic equipment. Within this type of application it is used to step down the voltage available from the mains supply to a value which is more easily converted to a low d.c. voltage via a rectification circuit. A circuit diagram for a simple low voltage d.c. supply is shown in Figure 5.3(b).

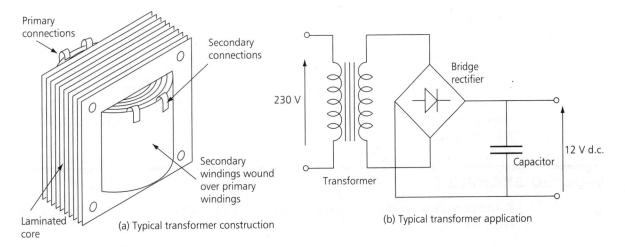

(a) Typical transformer construction

(b) Typical transformer application

Figure 5.3 Simple low voltage supply using single-phase, double-wound transformer

In this example the mains voltage is stepped down and rectified to produce a 12 V d.c. output.

Isolating transformer

One of the main advantages of this type of transformer construction is the fact that there is electrical isolation between the primary side of the circuit and the secondary side – the two coils are not connected physically, only through electromagnetic forces.

The double-wound, single-phase transformer can also be used as a so-called isolating transformer. In this application the number of turns on the primary is normally equal to the number of turns on the secondary. The transformer is also designed to exhibit good regulation characteristics. Isolating transformers are used to electrically isolate one electrical circuit from another and is often done for safety reasons. Typical examples include bench supplies in laboratories and equipment supplies in dental and doctors surgeries.

Matching transformer

In addition to stepping up or stepping down current or voltage a double-wound transformer can also be used to match dissimilar circuit impedances. This type of transformer can be found in say the output stage of an audio amplifier where the output impedance of the amplifier may need to be matched to the low impedance of a loudspeaker. The impedance matching can be achieved by considering the relationships between current, voltage and turns as follows:

$$Vs = Vp \frac{Ns}{Np} \quad \text{and} \quad Is = Ip \frac{Np}{Ns}$$

Therefore

$$Zp = \frac{Vp}{Ip} \quad \text{and} \quad Zs = \frac{Vs}{Is}$$

where Zp is the primary impedance and Zs is the secondary impedance. Note:

$$\frac{Zs}{Zp} = \left(\frac{Ns}{Np}\right)^2$$

and therefore

$$Zs = Zp \left(\frac{Ns}{Np}\right)^2$$

From this expression it can be seen that the secondary impedance is equal to the primary impedance multiplied by the turns ratio squared. To match the primary impedance to the secondary impedance is therefore simply a matter of choosing the correct turns ratio.

WORKED EXAMPLE 5.4

A transformer is required to match 600 Ω to a load of 8 Ω. Calculate the Ns/Np ratio required.

Solution

$$Zs = Zp(Ns/Np)^2$$

Therefore rearranging the above equation gives:

$$Zs/Zp = (Ns/Np)^2$$

The ratio is therefore

$$\sqrt{8/600} = 1:75 \ (0.114)$$

Auto-transformers

This type of transformer essentially has a tapped winding as shown in Figure 5.4(a) and therefore as can be seen there is no electrical isolation between primary and secondary circuits.

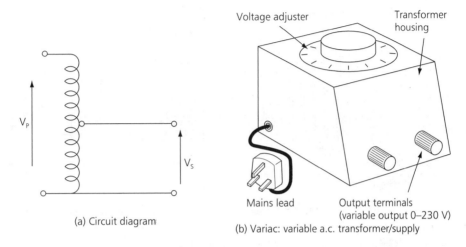

(a) Circuit diagram

(b) Variac: variable a.c. transformer/supply

Figure 5.4 Auto-transformer

This type of transformer is often used for high voltage generation and a typical example is the ignition coil used to produce the high voltage spark for the plugs in a car engine.

Another application for this type of transformer is to produce a variable output voltage by having the primary tap connected to a wiper which can be moved over the main winding therefore changing the turns ratio. This is often referred to as a **variac** and is useful for producing a variable AC supply for testing equipment. A typical example is shown in Figure 5.4(b).

Three-phase transformer

This type of transformer is often used in power transmission systems associated with the national grid. These transformers can be physically very large and are often oil cooled and are found in substations where the very high grid voltages are stepped down for onward feeding to consumers. Figure 5.5(a) shows a typical grid transformer with oil cooling.

The primary windings are often connected in a delta formation and the secondary in a star formation as shown in the circuit diagram in Figure 5.5(b). This is done so that only three cables are required to feed the primary of the transformer and on the secondary side there is a common or neutral connection providing a four-wire, three-phase connection.

Current transformer

A current transformer is a special case as the primary often consists of a single conductor. This type of transformer is often wound on a **toroidal core** so that a single conductor can be passed through the centre of the core as shown in Figure 5.6.

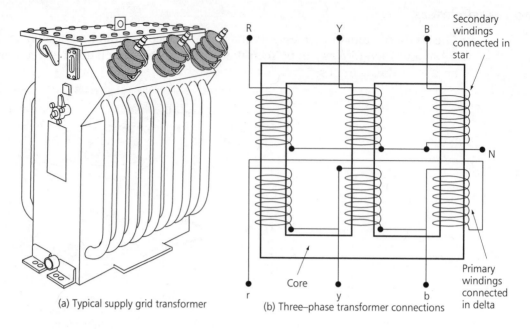

(a) Typical supply grid transformer   (b) Three-phase transformer connections

Figure 5.5 Oil-cooled three-phase transformer

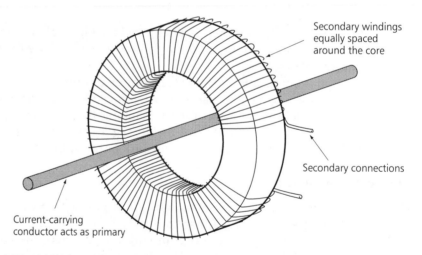

Figure 5.6 Toroidal core current transformer

When a large alternating current flows through the centre conductor it generates a magnetic field which cuts the windings wound around the toroid; this in turn induces an e.m.f. which appears as a voltage across the terminals of the winding. This voltage will in fact be proportional to the current flowing through the conductor. This type of transformer is therefore very useful for measuring large a.c. currents and is available in many different sizes to cater for a wide range of currents from a few amps to many 100's of amps.

Activity 5.1

Choose any four types of transformer and write a report describing their principles of operation and examples of applications.

The report should include a detailed sketch of the construction of each transformer and a description of its operation. Details of applications should be included with suitable circuit diagrams where appropriate.

Methods of determination

Measurements

Having considered the characteristics and applications of transformers we now turn our attention to testing transformers to ascertain their characteristics in terms of losses.

There are basically two tests which can be used on a transformer, the open circuit test and the short circuit test.

Open-circuit test

This test involves making measurements with the secondary side of the transformer open circuit. In this condition the transformer is effectively on no load and therefore the only losses present are core losses. Figure 5.7(a) shows a typical setup for an open-circuit test.

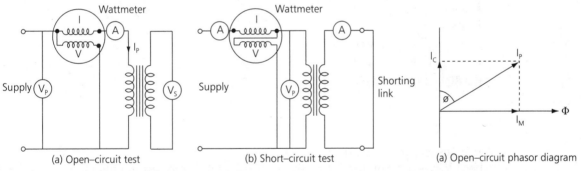

(a) Open–circuit test (b) Short–circuit test (a) Open–circuit phasor diagram

Figure 5.7 Open and short-circuit test

From this test a note is made of the input or primary current and the reading obtained from the wattmeter. From these two readings the value of the core losses can be obtained along with the no-load power factor.

Short-circuit test

This test involves making measurements with a short circuit on the secondary side of the transformer. Under these conditions full-load current is made to flow and from these a value for winding losses can be ascertained. Figure 5.7(b) shows the arrangement required for a short-circuit test.

The short-circuit test is more complex than the open-circuit test and requires access to a variable a.c. supply. The variable supply connected to the primary is slowly increased from zero until the rated full-load current of the transformer flows in the primary, via the shorting link. This will occur with a fairly small primary voltage. Under these conditions the wattmeter reading indicates the winding losses of the transformer. Also, by noting the value of primary voltage and current the actual value of the **winding resistance** can be obtained.

Calculations

Having carried out a short-circuit and an open-circuit test on a transformer there are a number of calculations which can be performed to ascertain its various characteristics. This is best illustrated by use of a case study related to measurements taken form an actual transformer.

Case study

A single phase transformer produced the following test results:

Open-circuit test: primary current 0.5 A
                       input power 40 W

Short-circuit test: primary voltage 24 V
                       primary current 35 A
                       input power 55 W

Note that the primary voltage in the short-circuit test was the value required to give full-load current.

From the results of the tests we are asked to find the following:

- core losses
- winding losses
- efficiency
- magnetising current
- no load power factor
- winding resistance.

Core losses and winding losses

These losses are obtained directly from the power measurements taken in the two tests. The power measurement taken in the open-circuit test gives the value of the core loss, which in this case study is 40 W. Therefore we can state that:

Core losses = 40 W

The power measurement taken in the short-circuit test gives the value for the winding loss, which in this case is 55 W, therefore we can state that:

Winding losses = 55 W

Efficiency

From these two figures the efficiency of the transformer can be calculated. However there is small piece of information missing. To actually calculate the efficiency, the maximum rated output power for the transformer is required. This would normally be supplied by the transformer manufacturer and in fact would normally be stamped on the structure of the transformer or printed on a label attached to it. This figure is often referred to as the transformer rating and would be stated in terms of volt-amperes or kilovolt-amperes. The transformer used in this case study has a rating of 1 kVA or 1000 VA. To calculate the efficiency the following expression is used:

$$\text{Efficiency} = \frac{\text{Output power}}{(\text{Output power} + \text{core loss} + \text{winding loss})} \times 100\%$$

From the figures obtained from the open/short-circuit tests the efficiency is given by:

$$\text{Efficiency} = \frac{1000}{1000 + 40 + 55} \times 100\% = 91.3\%$$

Magnetising current

The voltage applied to the primary of the transformer in the open-circuit test was 240 V, the power measured was 40 W. From these two figures the core loss component of the primary current can be calculated as follows:

$$\text{Ic} = \text{input power}/\text{primary voltage} = 40/240 = 0.167 \text{ A}$$

where Ic is the core loss component of current.

From our earlier work on transformer phasor diagrams the primary current is the phasor sum of the magnetising current component and the core loss current component. The open-circuit test gave us the primary current value, which was 0.5 A, and the core loss component has just been calculated. To find a value for the magnetising current the following expressions are used:

$$\text{Ip} = \sqrt{(\text{Im}^2) + (\text{Ic}^2)}$$

By rearranging the expression we get:

$$\text{Im} = \sqrt{(\text{Ip}^2) - (\text{Ic}^2)} = \sqrt{(0.5)^2 - (0.167)^2} = 0.47 \text{ A}$$

The phase relationships are clearly shown on the phasor diagram in Figure 5.7(c).

No-load power factor

The no-load power factor can easily be obtained form the phasor diagram. The angle between Ic and Ip shown in Figure 5.7(c) is in fact the power factor angle. To calculate the power factor we simply find the cosine of this angle. Measuring this angle from the phasor diagram gives an angle of 70.7°. The power factor is then given by cos (70.7) = 0.33 lagging.

Winding resistance

The final calculation required is that related to finding a value for the winding resistance of the transformer. For this calculation, measurements from the short-circuit test are used as follows. The power measured in the short-circuit test relates purely to winding losses and hence the winding resistance:

$$\text{Pw} = \text{Ip}^2\text{Rw}$$

where Pw = winding power loss, Ip = primary current and Rw = total winding resistance referred to the primary. Rearranging the expression gives:

$$Rw = \frac{Pw}{Ip^2} = \frac{55}{35} = 0.45 \ \Omega$$

Assignment 5

This assignment provides evidence for:
Element 12.2: Investigate transformers
and the following key skills:
Communication 3.2, 3.3
Information Technology 3.1, 3.2, 3.3
Application of Number 3.1, 3.2, 3.3

You are required to perform a short-circuit and open-circuit test on a transformer, observing all the necessary safety precautions. From the results obtained you are required to calculate the following parameters:

- Core losses
- Winding losses
- Efficiency
- Magnetising current
- No load power factor
- Winding resistance
- Regulation

Report

A report is to be prepared and is to include diagrams showing connections of equipment used to perform the two tests, details of your calculations including the drawing of phasor diagrams where appropriate, observations and comments relating to your findings.
The evidence indicators for this element require you to:

- describe the operation of an ideal transformer, using appropriate laws and principles
- describe the construction of four different types of transformer
- describe applications of four types of transformer
- provide a record of a practical investigation to determine the characteristics of one type of transformer including supporting calculations of efficiency and regulation.

There are opportunities for students to give presentations and to use IT to collect information and present reports and thereby cover key skills in Communications and Information Technology.

Chapter 6: Investigating d.c. machines

D.C. machines can be used as generators and motors and are used in a wide variety of applications. One of the main reasons for their popularity is the fact that they are easy to use and that their speed can be controlled using relatively simple methods. This chapter is concerned with investigating the constructional features and characteristics of d.c. machines.

When you complete this chapter you should be able to:

- describe the construction features of d.c. machines
- describe the operation of d.c. machines using appropriate laws and principles
- identify control techniques for d.c. machines
- determine the characteristics of d.c. motors by measurement.

D.C. machines

Constructional features

In general, d.c. generators and motors have the same basic construction, as shown in Figure 6.1. The main component parts are the **armature** or rotor which consists of windings which lie in slots on an iron based former or core and a stationary magnetic field generating system called the **field windings** or stator. The armature sometimes has a fan attached for cooling purposes.

The stationary magnetic field is usually generated by a separate set of windings supplied with a d.c. current but can be generated by a permanent magnet. The armature windings are connected to a **commutator**, the segments of which are connected to the outside world by carbon brushes.

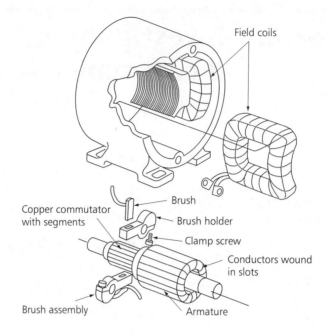

Figure 6.1 General construction of a d.c. machine

Basic operating principles

The basic underlying principle is that of **Fleming's left-hand rule**, which states that a current-carrying conductor placed in a magnetic field will experience a force, the strength of which is given by the expression:

$$F = BLI \sin \phi$$

where F is force, B is magnetic flux density, I is current, L is the length of the conductor and ϕ is the angle between the conductor and the direction of the magnetic field.

Thus if a d.c. supply source where connected to the brushes of the commutator, a d.c. current would flow through the armature windings which in turn would generate a turning force and hence cause the armature to rotate, therefore turning the d.c. machine into a motor.

In addition to this when a conductor cuts through a magnetic field a **back e.m.f.** is generated. This effect is described by the laws of Faraday and Lenz, see Chapter 2. This implies that if the armature of a d.c. machine is mechanically rotated within a magnetic field a voltage will appear across the terminals connected to the commutator. In this mode the machine is acting as a generator.

Progress check 6.1

What distinguishes a generator from a motor?

Generators

Having established that a d.c. machine can be used as either a generator or a motor, we shall start by considering its characteristics as a generator. Basically, Lenz's law states that e.m.f (E) is proportional to magnetic flux (Φ) multiplied by angular velocity (ω), hence for a given machine the output voltage under open circuit conditions will be

$$E = K \Phi \omega$$

If the field winding is connected to a variable d.c. supply as shown in Figure 6.2(a) and the the armature is rotated at a linearly increasing speed then the open circuit voltage for a given value of field current (If) will be as shown in Figure 6.2(b).

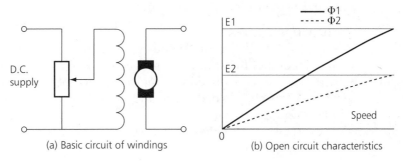

(a) Basic circuit of windings   (b) Open circuit characteristics

Figure 6.2 Field winding circuit of a d.c. generator

When an electrical load is connected to the output terminals of the generator the voltage will be less than that which is obtained under no-load conditions. This is due to the resistance of the armature windings and therefore the loaded output voltage (V) will be given by:

$$V = E - (Ia \times Ra)$$

where Ia is armature current and Ra is armature resistance.

The generator type just described is known as a **separately excited d.c. generator** because the field current is provided by a separate supply. This is clearly inconvenient, and it would be far better to power the field windings from the generator's own output. There are three ways in which this may be done as shown in Figure 6.3.

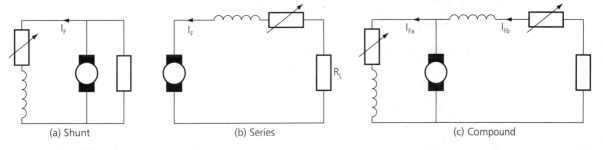

(a) Shunt   (b) Series   (c) Compound

Figure 6.3 Field winding configurations of a separately excited d.c. generator

Progress check 6.2

Why is the generator output voltage less under loaded conditions?

Both the shunt and series generators have one field winding whereas the compound generator has one series and one compound winding.

Operating characteristics

Each of these arrangements gives the generator a particular characteristic which will be investigated later. Firstly we need to ask an important question. How will these generators start? The answer is that a small e.m.f will be produced with no field current due to the remnant magnetism of the field winding iron former. This will in turn create a small field current which generates a larger e.m.f and so on. Having established how the generator starts, each field winding configuration will now be considered in some detail.

Shunt wound generator

This is probably the most obvious arrangement for the field coils.

The diagram of the shunt wound generator shows a variable resistance in series with the field coil. This allows the resistance of the field coil to be varied and hence controls the field current and generator output voltage. To compare the shunt wound generator to the separately excited generator the loaded characteristics need to be considered. The shunt wound generator load characteristic is shown in Figure 6.4(a).

As the load current is increased the generator terminal voltage reduces but the voltage does not smoothly reduce to zero. There is a point X on the curve where the terminal voltage falls below a level which can no longer sustain sufficient field current to maintain correct generator operation and the terminal voltage rapidly falls to zero. The term used to describe the variation in terminal voltage with increasing load current is **regulation** and is defined by the expression:

$$\text{Regulation} = \frac{\text{no-load voltage} - \text{full load voltage}}{\text{no-load voltage}} \times 100 \ (\%)$$

WORKED EXAMPLE 6.1

A generator is found to have an open circuit voltage of 50 V and a full-load terminal voltage of 48 V. Calculate the percentage regulation of the generator.

Solution

$$\text{Regulation} = \frac{\text{no-load voltage} - \text{full-load voltage}}{\text{no-load voltage}} \times 100\%$$

$$= \frac{50 - 48}{50} \times 100\% = 4\%$$

As can be seen from the load characteristic of the shunt wound generator it has very poor voltage regulation compared with the separately excited generator and is therefore not the ideal configuration. See Figure 6.4(a).

Another possibility would be to connect the field winding in series with the armature. This configuration is known a the series wound generator.

Progress check 6.3

On open circuit it is found that the output voltage of a generator is 24 V; on full load this reduces to 22 V. From these figures calculate the percentage regulation of the generator.

Series wound generator

This configuration is shown in Figure 6.3(b) and as can be seen the load current flows through the field windings and therefore they have to be constructed from much heavier gauge wire but will have fewer windings compared with that of the shunt wound generator.

The load characteristic of the series wound generator is shown in Figure 6.4(b) and as can be seen the terminal voltage increases with increasing load current.

Effectively the terminal voltage would continue to increase without limit if it were not for the eventual saturation of the iron former on which the field windings are wound. This limits the actual maximum level of magnetic flux which can be produced. Like the shunt wound generator the voltage regulation is far from ideal and in actual fact this configuration is very rarely used in practice.

Compound wound generator

To achieve the best voltage regulation two field windings are required as used in the compound wound generator. In this case both series and shunt windings are used as shown in Figure 6.3(c).

By using this configuration the characteristics of both the shunt and series generators add together to produce a load characteristic with excellent voltage regulation as shown in Figure 6.4(c).

As the load current increases the armature losses will increase and the terminal voltage will try to fall. However, the increased current in the series field winding will compensate by increasing the field flux and hence the output voltage.

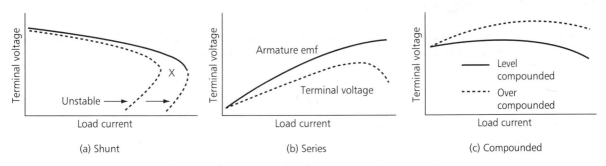

Figure 6.4 Generator load characteristics

Activity 6.1

Explain the different methods for exciting the field windings of a generator and explain how the different methods affect the generator characteristics.

Motors

As was mentioned earlier, generator principles can be applied to motors and therefore it is not surprising to find that there are also three different types of d.c motor: shunt wound, series wound and compound wound. To understand the characteristics of each of the three motors consideration must first be given to the basic principles of motor operation.

Basic principles

Starting with the separately excited field motor we can draw a direct comparison with the separately excited generator as shown in Figure 6.2(a).

By connecting a voltage source across the armature windings a current will flow generating a magnetic flux which will react with the flux produced by the field windings to generate a turning force, as was stated earlier in this chapter. This turning force is known as **torque** (T) and is given by the equation:

$$T = K \, \Phi \, Ia$$

where K is a constant for a given motor, Ia is the armature current and the unit of torque is the newton-metre (Nm). The mechanical power output of the motor is given by:

$$P_o = T\omega$$

where the unit of angular velocity is rads/sec or 2π revs/sec.

Considering the electrical side of the motor the e.m.f. equation used for the generator can be modified for use with the motor as follows:

The armature back e.m.f. $Eb = V - (Ia \times Ra)$

where V is applied terminal voltage and Ia is the armature current and Ra is the armature resistance. If both sides of this equation are multiplied by (Ia) we will get:

$$EbIa = VIa - Ia^2Ra$$

where EbIa = armature power, VIa = input power and Ia^2Ra = machine losses. From this we can estimate the efficiency of the motor:

$$\text{Efficiency} = \frac{\text{Input power} - \text{losses}}{\text{Input power}} \times 100\%$$

WORKED EXAMPLE 6.2

Measurements taken on a motor indicate that the input power is 1 kW and that the losses are 150 W. Calculate the efficiency of the motor.

Solution

$$\text{Efficiency} = \frac{\text{Input power} - \text{losses}}{\text{Input power}} \times 100\%$$

$$= \frac{1000 - 150}{1000} \times 100\% = 85\%$$

Having now established the basic principles and relationships the three basic types of d.c. motor will be considered.

Shunt wound motor

The arrangement is the same as that used in the shunt wound generator and is shown in Figure 6.5(a).

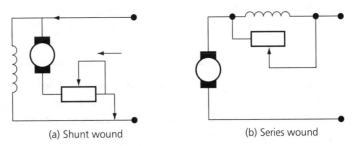

(a) Shunt wound   (b) Series wound

Figure 6.5 Shunt and series wound motors

As a constant terminal voltage is applied to the motor a constant field current will flow and therefore the motor will tend to run at a constant speed for a range of load conditions. Obviously if the load is increased above a certain level the motor will start to slow down as the armature current tries to increase until ultimately the motor will stall. The shunt wound motor is therefore ideal for constant speed load applications such as driving an extraction fan.

Series wound motor

The field windings are connected in the same way as for the series wound generator as shown in Figure 6.5(b).

As for the generator, the field windings have to be constructed from heavy gauge wire and require fewer turns compared with the series wound motor. This arrangement produces radically different torque/speed characteristics as shown in Figure 6.6.

As can be seen from the torque/speed characteristics the available torque rapidly increases with armature current and high speeds are achieved with low values of armature current. This type of characteristic is useful in applications

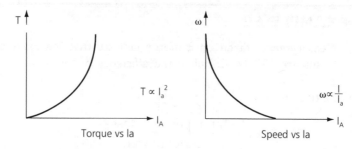

Figure 6.6 *Torque/speed characteristics for series wound motor*

where the load has a large amount of inertia or where initial friction levels are high. The main danger associated with this characteristic is that if the load is suddenly removed for some reason the motor speed would theoretically increase to infinity causing the motor to be destroyed. This type of motor therefore would need to be permanently loaded.

Compound wound motor

As with the compound wound generator the problems associated with series wound and shunt wound motors can be overcome by using both types of winding to form a compound wound motor. This gives a torque/speed characteristic somewhere between the shunt wound and series wound characteristics.

Activity 6.2

Produce a report describing the constructional features and characteristics of two d.c. machines. The report is to include diagrams showing the constructional details of the two machines with a written description of their characteristics and principles of operation.

Control techniques for d.c. machines

As we have seen the characteristics of a d.c. motor can be controlled by using different arrangements of field windings. In actual fact prior to the 1980's d.c. motors were controlled by switching in and out of circuit different field winding arrangements such as shunt, series and compound using a constant d.c. supply. However, since the 1980's there has been a move towards the use of separately excited d.c. motors using variable d.c. supplies.

Thyristor control of d.c. motors

The vast majority of d.c. motors are now controlled electronically using a controlled rectifier (thyristor) drive. The thyristor drive is supplied by either a single phase or a three phase supply depending on the size of the motor. A typical drive arrangement is shown in Figure 6.7.

The purpose of the **controlled rectifier** or thyristor is to generate a variable d.c. supply for the armature and field windings. The thyristor can be controlled

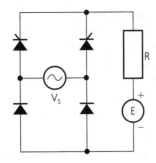

Figure 6.7 Thyristor-controlled d.c. motor drive

electronically and therefore control of both armature supply voltage and field winding current is possible. In this way both the torque and the speed of the motor can be controlled.

Control systems

A control system is required if the speed of the motor is to remain constant over a range of load torques. The most common form of control is the so-called two-loop system which is shown in Figure 6.8.

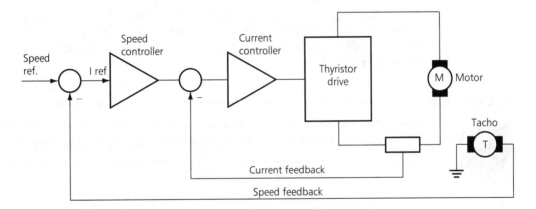

Figure 6.8 Two-loop control system

The system consists of two control loops, an inner current control loop and an outer speed control loop.

Torque control

The purpose of the current loop is to make the actual motor current follow the current reference signal (Iref). The current feedback signal is compared with the reference signal, the difference between these two signals is amplified and the resulting signal is used to drive the thyristor system which in turn controls the d.c. supply voltage. The current feedback signal is obtained from a small current transformer connected in series with the motor armature. The idea of this control system is to to keep the motor current constant regardless of the speed of the motor. The torque output of the motor will therefore remain constant regardless of motor speed.

Controller characteristics

In Chapter 18 the different types of control systems are discussed in detail, including proportional, proportional plus integral, proportional plus derivative and proportional plus integral plus derivative. To obtain the correct control characteristics for the d.c. motor controller the current error amplifier is usually of the proportional plus integral (PI) type. This means that the actual motor current will be exactly equal to the reference current value under steady-state conditions.

Speed control

The outer control loop shown in Figure 6.8 is used to control the motor speed. Feedback is normally supplied by a d.c. tacho-generator, discussed in Chapter 17. The actual and required speeds are then fed into the speed/error amplifier. Any difference is amplified and serves as the input to the current loop. This loop will therefore try to keep the speed of the motor constant under differing load conditions. As with the current/error amplifier PI control is used to ensure that under steady-state conditions the actual motor speed will exactly equal the required speed.

Industrial d.c. motor controllers

These are now in widespread use and have the characteristics discussed previously but in addition have other safety features built-in including field current weakening for finer speed control, tacho-loss protection circuitry to limit motor speed in the event of tacho failure and current limiting to protect the motor armature. Many industrial motor drives are programmable and come with a small operator panel which allows the motor torque and speed to be pre-programmed.

Activity 6.3

Using manufacturers' data investigate the facilities available on a typical modern electronic d.c. motor controller and write a short report covering your findings.

Assignment 6

This assignment provides evidence for:
Element 12.3: Investigate d.c. machines
and the following key skills:
Communication 3.1, 3.2, 3.3
Information Technology 3.1, 3.2, 3.3
Application of Number 3.1, 3.3

Using a motor test set similar to the one shown in Figure 6.9, carry out measurements on a compound wound d.c. motor and a compound wound d.c. generator

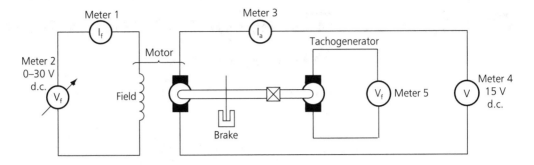

Figure 6.9 D.C. motor test set

to establish the torque/speed and load/speed characteristics of the motor and the open-circuit and loaded characteristics of the generator.

Report

A report is to be produced which is to include records of measurements taken, graphs showing torque/speed, load speed, open circuit and loaded characteristics.

The element indicators for this element require you to produce a report on one d.c. motor and generator including for each:

- description of constructional features
- description of operation, using appropriate laws and principles
- identification of control techniques
- measurements determining characteristics

This element offers opportunities to generate evidence towards key skills in Application of Numbers, Communication and where possible IT.

Chapter 7: Three-phase induction motors

This chapter covers:
Element 12.4:   Investigate three-phase induction motors.

... and is divided into the following sections:
- Three-phase induction motor types
- Operating principles
- Operating characteristics
- Starting methods
- Motor testing.

Generally speaking induction motors are superior to d.c. motors and are preferred for most industrial applications. They are more reliable due to the fact that they have fewer parts and therefore also need less maintenance. As the induction motor is the most widely used machine in industry it is important to understand the operational characteristics and the methods used for starting and indeed this is the purpose of this particular chapter.

When you complete this element you should be able to:

- describe types of three-phase induction motor
- explain the operating principles of three-phase induction motors
- describe and determine the operating characteristics of three-phase induction motors
- describe the methods of starting three-phase induction motors.

Three-phase induction motor types

Three-phase motors are preferred to single-phase motors due to their ease of starting and their quieter operation. There are two basic types of three-phase induction motor in use, the cage rotor and the wound rotor types. Each of these will be described in detail but it has to be said that by far the most widely used type is the cage rotor or squirrel cage motor as it is often called.

Wound rotor

There are two main parts to any motor, the **stator** and the **rotor**. As their names suggest the rotor is the part which rotates and the stator is the part which remains

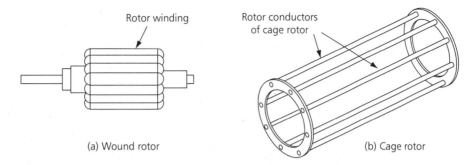

Figure 7.1 Types of rotor

stationary. The stator has the same construction for both wound and cage rotor machines and will be discussed a little later.

The wound rotor component construction is shown in Figure 7.1(a) and consists of a former over which are wound three star-connected windings, the free ends of which are connected to slip rings which are mounted on the end of the rotor shaft.

The main advantage of the wound rotor machine is that under start-up conditions it can be connected to its full rated mechanical load.

Cage rotor

The cage rotor is completely different to the wound rotor in that it is a solid construction with no windings as such. Figure 7.1(b) shows the construction of a typical cage rotor. It is made from a laminated steel core with copper bars the ends of which are shorted together by means of a copper ring.

Progress check 7.1	Describe the main constructional differences between the wound and the cage rotor motor.

Operating principles

One of the main considerations in terms of the operating principles of an induction motor is the generation of a rotating magnetic field within the stator assembly. As was mentioned earlier the stator is the same for both the wound and cage rotor machines and therefore the following principles apply to both.

Rotating magnetic field

The three-phase induction motor stator has three windings set 120° apart. When a three-phase supply is connected to these windings each will produce an alternating magnetic field. If we use Red, Blue and Yellow to represent the phases then Figure 7.2(a) shows how the magnetic field strength varies with time.

Effectively each winding is alternately producing a magnetic south pole and then a magnetic north pole. Due to the physical positioning of the windings the

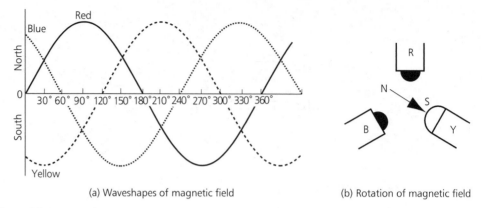

(a) Waveshapes of magnetic field   (b) Rotation of magnetic field

Figure 7.2 Magnetic field generation in a three-phase motor

net result of this is that a rotating magnetic field is generated at the centre of the rotor, this can be effectively represented by a rotating magnet as shown in Figure 7.2(b).

The speed of rotation of the magnetic field depends on the frequency of the three-phase supply and the number of stator windings or pole pairs. This relationship is shown by the following expression:

$$Ns = \frac{f}{p}$$

where (Ns) is the speed of the rotating magnetic field in revolutions per second, f is the three-phase supply frequency and p is the number of pole pairs.

Progress check 7.2

Briefly describe how a rotating magnetic field is generated.
Calculate the speed of the rotating magnetic field generated by six pole pairs with a supply frequency of 50 Hz.

Having established a rotating magnetic field in the stator we now need to consider the effects of this on the rotor.

Operating characteristics

Torque

For a motor to be of any use it must be able to produce torque, that is it must produce a *turning force about the centre* of the rotor shaft. This in turn can then be used to rotate a mechanical load usually via a pulley and belt arrangement.

The effect of the magnetic field on a cage rotor is to induce an e.m.f. in the bars of the rotor due to the fact that they are being continuously cut by the rotating magnetic field. As the bars are all connected together a currents flows and this in turn generates a magnetic field in opposition to the rotating magnetic field. As unlike magnetic poles attract and like poles repel the net result is that

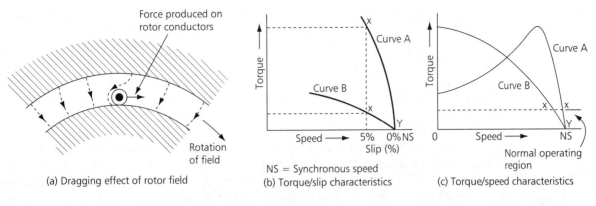

Figure 7.3 Rotor torque generation

the rotor magnetic field will be dragged around with the rotating magnetic field. As the rotor magnetic field is generated by the rotor itself the rotor will also be physically dragged around with the rotating magnetic field as shown in Figure 7.3(a).

Synchronous speed
The speed of rotation of the magnetic field is also known as the synchronous speed, as the rotor is actually being dragged by the rotating magnetic field the rotor lags slightly behind it. This effect which is known as **slip** means that the rotor speed is always less than the synchronous speed. Slip can be defined by the expression:

$$\text{slip} = \frac{\text{synchronous speed} - \text{rotor speed}}{\text{synchronous speed}}$$

In this form the answer is given in terms of per unit slip, sometimes it is more convenient to have the answer as a percentage in which case the expression is simply multiplied by 100%.

It is clear that as a mechanical load is applied to the motor the rotor will lag further and further behind the rotating magnetic field until eventually the motor will stall. An induction motor is therefore designed to operate with a particular value of mechanical load also known as **full load**. Typically the motor will be designed such that under full-load conditions the slip is approximately 5%. To understand the implications of this the torque/speed and torque/slip characteristics will be considered.

Torque/slip
The differences between wound rotor and cage rotor machines become apparent when considering their torque/speed and torque/slip characteristics as shown in Figure 7.3(b) and (c).

Curve A shows the typical cage rotor characteristic and curve B shows the typical wound rotor characteristic. The normal operating range for the induction motor is between points X and Y.

Torque/speed
When operated within the XY region of the torque/slip characteristic the motor operates almost as a constant speed machine as shown in Figure 7.3(c).

There is one disadvantage however; the starting torque is low and therefore the motor can only be started against moderate load conditions.

Curve B in Figure 7.3(c) is the characteristic of the wound rotor machine and as can be seen maximum torque is available from almost zero rotor speed. This means that this type of motor can be started against full load conditions. With this type of motor it is common practice to open-circuit the slip rings once the motor is running at the required speed. It then exhibits the same characteristics as the cage rotor machine.

Generally speaking induction motors are very efficient machines with typical figures approaching 90% and operate at high power factors typically between 0.8 and 0.9. This is only true of course under full-load conditions. It is therefore important to consider how this type of motor is controlled under start-up conditions to achieve *maximum efficiency and high power factors*.

Progress check 7.3

Calculate the percentage slip for a motor with a synchronous speed of 1000 r.p.m and a rotor speed of 700 r.p.m.

Briefly explain the differences in the operating characteristics of the cage and wound rotor motors.

Starting methods

There are basically four starting methods used with induction motors, on-line, star/delta, auto-transformer and wound rotor resistance. The on-line method is the most popular with star/delta and auto-transformer using current reduction techniques. None of these methods can be started against their rated loads; only the wound rotor resistance method can be used when starting with a rated load is required.

Switchgear

Each of these methods uses a number of switchgear components, the operation of which will need to be understood before these starting methods can be described in detail.

Contactor

This is the main switchgear component used in motor starter circuits; its function is to switch high current circuits ON and OFF. It has two basic forms, manually operated and solenoid operated with Figure 7.4(a) showing the construction of a mechanical type.

When used in a switching application a schematic diagram would be drawn showing its connections, as shown in Figure 7.4(b).

Notice the schematic symbols used for the contactor and that in a three-phase circuit it has three sets of contacts. The three-phase circuit would be turned ON by physically pressing the button on the manually operated contactor whereas in the case of the solenoid-operated contactor a switch in the low current 24 V circuit is operated.

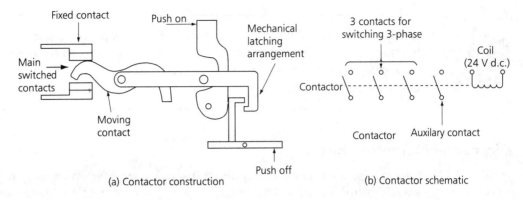

Figure 7.4 *Contactor arrangement*

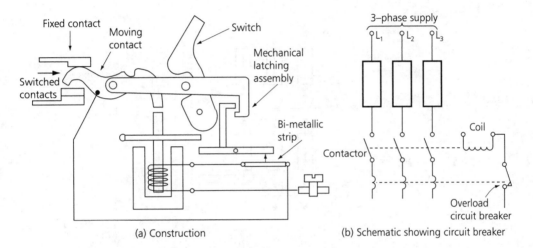

Figure 7.5 *Circuit breaker*

Circuit breaker

This device is used mainly for circuit protection and is available with different current ratings. The principle of operation is based on the bi-metalic strip connected to a mechanical latching arrangement as shown in Figure 7.5(a).

When the current flowing through the contacts exceeds a pre-defined value the heating effect of the current will cause the bi-metallic strip to bend until it eventually trips the mechanical latching arrangement. This will cause the contacts to latch open therefore turning the circuit OFF. The contacts can be closed again by physically pressing the reset button which latches the contacts ON again. This device can therefore be used to monitor the circuit current and when under fault conditions the current exceeds a predetermined value the circuit will automatically turn OFF, thereby protecting the circuit from overload. The circuit breaker is often used in conjunction with a contactor as shown in Figure 7.5(b).

Having briefly considered the operation of two switchgear elements we can now turn our attention to their application with regard to starter circuits for induction motors.

Activity 7.1

Using manufacturers' data investigate the different types of contactors and circuit breakers available. Produce a short report including sketches of three types of contactor and three types of circuit breaker with a short description of their main characteristics and features.

Direct-on-line motor starter

One of the most common methods used for starting an induction motor is the direct-on-line method. This method makes use of a solenoid-operated contactor with a circuit breaker for overload protection and a start and stop button as shown in Figure 7.6.

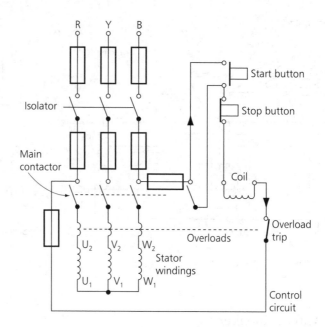

Figure 7.6 Direct-on-line method

The main contactor used in this application has four sets of contacts; three are used to switch the phases and the fourth is used as a retaining contact. When the start button is pressed the Red and Blue phases are connected across the coil of the main contactor activating the contactor and connecting the three phases to the induction motor. The fourth set of contacts effectively shorts out the start button and retains the circuit to the contactor coil. The only way to stop the motor is to press the stop button which breaks the retaining circuit, de-energising the contactor and hence stopping the motor. The motor will also stop if the current drawn by the motor exceeds the trip level of the overload circuit breaker. Once tripped the overload contacts will open causing the motor to stop. The overload circuit breaker has to be reset before the motor can be restarted.

This starting method tends to be used only for starting cage rotor induction motors. This method of starting has certain drawbacks in that the level of current at start-up is approximately six times higher than that at normal running speed; also the available starting torque is lower.

Star/delta starters

To reduce the current demand on start-up some form of current reducing technique is required. One such technique is to start the motor with its stator coils connected in star formation and then switch to delta when the motor is running. The reduction in starting current occurs due to the fact that the power available in each coil when connected in star is 1/3 of that when connected in delta.

When using this starting method the load has to be disconnected from the motor on start-up. Hence this starting method is not applicable in situations where the motor is permanently connected to a load.

The schematic diagram showing the switchgear connections is much more complicated than the direct-on-line method, as shown in Figure 7.7.

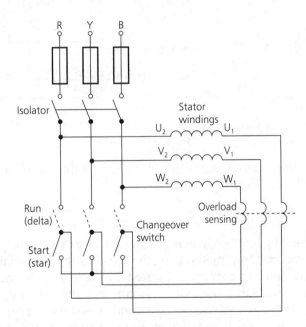

Figure 7.7 Star/delta schematic

The main contactor switches the mains supply, with the main switch in the start position a second contactor connects the stator coils in star. When the main switch is put into the run position a third contactor connects the stator coils in delta. For safety reasons mechanical and electrical interlocking is used to prevent the star/delta contactors from being activated simultaneous.

Auto-transformer starter

Where there is a requirement to start a motor against a partial load such as when a motor is driving a fan then the auto-transformer method can be used. Starting current is reduced by using tapped auto-transformers to reduce the voltage applied to the stator windings. A number of transformers could be used and would be switched out as the speed of the motor increases as shown in Figure 7.8.

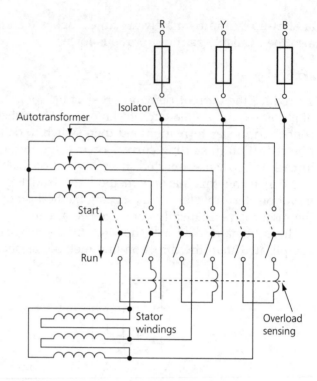

Figure 7.8 Auto-transformer starter

In modern systems the switching of the transformers would be done automatically via electronic means.

Wound-rotor-resistance

For applications where the motor is to be started against its full rated load the best starting method is the wound-rotor-resistance method. In this method the starting current is limited by adding resistance to the rotor windings. This also increases the power factor and the starting torque. The resistance is provided by variable wire-wound resistors connected to the rotor via carbon brushes which rub against the slip rings on the rotor shaft. The motor is started with the resistance set to its maximum value. As the speed of the motor increases the resistance is then slowly reduced until the motor reaches full speed at which point the resistance value is zero. The schematic diagram for this starter circuit is almost the same as for the direct-on-line method with the addition of external rotor resistors.

Motor testing

To establish torque/speed characteristics and efficiency of an induction motor certain tests will need to be carried out. These tests will require the use of a **motor test rig** which has facilities for braking the motor, measuring torque, electrical power, current and voltage which will allow measurements to be taken of power, current, voltage, speed and torque at different loads. From these results graphs may be plotted of torque, speed, current and efficiency.

Case study

In this case study a motor test set was used to establish the efficiency of a three-phase induction motor on full load. Under these conditions the test set returned the following results:

- line current = 10 A
- line voltage = 415 V
- wattmeter readings = 4500 W and 900 W
- speed = 1000 rev/min
- torque = 40 Nm

From the above readings the efficiency can be calculated by using the expression:

$$\text{Efficiency} = \frac{\text{output power}}{\text{input power}} \times 100\%$$

where input power = 4500 + 900 = 5400 W

Output power = $2\pi nT$

where n = speed (revs/sec) and T = torque (Nm).

Therefore:

Output power = 2 × 3.14 × (1000/60) × 40 = 4186.6 W

Efficiency = 4186.6/5400 × 100% = 77.5%

Assignment 7

This assignment provides evidence for:
Element 12.4: Investigate three-phase induction motors
and the following key skills:
Communication 3.1, 3.2
Application of Number 3.1, 3.3

For this assignment you will need access to a three-phase motor test set. Taking suitable safety precautions you will need to set-up the test set to enable sufficient readings to be taken to enable torque/speed and torque/slip characteristics to be plotted similar to those shown in Figure 7.3(b) and (c). You are also required to estimate the motor efficiency under full load conditions.

Report

A report is required containing details of your results, graphs, calculations and descriptions of motor operation and starting method used in the test.

The evidence indicators for this element suggest that a report should be produced on an investigation of one three-phase induction motor including:

- description of the type of three-phase induction motor
- description of the operating principles of the induction motor
- description and determination of how torque is produced on the rotor of an induction motor and the relationship between slip and speed
- description and determination of torque/slip characteristics
- description of starting method used

There are opportunities within this element to collect evidence towards key skills in Application of Numbers.

PART THREE: COMMUNICATIONS ENGINEERING

Chapter 8:   Signals and their conversion
Chapter 9:   The media used for signal transmission
Chapter 10: Communication techniques

- The ability to send and receive information electronically is now a vital part of our everyday lives. Watching a live TV broadcast via satellite or, perhaps, accessing the Internet from home or college is now almost taken for granted. It is only when one realises that it was not until the mid-1930s that the first domestic TV transmissions were made or that, even by 1975, the idea of a personal computer was still a dream – can one begin to appreciate how remarkable and rapid the advances in communication technology have been.
- The aim of Part 3 is to provide a broad introduction to the principles and practice of information signal transmission and to investigate some of the important techniques involved.

Chapter 8:   Signals and their conversion

This chapter covers:
Element 9.1:   Investigate signals and their conversion.

... and is divided into the following sections:
- Differences between information signals
- Analogue signals
- Digital signals
- Using decibels
- Noise and signal-to-noise ratio
- Digitising analogue signals – introduction to PCM.

In all communications systems the primary concern is to transmit information from one location to another as efficiently as possible. The information (e.g. the human voice, computer data etc.) will be converted to a voltage or current signal having a range of frequencies and is usually called the **baseband signal**. In general, signals can be of two types, either **analogue** or **digital**. Of the two, digital signals are the simplest and have many advantages over analogue signals. It is because of this relative simplicity that analogue signals are often converted into digital form before they are sent over communication systems.

After completing this element you should be able to:

- describe the properties of audio, video and digital baseband signals using logarithmic units where appropriate
- explain the reasons for converting analogue signals to digital form
- describe the stages in converting from analogue to digital and vice versa
- explain the requirements for signal conversion.

Differences between information signals

An **analogue signal** is one in which the voltage (or current) can take *any value within a given range*. The voltage is a copy of or an analogy of the physical quantity that it represents. As shown in Figure 8.1(a), the voltage produced by speaking into a microphone is an example of an analogue signal. The voltage produced by the microphone is a 'mirror image' of the sound wave variations produced by the voice. The signal varies from zero when there is silence up to a maximum level when the talker is producing maximum volume. The amplifier 'boosts' the

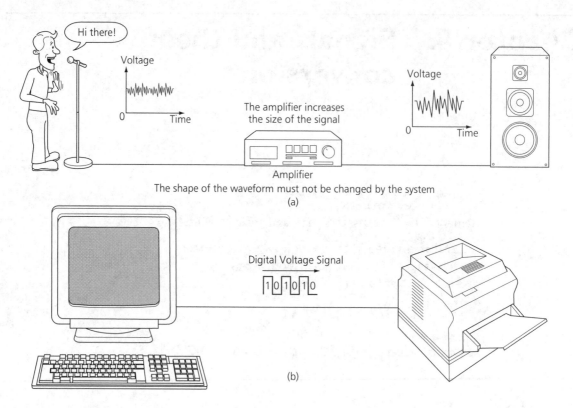

Figure 8.1 Analogue and digital signals

signal and the loudspeaker converts the voltage back into sound waves of higher intensity. It is important to realise that if the amplifier or the loudspeaker modifies the *shape* of the signal in any way, then the output will not be a faithful reproduction of the voice. Any change in shape of a waveform is called **distortion**. Thus, ideally, in analogue systems there should be no distortion whatsoever for perfect operation.

A **digital signal** is one in which the voltage *can only have two distinct values*. The signal sent from a computer to a printer is an example of a digital signal (see Figure 8.1(b)). Only two voltages are used and a 'stream' of voltage pulses representing the information is sent from the computer via the cable to the printer. You will see as you progress through this element that digital signals have many advantages when compared to analogue signals. In fact it is advantageous in many applications to *change* analogue signals into digital signals. Before we can examine this process it is necessary to know what frequencies are contained in the most common types of signals.

Analogue signals

When an information signal such as the human voice or computer data etc. is sent over a communication link, the voltage or current that represents it has a range of **frequencies** that characterise the information. The resulting signal voltage or current is called the **baseband signal**. In the case of voice, the range of frequencies will be quite 'narrow'. In contrast, a TV signal will have a relatively wide range of frequencies within it. The communication system carrying the signal must, of course, be able to transmit all of the important frequency components that the signal contains. The range of frequencies that the system can accommodate is called the **system bandwidth** and must be greater than or at least equal to the frequency range of the signal it is carrying. In order that the system bandwidth can be correctly chosen, the frequency content and characteristics of typical baseband signals need to be known. We'll start by investigating the frequencies contained in the human voice.

The human voice

Human voice reproduction is a complex operation involving the vocal chords, larynx, windpipe, mouth and nasal cavities. Air from the lungs causes the vocal chords to vibrate which alternately blocks and releases air into the mouth. This produces compressions and rarefactions of the air molecules which is known as a sound wave. The sound waves produced by speech have frequencies in the range from about 100 Hz up to about 10 kHz. The power, however, is not evenly distributed across the frequency range but peaks around 500 Hz to 600 Hz for men and about 800 Hz to 1 kHz for women.

The commercial speech bandwidth

In a telephone system it is not necessary to transmit all the frequencies produced by the human voice in order to have intelligible communication. By international agreement, the voice frequency band is restricted to the frequency range 300 Hz to 3.4 kHz, i.e. the bandwidth of the signal is (3400 − 300) = 3.1 kHz. Thus, when you are talking on the telephone you are hearing no frequencies above 3.4 kHz and none below 300 Hz. Even though the bandwidth has been limited, you will agree that most conversations are intelligible and enough frequency information is retained for you to recognise the speaker's voice.

The range of frequency from 300 Hz to 3.4 kHz is called the **commercial speech bandwidth** or *the bandwidth of a voice channel*. For reasons that we shall investigate later, the bandwidth allocated to a voice channel in a commercial telephone system is greater than this. Each voice channel is allocated a frequency 'slot' of 4 kHz. Thus a little 'frequency space' is allowed above and below the voice channel. This is shown in Figure 8.2(a). The frequency spaces are known as **guard bands**. Why they are needed will be examined in Chapter 10.

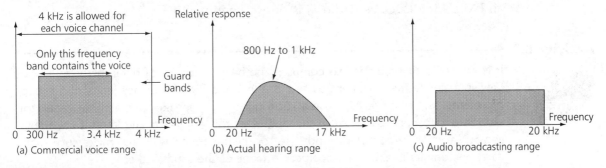

Figure 8.2 The commercial and actual speech bandwidth

Activity 8.1

A small loudspeaker can be used as a simple microphone. Connect the loudspeaker directly to channel 1 of an oscilloscope. Adjust the sensitivity of channel 1 until when you speak into the loudspeaker a waveform is displayed. The waveform will be quite complicated indicating that the human voice contains a range of frequencies. Connect another loudspeaker to a signal generator set to produce a 1 kHz sinewave. The 'scope trace from the first loudspeaker should now be a fairly good sinewave when the 'scope is adjusted correctly.

Human hearing

When we hear, sound waves cause the ear drum to vibrate and these vibrations are transferred to nerve fibres within the inner ear. The nerve fibres convert the vibrations into minute voltages which are then passed to the brain and interpreted as the sensation of sound.

The frequency range over which the average person is able to respond to sound is typically in the range of about 20 Hz–17 kHz, Figure 8.2(b). This range, however, varies considerably from one person to another and depends on such factors as age and past exposure to high levels of sound etc. For communication system purposes, this **audio frequency band** is defined as lying between 20 Hz–20 kHz. Thus the **audio frequency bandwidth** is almost 20 kHz, Figure 8.2(c).

The ear can respond to a wide range of sound intensities. If the sound intensity is too low, the sound will not be heard. The minimum intensity level is called the **threshold of hearing**. It varies considerably with frequency over the hearing range and from individual to individual. If sound levels are too large, then feelings of discomfort and pain will be experienced. This is described in terms of the **threshold of feeling**. Like the threshold of hearing, the threshold of feeling varies with frequency and from person to person.

The 'average ear' is most sensitive to frequencies within the range of about 800 Hz–2 kHz. For this reason the **standard test tone** used in telephone systems is either 800 Hz (Europe) or 1 kHz (USA). The test tone is used to check that the various circuits in the system are operating correctly.

Music

The range of frequencies produced by musical instruments is much wider than that produced by the human voice. Consequently a wider bandwidth must be provided for the transmission of music signals over commercial systems if the quality is to be acceptable. For FM broadcasts, the frequency range for music is from 40 Hz–15 kHz. You would probably agree that the sound quality from an FM station (such as Radio 1) is of excellent quality.

Activity 8.2

It is an interesting exercise to compare the hearing ranges of different people. Devise a simple measurement to test the low and high frequency limits of a group of volunteers. (Hint: Perhaps use a good quality loudspeaker and a signal generator. Use an oscilloscope to measure the output of the generator and keep it constant as you vary the frequency. Do not use too high a sound level and make sure that each person sits in the same position in front of the loudspeaker.

Television transmission

To receive TV signals, information must be transmitted containing a **luminance signal** which represents the image in black and white, a **chrominance signal** containing colour information and a **sound signal** (see Figure 8.3). In view of the large quantity of information you would expect that a fairly wide bandwidth would be required when compared with, say, a telephone channel. In practise, a TV signal has a bandwidth of 5.5 MHz.

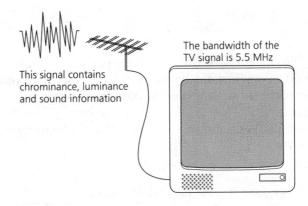

The bandwidth of the TV signal is 5.5 MHz

This signal contains chrominance, luminance and sound information

Figure 8.3 Bandwidth of TV signals

Digital signals

A digital signal is one in which the voltage (or current) can have *only two possible values*. Each value is known as a **bit** and are usually given the symbol '1' and '0'. Two examples of digital signals are shown in Figure 8.4.

In Figure 8.4(a) a '1' is represented by a positive voltage 'V' and a '0' is represented by zero volts. Such a digital signal is called unipolar because only one polarity of voltage is used. A bipolar digital signal is drawn in Figure 8.4(b). Here '+V' represents '1' and '-V' represents '0'. In both cases, each voltage value is called a bit. Both examples in Figure 8.4 are seven bits long.

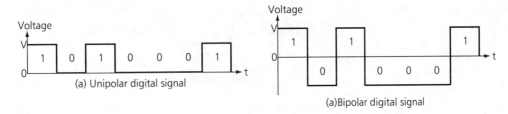

(a) Unipolar digital signal

(a)Bipolar digital signal

Figure 8.4 Digital signals

Bit rate

The number of bits that occur in one second is called the bit rate. If, for example, each bit in Figure 8.4 lasted 10 ms then the bit rate is:

$$\frac{1}{10 \times 10^{-3}} = 100 \text{ bits/s}$$

Communications engineering

The frequencies contained in a square wave

In order to determine the necessary bandwidth required of systems which carry digital signals, it is vital to know what frequency components are contained in the signals. Consider the square waveform in Figure 8.5(a). The waveform has a period T_F so its frequency:

$$f_F = \frac{1}{T_F} \quad \text{(hertz)}$$

Figure 8.5(b) shows what happens if a sinusoidal-shaped waveform or **sinusoid** of frequency f_F is added to a smaller sinusoid of frequency $3 \times f_F$. The first sinusoid is called the **fundamental wave** and the other is called the **third harmonic**. (This is why the subscript 'F' has been used.) As you can see, the result out of these waveforms is already beginning to look like the original square wave.

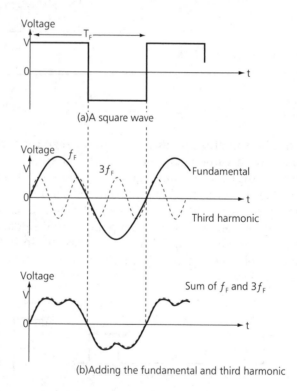

(a) A square wave

(b) Adding the fundamental and third harmonic

Figure 8.5 *The frequency components in a square wave*

If we continued by adding the fifth harmonic ($5 \times f_F$) and seventh harmonic ($7 \times f_F$) with progressively decreasing peak values of the correct size, the resulting waveform gets more and more like a square wave. (Theoretically, to get a perfect square wave All of the odd harmonics up to infinity would need to be included. In practice, the inclusion of up to the fifteenth is acceptable.) This result is very important because it means that we can make (or synthesise) square waves by

134

adding together odd harmonic sinusoids of the correct peak value or we can 'break down' (or analyse) square waves into its odd harmonic components.

Progress check 8.1

A square wave has a period of 1 ms. What are the first four sinusoidal frequencies contained within it?

The frequencies contained in a digital signal

The digital data stream transmitted from, say, a computer to a printer will consist of all sorts of combinations of bit patterns depending on the information being sent. Our knowledge of the frequency content of a square wave, however, will

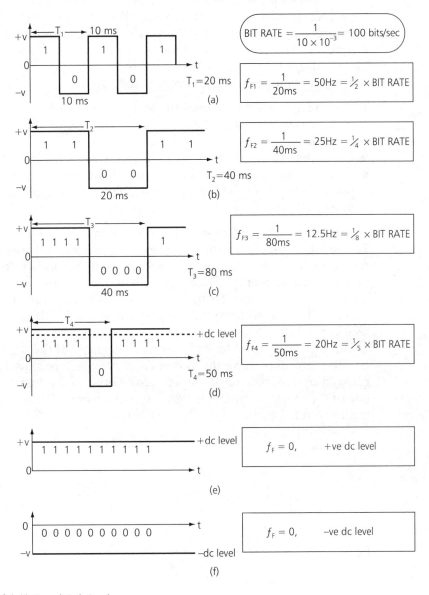

Figure 8.6 Various digital signals

help work out the frequency content of any digital signal. Suppose, for simplicity, that the duration of each bit is 10 ms. The bit rate is therefore:

$$\frac{1}{10 \text{ ms}} = 100 \text{ bits/s}$$

Figure 8.6 shows six different digital waveforms that *could* be possible. For each, the fundamental frequency f_F has been calculated.

In Figure 8.6(a) the most rapidly changing waveform possible is shown. It is a square wave consisting of alternate '1's' and '0's'. The fundamental frequency is 50 Hz which is *one half of the bit rate value 100*. The waveforms in (b) and (c) are other possible square waves. Their fundamental frequencies are one quarter and one eighth of the bit rate. Since each waveform in Figure 8.6(a), (b) and (c) are square waves, the average value (which is the d.c. value) is equal to zero. Figure 8.6(d) shows a possible waveform in which the d.c. level is NOT zero. The waveform spends more time at '1' than at '0' so the d.c. level is positive.

Figure 8.6(e) and (f) show the possible waveforms having **zero fundamental frequency**. They consist of 'streams' of '1's' or '0's'. They have average values (d.c. values) of +V and −V respectively.

We can summarise the findings as follows:

● The maximum fundamental frequency is one half of the bit rate. If the bit rate is increased (i.e. more bits are sent per second) the fundamental frequency increases.
● Generally, the 'bit stream' will contain a d.c. voltage level.
● To retain the rectangular shape of the digital signal through a system all of the odd harmonics of the fundamental frequency would have to be passed by the system. Theoretically, the system would need an infinite bandwidth.

But we do not need to send all of the harmonics. The shape is not important. To see why, refer to Figure 8.7. Here a digital signal is being sent from a transmitter to a receiver through a system whose bandwidth will only pass the fundamental frequency which equals half the bit rate. None of the harmonics are transmitted. Even though the digital signal is distorted the **regenerator circuit** senses a '1' or '0' and produces a brand new digital signal which is passed to the receiver.

The important point to realise is that the shape of the digital waveform is fairly unimportant. Providing the existence of voltage levels at '1' or '0' can be recognised, a new signal can be reproduced without any loss of information. This is one of the major advantages digital systems have over analogue systems. We will return to this point later in the element.

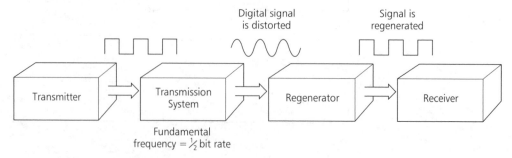

Figure 8.7 Regenerating a digital signal

Progress check 8.2

A digital signal has a bit rate of 9600 bits per second. What are the maximum and minimum fundamental frequencies in the signal?

Using decibels

The decibel unit (dB) and other units related to it are very common throughout communications and in engineering generally. Consider the system in Figure 8.8(a). This shows a system (it could be an amplifier, for example) which has an input power P_{in} and an output power P_{out}.

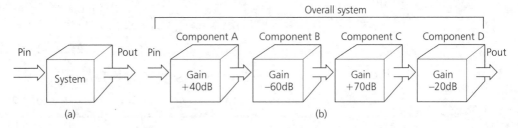

Figure 8.8 Defining power gain and loss

If P_{out} is bigger than P_{in}, we would say that the system has **power gain**. If P_{out} is less than P_{in} then the circuit has **power loss**. The ratio of P_{out} to P_{in} is the power gain or loss of the system. That is,

$$\text{Power gain (or loss)} = \frac{P_{out}}{P_{in}} \quad \text{(no units)}$$

This ratio has no units – it is just a number. For example, suppose $P_{out} = 100$ W and $P_{in} = 1$ mW. The power gain is:

$$\text{Power gain} = \frac{100}{1 \times 10^{-3}} = 100{,}000$$

Conversely, suppose $P_{in} = 10$ W and $P_{out} = 1$ nW, (n = nano = 10^{-9}). The power loss is:

$$\text{Power loss} = \frac{1 \times 10^{-9}}{10} = 10^{-10}$$

The power gain or loss is normally expressed in decibels as

$$\text{Power gain or loss} = 10 \log_{10}\left(\frac{P_{out}}{P_{in}}\right) \quad \text{(dB)}$$

If we repeat the two examples above using dBs we get:

Power gain = $10 \log_{10}(100{,}000) = +50$ dB

Power loss = $10 \log_{10}(10^{10}) = -100$ dB

Notice that the fairly large and small values in the two examples have been 'compressed' into more easily manageable numbers using the logarithmic unit. This is a major advantage of using the dB notation. Note that for a power gain, a positive dB value results; for a power loss, the dB value is negative.

WORKED EXAMPLE 8.1

Determine the dB gain or loss if $P_{in} = 1$ W when P_{out} is (a) 2 W, (b) 0.5 W, (c) 2 kW and (d) 0.1 W.

Solution

(a) $P_{in} = 1$ W, $P_{out} = 2$ W. Power gain $= 10\log_{10}(2/1) = +3$ dB

(b) $P_{in} = 1$ W, $P_{out} = 0.5$ W. Power loss $= 10\log_{10}(1/2) = -3$ dB

(c) $P_{in} = 1$ W, $P_{out} = 2000$ W. Power gain $= 10\log_{10}(2000) = +33$ dB

(d) $P_{in} = 1$ W, $P_{out} = 0.1$ W. Power loss $= 10\log_{10}(1/10) = -10$ dB

Progress check 8.3

If $P_{in} = 5$ W and $P_{out} = 50$ W calculate the power gain in dB. If P_{out} is doubled, by how much does the gain increase?

Another advantage of using dBs is that if the gains and losses of the components making up a system are expressed in dBs the overall gain or loss can be found simply by *adding* the individual gain and losses together. Consider the system in Figure 8.8(b).

The overall gain is:

$$10 \log_{10}\left(\frac{P_{out}}{P_{in}}\right) = +40 - 60 + 70 - 20 = +30 \text{ dB}$$

Now $\log_{10}\left(\dfrac{P_{out}}{P_{in}}\right) = \dfrac{30}{10} = +3$

So $\dfrac{P_{out}}{P_{in}} = 10^3 = 1000$

In this example, the overall system has a gain of 1000. If $P_{in} = 1$ mW then P_{out} would be 1 mW $\times 1000 = 1$ W.

Progress check 8.4

If the system components in Figure 8.8(b) have the following gains and losses, determine the overall gain or loss of the system. If $P_{in} = 1$ W calculate P_{out}.

Component A $= -10$ dB, B $= +20$ dB, C $= -20$ dB and D $= -30$ dB.

Expressing power levels in decibel units

The dB is only a measurement of a **power ratio**. It is not an absolute unit like the volt and amp etc. It is, however, possible to express powers in decibel units provided a **reference power level** is chosen. Therefore, to say that a power level is, for example, +20 dB is totally meaningless *but* to say that the power is +20 dB above a reference power of 1 W is perfectly acceptable. When this is done the power level is given the unit of dBW which means *'decibels relative to 1 W'*. For example, if power P = 100 W, then:

$$P \text{ in dBW} = 10 \log_{10}\left(\frac{100}{1}\right) = +20 \text{ dBW}$$

If, say, P = 1 mW then:

$$P \text{ (dBW)} = 10 \log_{10}\left(\frac{1 \times 10^{-3}}{1}\right) = -30 \text{ dBW}$$

If the reference level is chosen to be 1 milliwatt, then the unit that results is called the dBm which means *'decibels relative to 1 mW'*. If P = 100 W then,

$$P \text{ in dBm} = 10 \log_{10}\left(\frac{100}{1 \times 10^{-3}}\right) = +50 \text{ dBm}$$

WORKED EXAMPLE 8.2

Express the following power levels in dBW and dBm: (a) 1 W, (b) 1 mW, (c) 2 W, (d) 1 W.

Solution

(a) In dBW, $1 \text{ W} = 10 \log_{10}\left(\frac{1}{1}\right) = 0 \text{ dBW}$

In dBm, $1 \text{ W} = 10 \log_{10}\left(\frac{1}{1 \times 10^{-3}}\right) = 30 \text{ dBm}$

(b) In dBW, $1 \text{ mW} = 10 \log_{10}\left(\frac{1 \times 10^{-3}}{1}\right) = -30 \text{ dBW}$

In dBm, $1 \text{ W} = 10 \log_{10}\left(\frac{1 \times 10^{-3}}{1 \times 10^{-3}}\right) = 0 \text{ dBm}$

(c) In dBW, $2 \text{ W} = 10 \log_{10}\left(\frac{2}{1}\right) = +3 \text{ dBW}$

$$\text{In dBm,} \quad 2\,W = 10\,\log_{10}\left(\frac{2}{1 \times 10^{-3}}\right) = 33\,\text{dBm}$$

(d) $\text{In dBW,} \quad 1\,W = 10\,\log_{10}\left(\frac{1 \times 10^{-6}}{1}\right) = -60\,\text{dBW}$

$$\text{In dBm,} \quad 1\,W = 10\,\log_{10}\left(\frac{1 \times 10^{-6}}{1 \times 10^{-3}}\right) = -30\,\text{dBm}$$

Have you 'spotted' how to convert from one unit to another? *To change from dBW to dBm just add 30. To change from dBm to dBW just subtract 30.*

Using voltage and current ratios

The power dissipated in a resistance is given by:

$$P = \frac{v^2}{R} \quad \text{or} \quad P = i^2 \times R$$

Here V is the r.m.s. voltage across R and i is the r.m.s. current flowing through it. Consider the equation:

$$P = \frac{v^2}{R}$$

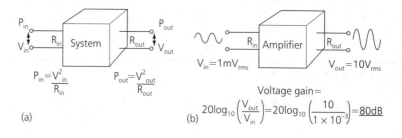

Figure 8.9 Using voltage ratios

Applying the decibel equation to the system in Figure 8.9(a) gives:

$$\text{Power gain or loss} = 10\,\log_{10}\left(\frac{V_{out}^2/R_{out}}{V_{in}^2/R_{in}}\right)$$

$$= 10\,\log_{10}\left(\frac{V_{out}^2}{V_{in}^2}\right) + \log_{10}\left(\frac{R_{in}}{R_{out}}\right)$$

$$\text{Power gain or loss} = 20\,\log_{10}\left(\frac{V_{out}}{V_{in}}\right) + 10\,\log_{10}\left(\frac{R_{in}}{R_{out}}\right)$$

If $R_{in} = R_{out}$ the second term in the equation is zero because $\log_{10}(1) = 0$, so:

$$\text{Power gain or loss} = 20\,\log_{10}\left(\frac{V_{out}}{V_{in}}\right) \quad \text{(dB)}$$

Even though the equation is a power gain or loss it is now common practice to describe this equation as being the *decibel voltage gain or loss of the system.* For example, the amplifier in Figure 8.9(b) would be described as having a voltage gain of 80 dB even though R_{in} and R_{out} would not be equal for a real amplifier circuit.

Progress check 8.5

Starting with $P = I^2 \times R$, determine an equation for the power gain or loss in terms of the currents i_{in} and i_{out}.

Noise and signal-to-noise ratio

Any signal that interferes with or obscures the desired signal is called **noise**. The 'hiss' you hear between tracks on a cassette tape is noise. The 'speckling' sometimes seen on a TV when a car passes is also noise. The ratio of the wanted signal power to the unwanted noise power is called the signal-to-noise-ratio (S/N) and is expressed in dBs as follows:

$$\frac{S}{N} = 10 \log_{10} \left(\frac{\text{wanted signal power}}{\text{unwanted noise power}} \right) \text{ (dB)}$$

The higher the signal relative to the noise, the higher will be the dB value. Generally, in communications systems we try to achieve the maximum value of S/N we can but the acceptable level will depend upon the system.

Noise will have many sources and may be produced within the system itself or 'picked up' from external sources. Any signal processed by a system, whether

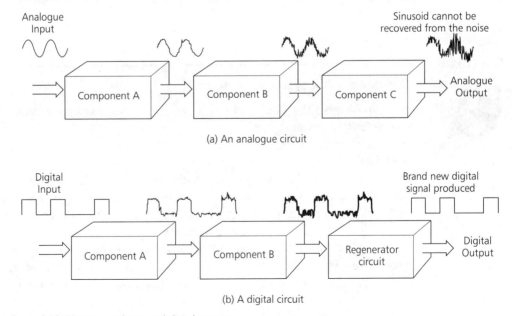

Figure 8.10 *Noise in analogue and digital systems*

analogue or digital, will be contaminated by noise. When the effects of noise are taken into account, another general advantage of digital signals over analogue signals can be seen. Consider the systems shown in Figure 8.10.

In Figure 8.10(a) an analogue signal (in this case a sinusoid) is being processed by a system. At each step noise is added to the sinusoid by the circuits within the 'blocks' and picked up from external sources. As the signal progresses through the system the signal-to-noise ratio continually decreases until at the output it is impossible to recognise the original sinusoid because it is so heavily contaminated with noise. For the digital system in Figure 8.10(b) on the other hand, providing the regenerator can still recognise a '1' or '0', a brand new digital signal can be regenerated. Of course, an occasional error may occur when the noise is so great that a '0' may be sensed as a '1' or vice-versa. In general though, a digital system has a much greater **noise immunity** than a comparable analogue system.

Benefits of digital systems

It should now be becoming apparent that, in many situations, digital signals have several advantages when compared to analogue signals. Two of these are:

- Digital signals only have two voltage levels which means that digital circuits are easier to design. Analogue signals can take up any voltage level within a given range making circuit design more complicated.
- The effect of noise and distortion on digital signals can be almost eliminated by using regenerators. Although it can be minimised, the effect of distortion and noise cannot be eliminated from analogue signals. .

Progress check 8.6

(a) A signal with a power level of 10 W is embedded in a noise signal of power level 1 mW. What is the signal-to-noise ratio?
(b) If the noise level is reduced to 0.5 mW, by how much does the signal-to-noise ratio increase?

Do this 8.1

Checking out a system for noise

You can observe the effects of noise and appreciate the importance of S/N ratio with your hi-fi system.

Procedure:
Switch on the amplifier but do not load a cassette or CD. Slowly turn the volume control from minimum to maximum and listen to the loudspeaker output. Depending on the quality of your equipment you might have to get quite close to the loudspeaker before you can detect any output. You will probably hear a 'hum' and a 'hissing' noise. Since you did not provide a wanted input signal but did get a discernible output, then there must be sources of noise within the system.

You can also illustrate the concept of signal-to-noise ratio quite easily. Adjust the amplifier to the normal level used and play a cassette or CD. Now listen for the noise output you could hear before. You will probably find that it is no longer distinguishable. In this case the wanted signal (music) is so much larger than the unwanted noise that the noise is inaudible, i.e. the signal-to-noise ratio is large enough for satisfactory operation.

If you compare the speaker output when the cassette and CD are 'between tracks' you should detect more noise being produced by the cassette than by the CD. Thus a digitally recorded CD will give a much higher S/N ratio than a cassette.

Digitising analogue signals – introduction to PCM

Before analogue signals such as speech and music etc. can be processed by digital systems, the analogue waveform must first be converted into digital form. The process is called **pulse code modulation** and the basic principles will now be described. Consider the system shown in Figure 8.11(a).

The analogue signal is first passed through a **low pass filter** which only allows frequencies up to a certain value to pass. For example, for speech in a telephone system the low pass filter 'cuts off' signals at 3.4 kHz so that no frequencies above this enter the system. The band-limited analogue waveform and another waveform called the **sampling pulse train** are fed into a circuit called a **sampler**. The output of the sampler is a stream of pulses, at the same frequency as the pulse train, whose amplitude is proportional to the amplitude of the analogue waveform at the instant of sampling. This waveform is called a **pulse amplitude modulated signal** (PAM signal). The question that now immediately 'springs' to mind is – how often must you sample the waveform? Figure 8.11(b) shows a waveform sampled at two different frequencies. Obviously, in the top diagram the sampling frequency is so low that details of the waveform shape have been lost. In the bottom diagram, the sampling frequency is such that the shape of the waveform has been retained. The minimum frequency that a waveform must be sampled at is given by the **sampling theorem**.

The sampling theorem

The sampling theorem states that a waveform containing frequencies up to a maximum of f_{MAX} must be sampled at least $(2 \times f_{MAX})$ times per second if the original waveform is to be successfully reconstructed from the pulse waveform. Thus, theoretically a voice waveform containing frequencies up to 3.4 kHz must be sampled at, at least, 6800 times per second. In practice, due to the requirements of the low pass filter, a sampling frequency of 8 kHz (2×4 kHz) is used. If you are speaking over a PCM telephone system your voice is being sampled 8000 times per second!

Progress check 8.7

What is the minimum frequency that an audio frequency music signal should be sampled at?

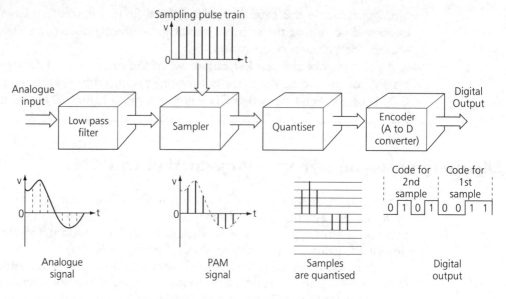

(a) Steps in digitising an analogue waveform

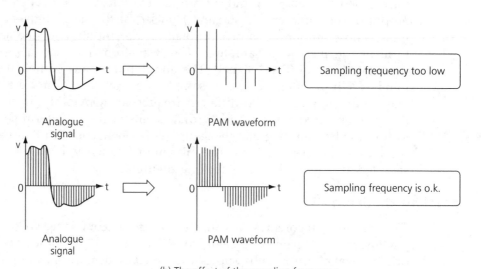

(b) The effect of the sampling frequency

Figure 8.11 Digitising an analogue signal

Quantisation and encoding

After sampling, the PAM waveform is passed to the **quantiser circuit**. The function of the quantiser is to 'round off' the pulse amplitudes to the nearest one of a range of allowed voltage levels. The process of approximating the sampled amplitudes is called **quantisation** and the allowed voltage levels are called **quantisation levels**. The quantised samples are then passed to the **encoder**, (**analogue to digital converter**), which converts the analogue sample into the equivalent digital form. For example, suppose the first and second samples are quantised to 3 V and 5 V respectively. The A to D converter will produce a digital output equivalent to these voltages as illustrated in Figure 8.11(a).

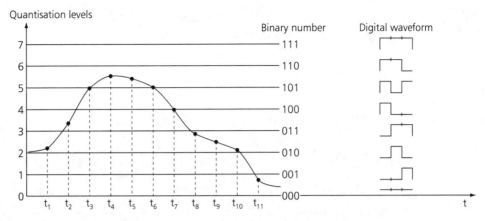

Figure 8.12 *Quantising a signal to 8 levels*

Obviously, the greater the number of levels allowed the less will be the error between the *actual sample and its quantised value*. The number of bits required in the encoding process, however, will increase. In general, the number of levels possible using 'n' bits is 2^n. So if 3 bits are used for the digital output then 8 levels are possible – 4 bits would allow 16 levels and so on.

Figure 8.12 shows the quantisation and encoding of a signal where 3 bits are used in the quantisation process. These levels have been labelled 0 to 7.

The equivalent binary number will range from 000 up to 111. The analogue waveform is sampled at instants t_1, t_2 etc. At t_1, the waveform is between levels 2 and 3 but because it is closer to level 2 it is approximated to this value. This quantised sample is encoded by the analogue to digital converter and the digital waveform shown is produced. At t_3, the analogue waveform coincides exactly with a quantisation level (5) so no error is produced by the quantisation process for this sample. At t_6, the sample is closer to level 5 than level 6 so the digital waveform for level 5 is sent from the analogue to digital converter. Thus the output from the encoder is a stream of pulses which is the digital representation of the original analogue waveform.

There are some important points which should be noted:

- the greater the number of levels used, the smaller will be the error between the PAM signal and the quantised signal
- increasing the number of levels increases the number of bits required to represent each level
- increasing the number of bits per sample means that more time is required to send each sample over a digital system.

In practice, of course, a compromise is made. It is found that very good quality speech can be transmitted over a PCM telephone system using *8 bits to code each sample*. This corresponds to $2^8 = 256$ levels.

Systems and standards for PCM are now laid down by international agreement. An 8 kHz sampling rate, with each sample encoded into 8 bits, means that each telephone coversation results in a continuous bit stream of 64000 bits per second (64K bits/sec).

Progress check 8.8

A compact disk stores its information in digital form. 16 bits are used for each sample. How many quantisation levels are used?

Recovering the analogue signal from the digital waveform

The main components of a system for receiving the digital waveform and reconstructing the original analogue signal are shown in Figure 8.13.

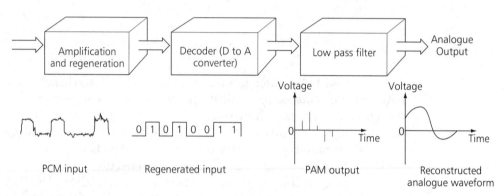

Figure 8.13 *The main components of a PCM receiver*

As previously explained, the digital signal will become contaminated with noise and be distorted by the system through which it passes. The first step is therefore to amplify and regenerate the signal and so produce a brand new digital waveform. The groups of bits representing each sample are then passed to the input of a **digital-to-analogue converter** (**decoder**). The decoder converts each digital sample into the corresponding PAM signal. The final recovery of the original analogue signal is achieved with a low pass filter with a bandwidth of 0 to f_{MAX} where f_{MAX} is the maximum frequency of the original analogue signal.

Quantisation noise

The analogue waveform at the output of the receiver in Figure 8.13 will not be an exact replica of the original analogue waveform. Rather, it is a quantised approximation to it. If we subtract the two waveforms then we have an error waveform which is called quantisation noise. It is given this name because it produces noise at the output of the receiver. The magnitude of the noise depends on the sampling rate and the number of quantisation levels. If either, or both, the sampling rate and number of quantisation levels is increased, the noise will be reduced.

Further considerations of digital signals

Once an analogue signal has been digitally encoded there is in effect no difference between it and data produced by computers etc. Thus the same systems can be used for the transmission of computer data as well as sound and TV signals providing the systems can accommodate the necessary data rates. There are other possibilities as well. For example, extra bit(s) can be added to the data to detect errors if they occur. Some systems allow not only for the detection of the error but also for its correction as well.

Case study

The **compact disk player** provides a good example of where analogue signals (e.g. music) are encoded in digital form for storage and finally decoded back into analogue form for replay over a hi-fi system. The process is, of course, quite complex but the general principles are fairly easy to understand. The following description is greatly simplified.

Encoding the information on disk

The analogue signal (e.g. music) is first recorded onto an analogue tape recorder to produce the master tape. (Alternatively, the recording is directly onto digital tape.) PCM is used to sample and encode each stereo channel. The sampling frequency is 44.1 kHz and each sample is encoded into 16 bits. This allows $2^{16} = 65,536$ separate levels per channel. The quantisation noise is so low that it corresponds to a signal-to-noise ratio of the recovered analogue signal of well over 90 dB. Before the master compact disk is made, extra bits are added which allows for error correction of the reconstructed analogue signal. More bits giving information about playing time, track and locational information are also included. This final digital signal is used to vary the intensity of a laser which 'burns' areas into the master disk called **pits** and leaves spaces between the pits which are called **lands**. The pits of varying lengths and the lands in between now contain the digital information which forms a spiral track on the master disk. The master disk is copied to produce the final CD.

Playback from the disk

The pits and lands are so small that they are invisible to the naked eye. Light from a laser in the CD player is directed towards the disk by a semi-transparent mirror. Lenses focus this laser light to a minute spot on the surface of the disk. As the disk rotates, light will be reflected from the lands back towards the semi-transparent mirror. If the laser light falls on a pit it will be scattered and not reflected back. Thus the digital information originally stored on the disk causes an 'ON' and 'OFF' flashing laser beam to return from the disk surface. The flashing beam passes through the semi-transparent mirror and falls on a photodetector which converts the light back into a digital voltage signal. The digital voltage signal is converted back into PAM form via a 16 bit digital-to-analogue converter. The final reconstructed analogue waveform is taken from the output of a low-pass filter.

Assignment 8

This assignment provides evidence for:
Element 9.1: Investigate signals and their conversion
and the following key skills:
Communication 3.2, 3.3
Information Technology 3.1, 3.2, 3.3

The evidence indicators for this element require you to investigate one piece of equipment which processes both analogue and digital signals. Your Centre will provide this equipment.

Report

Your report should include an explanation of the stages of signal processing and the effect each process has on the signal waveform. The results of your investigation should show clearly the types of signal and their magnitude, frequencies and bandwidth.

A record of an investigation of signals and their conversion should also be submitted. It should include:

● a description of the properties of audio, video and digital baseband signals
● an explanation of the reasons for converting analogue signals to digital form
● a description of the stages in conversion from analogue to digital signals
● an explanation of the requirements for signal conversion
● the use of logarithmic units where necessary.

Chapter 9: The media used for signal transmission

This chapter covers:
Element 9.2:   Evaluate media used for signal transmission.

. . . and is divided into the following sections:
● Transmission lines
● Using free space as a transmission medium
● Modes of propagation of radio waves
● Optical fibres.

There are two basic methods whereby an information signal (whether analogue or digital) can be sent from a transmitter to a receiver. The transmitter and receiver may be permanently connected together via copper cables or optical fibres. Alternatively, there may be no physical connection and the signal to be sent is broadcast through free space as a radio wave. Cables, optical fibres and free space are generally called **signal transmission media**. Each of the media has its own particular advantages and disadvantages and the choice of which is used will depend on many factors. In many communication systems, combinations of the various media are chosen to satisfy the ultimate objective of exchanging information between the transmitter or sender and receiver. In this chapter, the *properties of transmission media* will be investigated.

When you complete this element, you should be able to:

● describe signal transmission media
● describe the properties of signal transmission media
● evaluate signal transmission media for specific types of signal.

Transmission lines

Cables consisting of a pair of metal conductors separated from one another by an insulating material are called transmission lines. The insulating material serves to keep the distance between the two conductors constant and is usually called the **dielectric**. Polythene and polyethylene are typical dielectrics. The transmission lines shown in Figure 9.1 are the types most often used. The **coaxial cable** is probably familiar to you already because it connects your TV set to the antenna on the roof (Figure 9.1(a)). If you have satellite TV, coaxial cable connects the low noise block (LNB) in front of your receiving dish to the satellite receiver in your

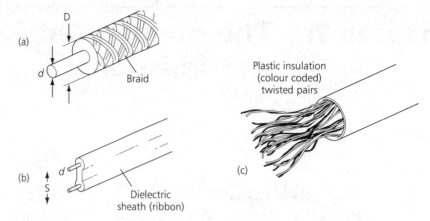

Figure 9.1 Coaxial cable, twin lead cable and twisted pairs

living room. It is often used in computer networks to wire computers together so that software can be shared.

Twisted pairs are used to connect individual telephone subscribers to their local exchanges. (The reason for twisting the pairs is to cancel out interference or cross-talk between the pairs.) Some computer networks also utilise twisted pairs. **Twin lead cables** are sometimes employed to connect low frequency antennas to transmitting and receiving equipment.

The choice of which type of line is best suited for a particular task will depend on many factors. Principally, however, it will be the frequency and bandwidth of the signals involved and the **attenuation** (loss of signal strength along the line) which will be the main considerations. Cost will also be a factor.

Activity 9.1

Your Centre will almost certainly have computers connected together as a network. Find out from the network technician what type of transmission medium is used to connect the computers together.

Waves on transmission lines

When an alternating voltage is connected between the conductors of a transmission line the voltage and the current it produces travels down the line in the form of a wave. If the line is long, it will take a finite time for the effect of the voltage and current waves to reach the far end. Understanding this wave behaviour is important and can be appreciated by using an analogy. Consider the diagram in Figure 9.2.

If the rope is moved continuously up and down, a wave motion will travel along the rope towards the far end. You would notice the following:

● Any position (e.g. X) on the rope would move up and down in the same way as that at the 'hand end' but not necessarily in step with it. There is a **time phase difference** between the motion of the hand and the points on the rope.

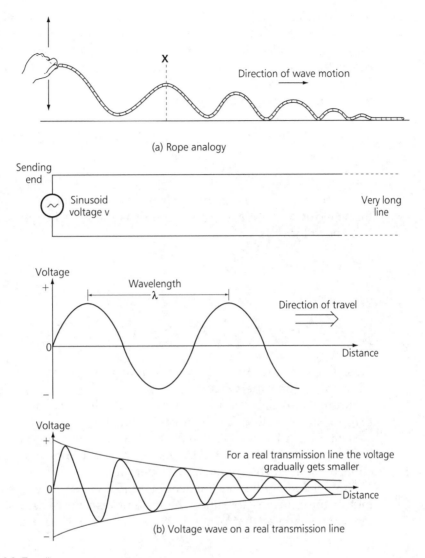

(a) Rope analogy

(b) Voltage wave on a real transmission line

Figure 9.2 Travelling waves

- The amplitude of the wave gets smaller as it travels along the rope. We say that the wave is **attenuated**, (i.e. loses energy). The further away we get from the source of the wave, the greater is the overall attenuation.
- The wave takes a certain time to reach the far end, i.e. it travels at a certain **speed**.
- Even though the amplitude of the wave decreases as it moves down the line, there is a characteristic distance between like displacements of the rope. This is called the **wavelength** and is given the symbol λ (lambda). It is the distance between those points over which the wave shape repeats itself.

A similar phenomenon occurs when an alternating voltage source is connected between the conductors of a transmission line. Consider the transmission line in Figure 9.2(b) where a sinusoidal voltage of frequency 'f' is connected to the sending end of the line. For the moment, we will assume that the line is very long (e.g. infinity) and it has no attenuation at all. As shown in the diagram, a voltage wave would travel down the line at a certain velocity. The distance between points on the line which have the same value of voltage (for example, the peaks) is called the wavelength λ. For example, suppose that a point d metres down the line lags behind the voltage at the sending end by 90°. At a distance of 2 d the voltage would lag by 2 × 90° = 180°. At a distance of 4d the voltage would be **in phase** with the sending end voltage because 4 × 90° = 360°. The distance 4d would be the wavelength, λ, of the voltage wave on the line.

Since the sinusoidal voltage has a frequency 'f' its period is:

$$T = \frac{1}{f}$$

In a time T, the voltage will travel a distance along the line of T multiplied by the velocity of the wave. This distance is the wavelength λ.

If the velocity is v then,

$$\lambda = v \times T = v \times \frac{1}{f}$$

or

$$V = f \times \lambda \ (\text{metres/second})$$

This is a very important equation because knowing any two of the quantities allows us to find the third.

Figure 9.2(b) shows how the wave gradually gets smaller as it travels along a real transmission line. This is because a real line has resistance and energy is lost as heat. The wave is **attenuated** as it travels down the line.

Obviously, we always try to ensure that real lines have the minimum attenuation possible at the frequency of operation.

The voltage applied at the sending end causes current to flow in the conductors of the transmission line. The current also flows as a sinusoidal wave, with the same shape as that of the voltage wave in Figure 9.2.

Progress check 9.1

Voltage and current waves on a coaxial line have a velocity of 0.3 × 10⁸ m/s. If the frequency is 100 MHz determine (a) the wavelength and (b) the time taken for a wave to travel along 30 km of line.

Characteristic impedance of a transmission line

Any transmission line, whatever its form, has a property which is known as its **characteristic impedance**, Z_0. Its value is the most important piece of information to have about a line. It is defined as follows. Consider the line in Figure 9.3(a) which is assumed to be infinite in length.

The ratio of voltage to current at the sending end is called the **input impedance**. Z_0 is defined as the input impedance of an infinite line so

$$Z_0 = \frac{V_s}{i_s}$$

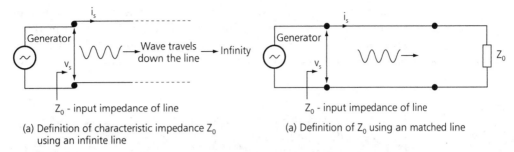

(a) Definition of characteristic impedance Z_0 using an infinite line

(a) Definition of Z_0 using an matched line

Figure 9.3 Characteristic impedance

If we measure *the ratio of voltage to current at any point* on the infinite line we would always get the value Z_0. Infinitely long lines are, of course, impractical but if we terminate a line of any length with an impedance equal to Z_0 then the ratio of voltage to current at any point, including its input, still turns out to be Z_0. This gives us a much more practical definition of characteristic impedance. Z_0 is defined as 'the input impedance of a transmission line which is terminated in its own characteristic impedance'. Such a line is said to be **impedance matched**. A matched line is shown in Figure 9.3(b).

Equation for Z_0

The characteristic impedance Z_0 we have said is the most important property of a transmission line. Its value depends on the dimensions of the conductors and the electrical properties of the insulating material between them. For the coaxial and twin lead lines in Figure 9.1 the equations are:

$$Z_0 = \frac{60}{\sqrt{\varepsilon_r}} \log_e \left(\frac{D}{d}\right) \quad \text{(coaxial line)}$$

$$Z_0 = \frac{120}{\sqrt{\varepsilon_r}} \log_e \left(\frac{2S}{d}\right) \quad \text{(twin lead line)}$$

In the equations ε_r is the 'relative permittivity' of the dielectric material separating the conductors.

WORKED EXAMPLE 9.1

A coaxial line has an outer copper braid diameter of 9 mm and the inner conductor has a diameter of 1.5 mm. If the relative permittivity of the polythene dielectric is 2.2, calculate the characteristic impedance.

Solution
For a coaxial line,

$$Z_0 = \frac{60}{\sqrt{\varepsilon_r}} \log_e \left(\frac{D}{d}\right) = \frac{60}{\sqrt{2.2}} \log_e \left(\frac{9}{1.5}\right) = \frac{60}{\sqrt{2.2}} \times 1.7918) = 72.48 \ \Omega$$

Commercial coaxial lines usually have characteristic impedances of less than about 100 Ω. This is mainly because of constructional considerations. The most common coaxial cables have impedances of 50 Ω and 75 Ω.

Progress check 9.2

If the coaxial line in Worked Example 9.1 has a dielectric of $\varepsilon_r = 3$, calculate its characteristic impedance.

Activity 9.2

Your TV antenna is connected to the TV by coaxial cable. What is its characteristic impedance? If the computers at your centre are connected by coaxial cable, investigate and determine the value of its characteristic impedance. (Hint: An RS Components catalogue will help you.)

Attenuation

As voltage and current waves travel down a real transmission line they gradually lose energy and their peak values decrease. This is indicated in Figure 9.2(b). Because the conductors have a resistance, power will be dissipated in them. Power will also be lost in the dielectric as well. In general, the losses increase as the frequency increases.

The amount of loss is described by the **attenuation coefficient** α of the line. α is usually quoted in dB/km. Figure 9.4 shows typical **attenuation/frequency characteristics** of a twisted pair and a coaxial cable used in telephone systems.

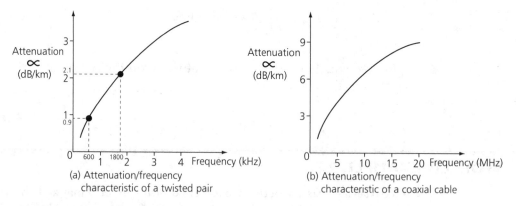

(a) Attenuation/frequency characteristic of a twisted pair

(b) Attenuation/frequency characteristic of a coaxial cable

Figure 9.4 Attenuation/frequency characteristics

Notice firstly the different frequency ranges in which the cables operate. **Twisted pairs** are used for audio frequencies such as the direct connection of a telephone handset to the exchange. Specially designed **coaxial cables** are able to carry many thousands of telephone channels. (How this is achieved will be investigated in the next element.) Secondly, since the attenuation changes with frequency, then signals containing a range of frequencies, for example voice signals, TV signals and data signals etc., will be distorted as they travel down the transmission line. This is because the frequency components in the signal suffer different amounts of attenuation. The effect is known as **amplitude distortion**.

WORKED EXAMPLE 9.2

A digital data signal of 1200 bits per second travels down a twisted pair whose attenuation/frequency characteristic is shown in Figure 9.4(a). The cable is 5 km long. Determine the attenuation suffered by the maximum fundamental frequency in the data signal and its third harmonic.

Solution
The maximum fundamental occurs when the data waveform consists of alternate '1's' and '0's'. This fundamental frequency is one half of the bit rate. Therefore,

$$\text{fundamental frequency} = \frac{1}{2} \times 1200 = 600 \text{ Hz}$$

The third harmonic has a frequency of $3 \times 600 = 1800$ Hz

From the curve in Figure 9.4(a) the attenuation suffered is: 0.9 dB/km at 600 Hz and 2.1 dB/km at 1800 Hz.

Since the cable is 5 km long the total attenuation is:

$5 \times 0.9 = 4.5$ dB at 600 Hz

$5 \times 2.1 = 10.5$ dB at 1800 Hz

Progress check 9.3

A coaxial cable is designed to have a useful bandwidth of 60 MHz. Theoretically, how many telephone channels could be accommodated by the cable?

Phase change coefficient
As the current and voltage wave propagate down a transmission line, the current and voltage at any point on the line **lags** behind the current and voltage at the sending end. The phase difference between two points 1 km apart is known as the **phase change coefficient**. It is measured in degrees (or radians) per km. For a transmission line, the phase change coefficient is not constant but depends on the frequency. Thus, if a signal containing a range of frequencies (such as voice) is sent down a line, the frequency components have different phase shifts when they reach the far end. This causes the signal to be distorted. This is in addition to the amplitude distortion caused by the change in attenuation with frequency. The net result is that the signal suffers a change in shape as it travels down the line. As you saw in the previous chapter, this produces problems with analogue signals. The effect is less serious with digital signals because, as has been seen, regenerators can be used to produce a new digital waveform whenever the overall distortion becomes too great.

Matched and mismatched lines

When a transmission line is terminated in its own characteristic impedance Z_0, it is said to be matched. All of the power in the wave travelling from the sending end, which we will now call the **incident wave**, will be absorbed by the load of value Z_0. If the load, Z_L, terminating the line is not equal to Z_0, then not all of the power in the incident wave will be absorbed by Z_L. Some power will be reflected back towards the sending end in the form of a **reflected wave**. As indicated in Figure 9.5, the incident and reflected waves combine to produce a resultant waveform called a **standing wave**.

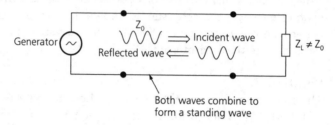

Figure 9.5 *Standing waves on a transmission line*

Generally, standing waves on transmission lines should be avoided since it means that only a portion of the incident power reaches the load and cable breakdown can occur in high power systems. Thus, whenever possible, a transmission should always be terminated in a load which is equal to its characteristic impedance.

Activity 9.3

The computer network at your Centre may use a system called Ethernet to connect the PCs together. Investigate how the ends of the Ethernet cable are terminated.

Using free space as a transmission medium

Instead of using transmission lines, information can be sent through free space using electromagnetic waves. The overall system is then usually called a **telecommunications system**. At the transmitter, the information signal is superimposed on the **electromagnetic wave** (e.m. wave) and 'launched' into free space by the transmitting antenna. At the receiver, the wave is received by the receiving antenna and the information signal is recovered from the wave by the receiving equipment. The techniques of superimposing the signal on the e.m. wave and recovering it are known as **modulation** and **demodulation** respectively. We will investigate this in the next element.

The propagation of the e.m. wave through space depends on many factors but the most critical one will be the frequency of the wave.

Electromagnetic waves and frequency bands

The nature of an electromagnetic wave is illustrated in Figure 9.6(a). The wave consists of electric and magnetic fields oscillating at right angles to one another. The wave does NOT require a medium through which to propagate because once 'launched' the electric and magnetic fields 'keep themselves going'. The changing electric field produces a changing magnetic field and vice-versa. Visible light is an e.m. wave and we know that no medium is required for propagation because light reaches the earth from the sun which travels across free space, (i.e. a vacuum).

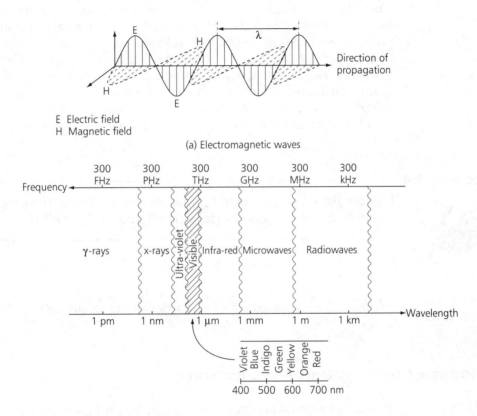

E Electric field
H Magnetic field

(a) Electromagnetic waves

Frequency range	Classification
3 – 30 kHz	Very low frequency (VLF)
30 – 300 kHz	Low frequency (LF)
300 – 3 MHz	Medium frequency (MF)
3 – 30 MHz	High frequency (HF)
30 – 300 MHz	Very high frequency (VHF)
300 – 3 GHz	Ultra high frequency (UHF)
3 – 30 GHz	Super high frequency (SHF) ⎫ (microwave band)
30 – 300 GHz	Extra high frequency (EHF) ⎭

Figure 9.6 Electromagnetic waves and frequency bands

All electromagnetic waves have the same nature. Thus, visible light, radio waves, x-rays and gamma rays are all e.m. waves. They only differ from one another in frequency f (and hence wavelength λ). In free space, they all travel at the same speed of 3×10^8 m/sec (or 186,000 miles/second). This is called the speed of light and the symbol c is used. (In air the speed is just slightly less than c but is usually assumed to be c.) The speed c is related to the frequency and wavelength by the following equation:

$$c = f \times \lambda \text{ (metres/second)}$$

Notice that this is the same form as the equation for waves on a transmission line.

The most important parts of the e.m. wave spectrum are shown in Figure 9.6(b). Note the tiny range of frequencies to which our eyes are sensitive. Each of the bands indicated on the diagram (radio waves, x-rays etc.) all have particular properties which are useful to mankind. An important observation to make is that as the frequency gets higher and approaches that of visible light (high frequency radio waves, microwaves etc.) we would expect the waves to have characteristics typical of light. We know, for example, that visible light travels in straight lines. Thus, high frequency radio waves and microwaves will tend to travel in straight lines. We will investigate the consequences of this a little later.

Activity 9.4

Find out the frequencies used by your local radio station. How are the frequencies classified using the frequency range classification in Figure 9.6?

Progress check 9.4

What is the wavelength of an e.m. wave whose frequency is 10 GHz?

Modes of propagation of radio waves

The use of radio waves for communication purposes has had a profound effect on mankind over the last 90 years or so and the techniques are continually being developed and refined. All systems that use radio waves are fundamentally the same in that a transmitting antenna will be fed with power at a particular frequency. An e.m. wave of the same frequency will then be radiated from the antenna and propagate through space. A receiving antenna then 'picks up' the wave and the receiver recovers the information originally transmitted. What distinguishes one system from another is the frequency used and hence the **mode of propagation** of the e.m. wave.

There are three main modes of propagation called the ground wave, sky wave and space wave. They are illustrated in Figure 9.7

The ground wave
The ground wave is used in the VLF and LF bands for worldwide communications, for example the BBC World Service. Sound broadcasting in the medium waveband also utilises the ground wave. It is found that at these frequencies the waves tend to 'hug' the earth's surface and so follow the curvature of the earth.

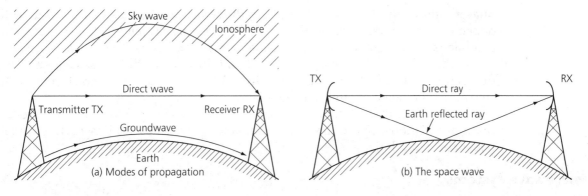

Figure 9.7 Modes of propagation

For efficient communication, high power transmitters are required and the range will depend on the frequency used. In the medium frequency band, coverage up to about 500 km is possible. At VLF and LF many thousands of km is easily achieved.

The sky wave

Between about 50 km to 500 km above the surface of the earth is a region which is known as the **ionosphere**. You can imagine the ionosphere as being like a belt which surrounds the earth. When ultra-violet radiation from the sun enters the atmosphere it causes electrons in some of the molecules to attain enough energy to break free from their atoms. Thus the negative electrons wander freely and a positive charge is left with the atoms from which they came. This process is known as **ionisation**. It is found that the density of electrons gradually increases with increasing height within the ionosphere. The electron density, however, is not constant but depends on the time of day. At night, for example, the free electrons closest to the earth recombine with the ionised atoms and the lower part of the ionosphere disappears altogether.

When a radio wave of a suitable frequency passes into the ionosphere it is bent or **refracted** by it back towards the earth's surface. This is particularly noticeable below about 30 MHz (i.e. the HF band) but depends on many factors. If the refraction is sufficient, the wave will be returned to earth and can be received by a suitable receiver. The distance between the transmitter and receiver is called the **skip distance**. In the HF band, the maximum skip distance is around 4000 km and a wide range of services use the technique. Sound broadcasting, some international telephone systems, as well as communication with ships and aircraft, are typical examples.

The space wave

In the VHF band and above (greater than 30 MHz) the space wave is used for communication between the transmitter and receiver. At these frequencies, the e.m. wave travels almost in a straight line although there is a slight refraction which gets less as the frequency gets higher. As indicated in Figure 9.7(b), the space wave actually consists of two waves, one called the **direct ray** and the other reflected from the earth's surface called the **reflected ray**. Notice the terminology used. The adoption of the word 'ray' is to highlight the fact that waves in this frequency band and above behave very like visible light rays, i.e. they can be reflected and refracted etc. At the receiver, the total received signal will be the phasor sum of the two rays. The phase difference will not be constant but will depend on the terrain between the transmitter and receiver and propagation conditions, (rain, temperature changes at sunrise and sunset etc.). The net result is that the strength of the received signal will fluctuate as the phase changes. The fluctuation in signal strength is generally called **fading** and in a practical system steps must be taken to minimise its effects.

In the VHF and UHF bands, the space wave is used for FM radio, TV broadcast transmissions, police, fire and ambulance services as well as multi-channel telephone radio links.

Microwave systems

The microwave band is defined as using frequencies above 3 GHz. (Gigahertz – GHz – is 10^9 Hz or 1000 MHz.)

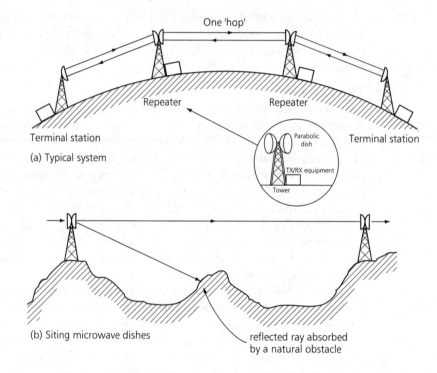

(a) Typical system

(b) Siting microwave dishes

reflected ray absorbed by a natural obstacle

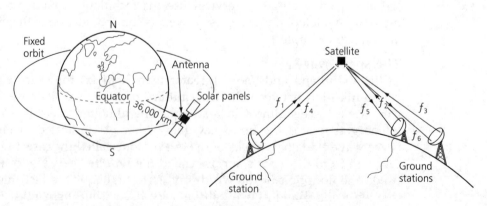

Figure 9.8 *Microwave line-of-sight and satellite systems*

(c) Communications via satellite

Here, to allow for the curvature of the earth, repeater stations are necessary between the main transmitting and receiving stations as shown in Figure 9.8(a). At these frequencies, the microwaves behave very much like light rays and travel in straight lines.

The antennas used are **parabolic dishes**, familiar to those of you with satellite TV. Essentially, the dish operates in the same way as a car headlamp which focuses the light from a bulb into a beam. The microwave dish focuses the microwaves into a narrow 'pencil beam' that is pointed directly towards the receiving dish. There is, therefore '**line-of-sight**' between transmitting and receiving dishes.

To allow for the curvature of the earth, the spacing between repeaters is limited to about 30 km–40 km. The actual position is important and depends on many factors. Usually the stations are sited on hills with the dishes mounted on towers. In the planning phase, the sites are chosen so that the earth-reflected ray is absorbed by a natural obstacle such as a hill, as shown in Figure 9.8(b). Thus fading is very much reduced. When the terrain in between is flat or water or desert, special techniques are used to combat the effects of fading.

You can think of a microwave system as a chain with many links. The signal is transmitted from the transmitting terminal station towards the first repeater, where it is received, amplified and then retransmitted towards the next repeater and so on. A point to remember is that 'a chain is only as strong as any of its links'. If the receiving or transmitting equipment at a repeater malfunctions, the whole communications link is broken. Microwave links in the U.K. carry telephone, TV and digital signals. A typical link might have the capacity for about 1800 individual telephone channels. Transmitted powers are quite low; a few watts is typical.

Activity 9.5

Use your Centre library (or possibly the Internet) to locate a photograph of the British Telecom tower in London. Note the array of dishes on the tower. Contact BT in your area and find out whether there is a microwave tower close to you. If possible, visit the site.

Communication satellites

Essentially, a communications satellite is a microwave repeater that is placed in orbit around the earth. They now form a vital link in worldwide communications as well as providing TV transmission and other services directly to the home and business enterprises. As shown in Figure 9.8(c), most satellites are placed in orbit above the equator at such a distance that they move at the same rotational speed as the earth. Thus, with respect to the earth the satellite appears to be stationary. An orbit of this type is called a **geostationary orbit**. (Other types of orbits are possible that are not around the equator and at a lower altitude. These satellites are mainly used for weather forecasting, surveillance and mapping, etc.)

The information to be sent over a satellite link, such as telephone traffic, TV signals and data, is first transmitted over normal networks to a ground station. The transmitting equipment at the station superimposes the information signals on a microwave frequency (called the **uplink carrier**) of typically about 6 GHz or 14 GHz. The earth station dish antenna is very large (30 m diameter) and this focuses the microwaves into a very narrow beam which is pointed directly at the satellite. At the satellite, the incoming signal is received, amplified and then retransmitted on a microwave **downlink carrier** of a different frequency towards the receiving earth station. TV signals are one-way and so at the receiving earth station the TV information is extracted from the downlink carrier for onward transmission over the national system. Telephone traffic (and data) is two-way so the person receiving the call will use the same satellite for the reply. This is carried on a different uplink frequency, amplified at the satellite and retransmitted towards the first earth station on another microwave downlink carrier. In this way, two-way transmission is established.

Many satellites are now used for the direct broadcasting of TV programmes and other services, which can be received by small, inexpensive domestic dishes. Such systems are called **direct broadcast satellites** or DBS for short. The ASTRA satellites positioned at 19.2° east over the equator are an example of a DBS system. The four satellites have the capacity to transmit 64 separate TV channels.

The TV signals are uplinked to the satellites from several locations across Europe, shifted in frequency and retransmitted back to earth. The antennas aboard the satellites are designed to focus the transmitted power over a particular area of the earth's surface. This area is called the **satellite footprint** and providing you lie within the footprint, excellent reception is possible with the correct size of receiving dish. A typical DBS receiving setup is shown in Figure 9.9.

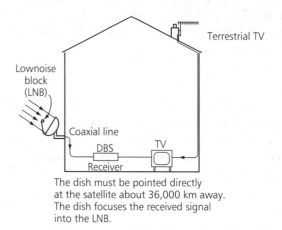

Figure 9.9 Receiving satellite TV

Activity 9.6

Find out the names of at least three satellites from which TV programmes can be received in the UK. Produce a short report briefly describing the position of each and list the frequencies on which they transmit. (Hint: Publications such as *What Satellite TV* will be useful.)

WORKED EXAMPLE 9.3

Estimate the time taken for a signal to be transmitted from a transmitting earth station to a receiving earth station.

Solution

We will assume that the distance to the satellite from the station is 36,000 km. The total time taken T is:

$$T = \frac{\text{total distance travelled}}{\text{speed of light}} = \frac{2 \times 36,000 \times 10^3}{3 \times 10^8} = \frac{2 \times 36 \times 10^{-2}}{3} = 0.24 \text{ s}$$

This is a quite noticeable amount and can lead to problems with data transmission. You may have noticed the effect yourself when watching TV live interviews conducted by satellite.

Optical fibres

In the last two decades or so, there has been a revolution in the technology of transmitting information from one point to another over land and under sea. The traditional electronic methods of using voltage and current along copper wires are gradually being replaced by **optoelectronic technology**. Here, a thread of extremely pure glass, no thicker than a human hair, guides pulses of laser light from the transmitter to receiver. The frequencies used are in the infrared band and because the frequencies are so high, it is usual to quote wavelengths rather than frequency. Currently, wavelengths in the range of about 1.6 μm to 0.8 μm are used in commercial systems.

Progress check 9.5

What is the frequency of light which has a wavelength of 1.6 μm?

The use of optical fibres, as they are called, has many advantages over copper cables. Some of these are:

- Copper is expensive whereas the raw material for glass is sand which is cheap and plentiful.
- Optical fibres are very small and light in weight. A km of coaxial cable uses about 30 kg of copper. A km of optical fibre uses around 14 g of glass.
- The attenuation of optical fibres is very low, typically less than 0.5 dB/km.
- The frequency bandwidth is very wide. Thus, very large numbers of data or telephone channels can be carried by a single fibre.
- There is no electromagnetic interference between closely spaced fibres (called **crosstalk**), allowing many fibres to be wrapped in the same cable.
- There is freedom from other forms of electrical noise.

A cable with two fibres is shown in Figure 9.10(a).

It appears that optical fibre technology will replace coaxial cables in long distance systems. Fibre is very suited for the transmission of digital signals and

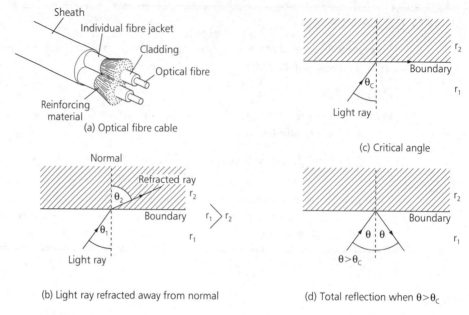

Figure 9.10 An optical fibre cable and properties of light

finds its greatest use in PCM telephone systems. Many computer networks now use optical fibres rather than coaxial or twisted pairs to connect computers and other devices together.

Types of optical fibre

Optical fibres operate on the principle that when light travels from one medium to another which have different **refractive indices**, the light will be bent or **refracted**. Figure 9.10(b) shows what happens when the light moves from a higher refractive index (n_1) medium to a medium of lower refractive index (n_2). The refracted light ray is bent away from the normal line at the boundary.

If the angle of incidence θ_1 is gradually increased, the refracted ray will begin to move along the boundary. The angle of incidence at which this occurs is called the **critical angle**, θ_C and is shown in Figure 9.10(c). For 'normal' glass θ_C is around 42°.

If the angle of incidence is increased further, then the light ray will be totally **reflected** at the boundary as shown in Figure 9.10(d). In effect, the boundary acts like a mirror and reflects the light rays back into the material with the higher refractive index.

In fibre optic technology, the glass with refractive index n_1, is called the **core** and the glass with refractive index n_2 is called the **cladding**, Figure 9.10(a). Figure 9.11(a) shows a typical fibre cable. Note in particular the dimensions involved. Light rays incident on the boundary at an angle greater than θ_C will be totally reflected and propagate down the fibre. Those whose angle of incidence is less than θ_C will be 'lost'. Since light rays taking many different paths can propagate down the core, we say that the light has many different **modes of propagation**. For this reason, optical fibre of this type is called **multimode fibre**.

There is a serious problem with multimode fibre because, since each mode takes a different path, they will arrive at the end of the fibre at different times. Thus

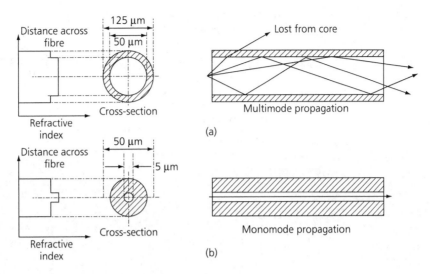

Figure 9.11 Multimode and monomode fibre cables

a light pulse transmitted at one end will spread out in time when it reaches the far end. The longer the fibre, the worse the effect will be. This limits the maximum bit rate that is possible. This pulse spreading effect is called **multipath dispersion**.

The obvious way of eliminating multipath dispersion is to ensure that only one mode can propagate down the fibre. It has been found that if the core diameter is decreased until it is about the wavelength of the light used, (about 2 to 8 μm), only one mode can propagate. This mode travels straight down the centre of the core and so multipath dispersion is eliminated. This type of fibre is called **monomode fibre** and is shown in Figure 9.11(b).

The glass used in optical fibre technology is not ordinary glass but is extremely pure. In fact, attenuations of around 0.1 dB/km are now possible. (The first optical fibres had attenuations of about 1000 dB/km.) A good analogy to remember is that if sea water was as transparent as the glass used in optical fibres, you could see the ocean floor 6–7 miles below the surface! The manufacture of optical fibre is a specialised technique and fibre producers closely guard the details of their manufacturing techniques from competitors.

Optical fibre telephone links

Optical fibre telephone links are steadily replacing the traditional coaxial telephone links throughout the UK and in most of the developed world. The basic elements of a system are shown in Figure 9.12.

Although only one telephone channel is shown in the diagram, in a practical system many telephone channels would be combined together to be transmitted over the single fibre link. The technique of combining the individual channels is called **time division multiplexing** and will be investigated in the next element.

The analogue voice signal is transformed into a PCM signal using the techniques described earlier in Element 9.1 (chapter 8). The PCM signal causes a **laser diode** to be switched ON and OFF in the optical transmitter circuit. The flashes of laser light, with a typical wavelength of 1.3 μm, are coupled into the optical fibre via an optical coupler. Depending on the length of the link, it may

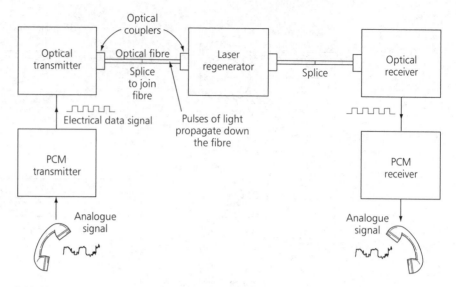

Figure 9.12 Main elements of a fibre optic telephone link

be necessary to **splice** lengths of fibre together. Obviously, perfect alignment of the fibres to be joined is essential if losses due to misalignment are to be kept to a minimum. The splicing procedure is a precision engineering task but it is possible in the field to produce splices with losses of only fractions of a dB. For long links, regeneration of the laser signal may be necessary. The regenerator is an optical receiver directly connected to a laser transmitter. A new stream of pulses is generated for onward transmission. When compared to a coaxial system, regenerator spacing is much greater in a fibre optic system. At the final optical receiver, the laser pulses are reconverted back into the original PCM signal. The analogue output signal is recovered at the output of the PCM receiver using the techniques previously described.

The bandwidth of the system is very wide. Typical bit rates of commercial systems are 1920 channels at 140 Mbits/sec and 8000 channels at 565 Mbits/sec.

Case study

It is useful to have some idea of the historical development of **fibre optic telephone links** within the UK. The technology is quite 'young' and by 1977 only a few experimental routes were installed by British Telecom to assess their viability and develop the technology. These early systems used a type of fibre called **graded index** in which the refractive index gradually decreased from the centre of the core outwards. Each cable contained 8 fibres. Some details of later links are tabulated (facing page).

Experimentation and development continues and some notable achievements to date are:

● The longest submarine fibre optic cable runs from the UK to Japan (27,000 km) and links 11 countries. It can support 600,000 telephone channels.

- In 1994 BT, at its research laboratories, achieved the longest transmission distance over a fibre link with a link 125,000 km long. The data rate was 20 Gigabits/sec.
- In 1996 in Japan, the highest transmission rate of 1.1 terabits/sec was recorded. This is equivalent to about 17 million telephone channels.

	London–Birmingham	Nottingham–Sheffield
Operational date	1984	1985
Bit rate	140 Mbits/sec	565 Mbits/sec
Radiation wavelength	1285 nm	1300 nm
Fibre type	Graded index	Monomode
No. of fibres in cable	8	8
System length	205 km	72 km
Repeater spacing	10 km (approx.)	26 km (maximum)
Attenuation	1.5 dB/km (approx.)	0.5 dB/km
Transmitter	Light emitting diode	Laser diode
Transmitter power	–9 dBm	–3 dBm

Case study

The UK has one of the most advanced national telephone networks in the world. It is called the **public switched telephone network** (PSTN). It has evolved over many years and is continuously being upgraded and refined. It consists of a number of automatic switching exchanges connected to one another by **trunk circuits**. These trunk circuits are coaxial cables and fibre optic cables. Fibre optic connections are now favoured and are replacing coaxial lines. The exchanges are linked together in such a way that any subscriber (i.e. you, with your home telephone) can be switched through and connected to any other subscriber on the network. A simplified diagram of the system is shown in Figure 9.13. Included in the system will be a number of exchanges which allow connection to subscribers in other countries via that country's telephone network. These international **gateway exchanges** are connected via satellite links, terrestrial microwave line-of-sight systems and undersea coaxial and fibre optic cables.

The exchanges are a key part of the system and in the UK, a large number are digital electronic telephone exchanges called **stored program control** (SPC) digital exchanges. Essentially, computers are used to control the switching processes necessary to make call connections. In the UK, it is called **System X** and it offers many advantages over the older exchanges which used electromechanical devices to perform the switching operations. A major advantage is that since the operations are controlled by computers, changes and improvements can be made by altering the computer software with little or no rewiring.

The sequence of events that occur when you make a telephone call will now be described. We will assume that you are using a modern push-button telephone.

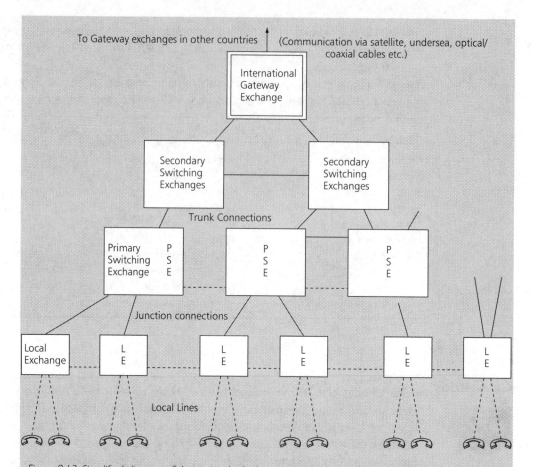

To Gateway exchanges in other countries ↑ (Communication via satellite, undersea, optical/coaxial cables etc.)

Figure 9.13 *Simplified diagram of the national telephone network*

- Lifting the handset (known as 'going off hook') sends a signal to the local exchange via a pair of wires in a telephone cable. The local exchange will then respond by sending a **dial tone** back to you.
- The number of the subscriber you desire is punched in sequence into the key pad. Each depression of a key sends two audio tones for each digit of the number and is called **dual tone multiple frequency signalling**.
- Depending on the number dialled, the system will route the call to the local exchange at the destination. If it is an international call, the route will include an international exchange, satellite or cable link(s) etc. to the international exchange in the destination country. Final **routing** is via that country's network to the local exchange at the destination.
- The local exchange will send 'ring current' to activate the bell at the destination. At the same time, your local exchange sends a **ring tone** to your telephone to indicate that a connection has been made. If the telephone is engaged, **busy tone** is sent instead.
- When the destination's telephone is lifted off hook, ring current and tone ceases and a **conversation** can take place.
- When the conversation is complete, replacing your handset **terminates** the connection.

Communication networks

Activities 9.1 and 9.3 asked you to investigate various aspects of the **data communication network** of computers at your centre. Technically, the connection of the PCs and other devices (printers, file servers etc.) is known as a **local area network** (LAN) since it is confined to one building or a few buildings close together. The devices that make up the LAN are wired together by copper cables or optical fibres and the interchange of data between the devices is restricted to the LAN itself. Unless other facilities and communication techniques are used, there can be no information interchange with other remote PCs or other LANs.

Obviously, there is a requirement for LANs and individual PCs to be able to exchange digital information even though they may be geographically far apart – perhaps in different countries. An example may be a High Street bank or other enterprise which needs regular and reliable data communications between many sites which are physically far apart. Computers connected as a LAN at one branch might, for example, need to communicate with mainframe computers held at head office. Data networks which allow communications over a large geographical area, nationwide or worldwide, are called **wide area networks** (WANs). Since the first WANs were introduced during the late 1960's and early 1970's there has been an incredible growth in demand for these services which is set to continue. It is natural that to satisfy this demand use has been made of the largest communication network in the world, which is the global telephone network.

Public telephone networks

One way of establishing a data link is to set up a connection via the **public switched telephone network** (PSTN). Even though the telephone systems in most advanced countries operate digitally internally (i.e. voice signals are converted to PCM signals) the local connection is analogue and is intended for analogue voice transmission. To send digital data over such a link, a device called a **modem** (meaning *MOD*ulator-*DEM*odulator) is required at each end of the link. The modem converts the 1's and 0's of the data waveform into different analogue voice frequencies which can be transmitted by the PSTN. (The technique used will be described in the next element.) Because the connection is created by dialling the telephone number of the other data terminal, the actual circuits used in the PSTN are chosen at random. The electrical characteristics of the circuits may not be adjusted for best performance and high error rates may occur.

An alternative is to lease a permanent link between the two locations from the public telecommunications operator (e.g. British Telecom) so that exclusive use of the link is provided. The characteristics of the link such as attenuation etc. can be adjusted for best performance ensuring higher speeds and lower transmission errors. Cost must always be taken into account of course and leasing might only be justified if the data traffic on the link is high. A good example would be the connection of a bank's cashpoint terminals to a central computer which can check accounts etc. A leased line in this case is very cost effective.

The use of digital technology and computers generally in telephone systems has allowed the development of fully digital systems for public use. This allows the transmission of integrated voice, computer data and other digital information from one user to another. The service is known as the **integrated services digital network** or ISDN for short. In its basic form, ISDN offers two types of communication channels called B-channels and D-channels. The B-channels are used for the transmission of voice, data and other digitised information (or a mixture of all three) at 64 Kbits/sec. This is the bit rate for the normal PCM transmission of voice. The D-channel is used for signalling purposes, e.g. setting up calls and terminating them. The basic type of service offered by ISDN is called **basic rate access** (BRA). BRA provides the user with two B-channels of 64 Kbits/sec and a D-channel of 16 Kbits/sec. The D-channel controls the calls over the two B-channels. ISDN is continually being developed. The newer **broadband-ISDN** offers much higher data rates as well as other services.

Case study

A **mobile telephone system** provides the subscriber with all the advantages of a normal telephone but does not require the telephone set to be permanently wired to a telephone exchange. Instead, free space is the communications medium with transmission and reception via radio waves. The basic principle behind older systems is illustrated in Figure 9.14 and the technique has been used for many years. A mobile user and a fixed telephone subscriber communicate via a **base station** which is interfaced to the PSTN. Additionally, private mobile radio networks used by the police, emergency services, taxis etc. have long been established but do not have a connection to the PSTN. The former system has been in operation for over 30 years using frequencies in the VHF band around 160 MHz. Their capacity was limited, however, and so in 1979 the frequency range from 862 MHz to 960 MHz was allocated for the development of a mobile system called **cellular radio**.

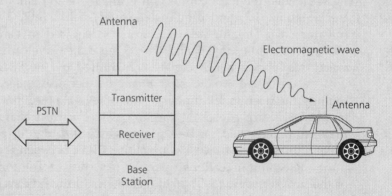

Figure 9.14 Principle of a mobile radio telephone system

The cellular radio system is a very clever way of using the available radio frequency bandwidth efficiently. In a cellular system, instead of having one high-power base station covering the area to be served, the area is divided into a number of smaller areas called **cells**. Each cell has its own relatively low power base station which provides coverage of that cell only. The number of radio frequency channels available is split into a number (n) of **channel sets**.

Each cell is allocated one channel set of frequencies according to a regular pattern. The pattern repeats itself to fill the number of cells in the system. Therefore, each set of channel frequencies is **reused** many times throughout the area covered by the system. A system using a cluster of seven cells (n = 7) is shown in Figure 9.15. Cells, although roughly circular in their area of coverage, are drawn as hexagons.

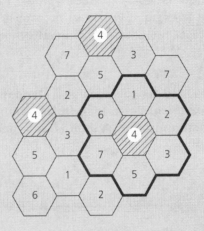

The heavy line shows the cluster of
seven cells.
Each different numbered cell has a
different set of frequencies associated
with it.

Figure 9.15 A seven-cell cluster

Each number (1 to 7) denotes that seven different channel sets are used. Notice that each cell is always spaced away from another cell of the same number which uses the same set of channel frequencies. This ensures that **co-channel interference** is kept to a minimum.

The base stations are connected to digital exchanges which are themselves linked together and interfaced to the PSTN. Digital lines are used with digital connection to the PSTN.

If a mobile phone user makes a call then the system must route that call to the cell where the mobile is so that connection can be made. To facilitate this, the cellular network is split into a number of areas each with its own identity number. This number is transmitted regularly from the base stations within that area. A mobile station, when not in use, will check the identity number sent by the base station. If, as it moves about, it detects a change in the identity number (i.e. it has moved to a different area), it automatically signals the base station to inform the system of its new location. Thus the cellular network can keep track of the area in which each mobile station is located.

Many mobile calls are made while the user is moving (e.g. in a car) so movement from one cell to another takes place. Unless the call is passed on to the base station in the next cell, the link would be broken. To overcome this, the signal strength from mobiles engaged in calls is constantly monitored. When the signal strength falls below a certain level, the system decides whether another base station can receive the signal at a stronger level. If it can, the mobile is commanded by a signal to switch to a frequency used by the new base station. This process is called **in-call handover**. The process takes a fraction of a second and is hardly noticeable. The system also controls the amount of power transmitted by the mobile. If the base station is receiving a signal of sufficient strength, it will signal the mobile and command it to reduce power. This reduces the possibility of interference.

Cellular radio frequencies and standards

The standard laid down for cellular communications within the UK is called **total access communication system** (TACS) and the frequency allocation is shown in Figure 9.16.

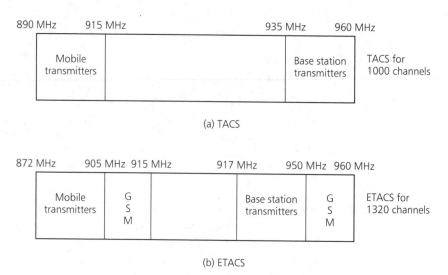

Figure 9.16 Frequency bands for cellular radio and GSM

The number of users rapidly increased, however, and so in 1987 more frequencies were allocated for cellular use as shown in Figure 9.16(b). This is known as **extended TACS** or ETACS. Notice that all cellular telephones can use the TACS channels but *not* all 'phones can access the ETACS frequency channels. In both systems, channels are spaced by 25 kHz.

In normal cellular communications, transmission of the voice is by analogue methods so when ETACS was introduced, a portion of the spectrum was allocated for a **pan-European cellular system** that is purely digital. This new cellular system is known as **group speciale mobile** or GSM for short. GSM is very complex, but being totally digital gives very high quality and reliable service. Estimates are that by 1999, over 10 million subscribers will use the system.

Assignment 9

This assignment provides evidence for:
Element 9.2: Evaluate media used for signal transmission
and the following key skills:
Communication 3.2, 3.3
Information Technology 3.1, 3.2, 3.3

The evidence indicators for this element require you to evaluate the transmission media used for two types of signal. Examples include computer data, broadcast signals such as radio and TV from satellite or terrestrial systems, or point-to-point signals such as telephone or facsimile.

Report

Your report should include:

- a description of the signal transmission media used; i.e. transmission lines, free space or optical fibre
- a description of the properties of the media; i.e. mode of propagation, bandwidth, noise, attenuation, phase change
- the criteria used for evaluation; i.e. properties, purpose and length of communication link
- a description of the types of signal used.

Chapter 10: Communication techniques

You have discovered in the two previous elements that signals have associated with them a characteristic range of frequencies that make up the total signal content. For example, the bandwidth of the human voice that is transmitted over a telephone system has a range 300 Hz to 3.4 kHz. Additionally, since the available frequency bandwidth for information transmission is limited, efficient and cost-effective use must be made of this bandwidth.

In this element we will investigate some communication techniques and examine the methods used to transmit many different channels over a common transmission system. Fundamental to this is the process of **modulation** which allows signals with a particular frequency content to be shifted (or translated) to different parts of the usable frequency spectrum.

When you complete this element you should be able to:

- describe baseband transmission of audio signals
- describe the principles of frequency translation
- carry out modulation and demodulation
- describe frequency spectrum sharing
- describe multiplexing and demultiplexing techniques

We will begin this element by describing in simple terms the two methods used to send many different information signals over a common transmission medium. This technique is called **multiplexing** and is fundamental to modern communication systems.

Introduction to multiplexing

To help you understand the principles behind multiplexing, we will investigate the options available to us to *transmit two telephone channels simultaneously* between

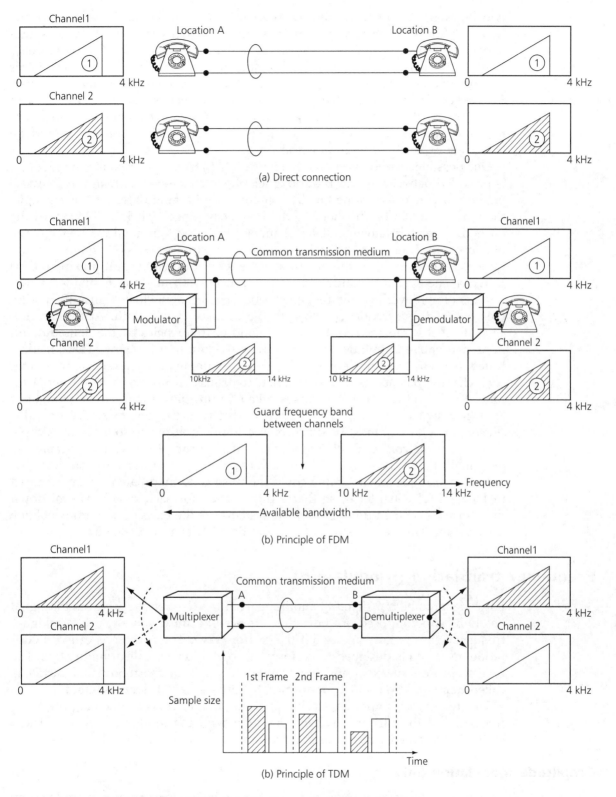

(a) Direct connection

Guard frequency band
between channels

Available bandwidth

(b) Principle of FDM

(b) Principle of TDM

Figure 10.1 The principle of multiplexing

two locations A and B. Obviously, the simplest solution would be to wire the two telephones together with a pair of wires so that each telephone is individually connected, as shown in Figure 10.1(a). Notice the way the frequency range of each channel (300 Hz to 3400 Hz) is shown as a triangle. The reason for this will become clear when you study modulation. This solution is fine, of course, for two channels and a cable pair is used to connect individual subscribers to their local exchange. But suppose many thousands of telephone channels are involved and the distance between them is large. Such a solution would be extremely expensive and impractical.

One possible solution is shown in Figure 10.1(b). A single transmission medium is provided between A and B but one telephone channel is *translated in frequency* before it is connected to the line by a circuit called a **modulator**. In this example, the modulator 'shifts' channel 2 to the frequency range 10 kHz to 14 kHz. At the far end, a **demodulator** translates channel 2 back to its original frequency range of 0 to 4 kHz. Thus each channel occupies its own frequency 'slot' on the common transmission line. This is the principle of **frequency division multiplexing** (FDM) in which the available bandwidth of the transmission medium is divided into a number of non-overlapping frequency 'slots'. Each channel to be transmitted occupies one of these frequency 'slots' in the available bandwidth. At the far end, each channel is returned to the normal band of frequencies by the **demodulator**.

An alternative technique, known as **time division multiplexing** (TDM) is illustrated in Figure 10.1(b). Here, the channels are sampled in turn repeatedly so that each channel has access to the common transmission line for a given period of time. The circuit that achieves this is called a **multiplexer** and can be thought of as a 'fast switch' that repeatedly switches between the channels. The **demultiplexer** at the far end must be synchronised to the multiplexer so that each sample is routed to its correct destination. Each batch of samples is called a **frame**. As you might have suspected, TDM is ideally suited for the transmission of digital data. For example, each sample could be pulse code modulated (as investigated in Element 9.1, Chapter 8) so that a digital code for each sample is sent down the transmission medium. TDM/PCM systems are the basis of modern telephone systems and will be investigated in more detail later in this element.

Frequency translation (modulation)

For a signal to be translated to another part of the frequency spectrum, the signal has to be superimposed on the characteristics of a third wave which is a higher frequency sinusoidal wave called the **carrier wave**. The process is called **modulation** and circuits designed to perform this operation are called **modulators**. The reverse process of extracting the original signal from the modulated carrier is called **demodulation** and is performed by circuits called **demodulators**.

The two most important techniques are to either vary the amplitude or frequency of the carrier wave in sympathy with the modulating signal. These methods are called **amplitude modulation** and **frequency modulation**.

Amplitude modulation (AM)

When a carrier wave of frequency f_c in Figure 10.2(a) is amplitude modulated by the signal frequency f_m (Figure 10.2(b)), its amplitude is made to vary in sympathy with the signal frequency f_m.

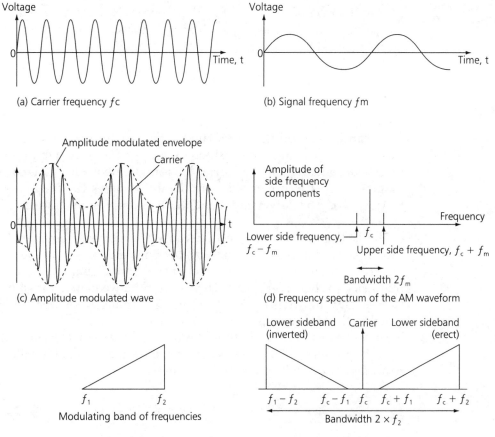

(a) Carrier frequency fc

(b) Signal frequency fm

(c) Amplitude modulated wave

(d) Frequency spectrum of the AM waveform

Modulating band of frequencies

(e) Frequency spectrum where f_c is modulated with a band of frequencies

Figure 10.2 Amplitude modulation

Notice in Figure 10.2(c) that the amplitude modulated wave varies in amplitude between a maximum value which is greater than its unmodulated amplitude and a minimum value less than the unmodulated amplitude. The waveform shape is called the **envelope** and is identical to the shape of the original modulating signal. It is this **shape** of the AM waveform which is 'carrying' or conveying the information in signal f_m.

Figure 10.2(d) shows the **frequency spectrum** of the AM wave. It contains three frequencies:

- the **carrier frequency** f_c
- an **upper side frequency** $(f_c + f_m)$
- a **lower side frequency** $(f_c - f_m)$

Each side frequency has exactly the same amplitude. The bandwidth of the AM signal is: $(f_c + f_m) - (f_c - f_m)$. Hence:

$$\text{Bandwidth} = 2 \times f_m \text{ (hertz)}$$

Notice that the bandwidth does not depend on f_c. It only depends on the frequency of the modulating signal.

177

<table>
<tr><td>Progress
check 10.1</td><td>A carrier wave of frequency 60 kHz is amplitude modulated with a signal of frequency 5 kHz. What frequencies are contained in the AM waveform and what is its bandwidth?</td></tr>
</table>

Sidebands

If the carrier is amplitude modulated with a signal consisting of a band of frequencies, then upper and lower **sidebands** are produced. Figure 10.2(e) illustrates these sidebands when the carrier f_c is modulated with a signal with frequency range f_1 to f_2. The **upper sideband** (USB) occupies the frequency range $(f_c + f_1)$ to $(f_c + f_2)$. The **lower sideband** (LSB) occupies the range $(f_c - f_1)$ down to $(f_c - f_2)$.

The USB is said to be 'erect', because the highest frequency in the signal (f_2) produces the highest frequency in the sideband $(f_c + f_2)$. The LSB is said to be 'inverted' because the highest frequency in the signal (f_2) appears as the lowest frequency in the LSB $(f_c - f_2)$. This is why the frequency range of the signal f_1 to f_2 is usually shown as a triangle. Using the 'triangle notation' shows immediately whether the sideband is erect or inverted.

Since the result of amplitude modulating a carrier (f_c) with a band of signals $(f_1$ to $f_2)$ produces two sidebands then the resultant signal is called a **double sideband signal** or DSB for short. The bandwidth is:

$$\text{Bandwidth} = 2 \times \text{highest modulating frequency (hertz)}$$

The highest modulating frequency in our example is f_2, so bandwidth $= 2 \times f_2$.

<table>
<tr><td>Progress
check 10.2</td><td>A carrier of 108 kHz is amplitude modulated with a voice signal occupying the commercial speech bandwidth. Sketch and label the spectrum of the DSB signal. What does a voice frequency of 1 kHz appear as in each sideband? What is the bandwidth of the DSB signal?</td></tr>
</table>

Modulation factor m

Figure 10.3 shows a sinusoidal carrier wave of peak value V_c amplitude modulated with a sinusoidal signal of peak value V_m. The amplitude modulated signal varies between a maximum value of $(V_c + V_m)$ and a minimum value of $(V_c - V_m)$.

The **modulation factor** m is defined as:

$$m = \frac{\text{maximum peak value} - \text{minimum peak value}}{\text{maximum peak value} + \text{minimum peak value}}$$

For the waveform in Figure 10.3, we have:

$$m = \frac{(V_c + V_m) - (V_c - V_m)}{(V_c + V_m) + (V_c - V_m)} = \frac{2V_m}{2V_c}$$

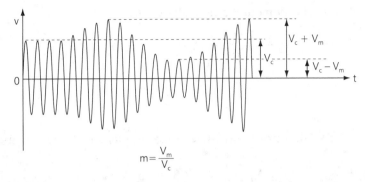

Figure 10.3 Modulation factor m

So

$$m = \frac{V_m}{V_c} \text{ (no units)}$$

It is also found that the peak value of the upper and lower side frequencies is equal to $\frac{1}{2}m \times V_c$. Usually, m is expressed as a percentage and is then called the **percentage modulation** or **depth of modulation**.

WORKED EXAMPLE 10.1

When displayed on an oscilloscope, a carrier wave modulated with a sinusoid varies between a maximum value of 4 V and a minimum value of 2 V. Calculate:

(a) the peak value of the carrier
(b) the peak value of the modulating signal
(c) the percentage modulation
(d) the peak value of the upper and lower side frequencies.

Solution

(a) The waveform would look like the diagram in Figure 10.3. In this case:

$$V_c + V_m = 4 \text{ V}$$
$$V_c - V_m = 2 \text{ V}$$

Adding these two equations gives,

$$2V_c = 6, \quad V_c = 3 \text{ V}$$

So the carrier has a peak value of $V_c = 3$ V.

(b) Subtracting the two equations gives:

$$2V_m = 2V, \quad V_m = 1V$$

The modulating sine wave has a peak value of $V_m = 1$ V.

(c) The percentage modulation is:

$$m\% = \frac{V_m}{V_c} \times 100\%$$

$$m\% = \frac{1}{3} \times 100\% = 33\tfrac{1}{3}\%$$

(d) The upper and lower side frequencies have a peak value of:

$$\frac{1}{2}mV_c = \frac{1}{2} \times \frac{1}{3} \times 3 = 0.5 \text{ V}$$

Maximum percentage modulation

Remember, the envelope of the modulated wave is a replica of the modulating signal. If the depth of modulation is increased to 100%, the maximum value of the waveform is $2V_c$ and the minimum value is zero *but* the envelope is still an exact replica of the modulating signal. The depth of modulation can be increased beyond 100% but the envelope is then distorted and is not an exact replica of the modulating signal. Thus, in practice depths of modulation exceeding 100% are never used.

Activity 10.1

You will almost certainly have a radio receiver at home. Study the tuning scale and write down the frequency range for the medium wave AM band.

Case study

If you completed Activity 10.2 you should have found that the frequency range for AM radio broadcasts was about 540 kHz up to about 1640 kHz. In this case study we will investigate the signal flow through a **typical AM receiver**.

In Europe, the stations use DSB amplitude modulation with carriers in the range 647 kHz up to 1546 kHz. The bandwidth of each channel is about 9 kHz and since DSB is used, the highest audio frequency transmitted is 4.5 kHz. This is adequate for voice but it means that music reproduction is of poor quality when compared to, say, the quality of reproduction from your CD player.

AM reception

A simplified block diagram of an AM receiver is drawn in Figure 10.4.

The **antenna** of the receiver picks up all the signals (including noise) and passes them to the **radio frequency amplifier**. Generally, the signals' strengths are quite weak and are typically only a few microvolts. The r.f. amplifier, mixer and local oscillator circuits contain resonant circuits with variable capacitors like the ones investigated in Element 11.2, chapter 2. We normally call them **tuned circuits**.

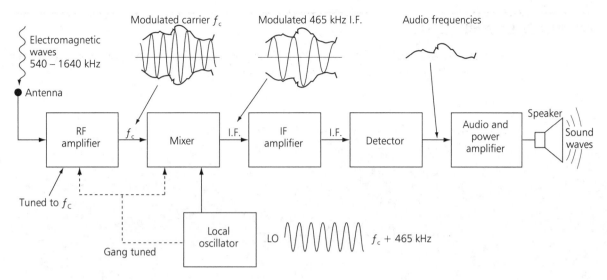

Figure 10.4 Block diagram of an AM receiver

When the tuning knob on the receiver is varied, the capacitors are changed simultaneously so that the frequency that leaves the mixer circuit is always 465 kHz, with the audio information amplitude modulated on it. This is called **ganged tuning**. The process is as follows.

Movement of the tuning knob causes the r.f. amplifier to select and amplify the desired station (carrier frequency f_c) and allows f_c and its upper and lower sidebands to be outputted from the amplifier to the **mixer**. The mixer has two inputs, the amplitude modulated signal and a sinusoidal signal from the local oscillator (LO) circuit. The ganged tuning ensures that the LO frequency is $(f_c + 465)$ kHz. The mixer produces sum and difference frequencies of f_c and $(f_c + 465)$. The **difference frequency**, i.e. $(f_c + 465) - f_c = 465$ kHz, is selected by a tuned circuit and appears at the output of the mixer. Thus, the original AM signal has been translated down to a frequency of 465 kHz and the voice or music signal is contained in the upper and lower sidebands of the 465 kHz signal. The 465 kHz signal is called the **intermediate frequency** or i.f.

The i.f. amplifier 'boosts' this signal and passes it to a circuit called a **detector** or **demodulator**. The detector recovers the envelope of the AM signal (which is the original audio information) and removes the i.f. signal of 465 kHz. After further amplification, the audio signal produces sound out of a loudspeaker.

Power in an amplitude modulated waveform

The power contained in an AM waveform will be distributed between the upper and lower sidebands and the carrier. For amplitude modulation with a single frequency it can be worked out mathematically that the total power P_T is equal to:

$$P_T = P_C \left(1 + \frac{m^2}{2}\right)$$

when P_c is the power in the carrier and m is the modulation factor. Using this equation to make some simple calculations will show that amplitude modulation is not a very efficient use of power.

WORKED EXAMPLE 10.2

The total power contained in an AM wave is 900 W. Calculate the power in the carrier and the side frequencies for percentage modulations of (a) 50% and (b) 100%.

Solution

(a) Since $P_T = 900$ W then using the equation above gives

$$900 = P_C \left(1 + \frac{m^2}{2}\right)$$

For $m = 0.5$,

$$900 = P_C \left(1 + \frac{(0.5)^2}{2}\right) = P_C (1.125)$$

So $P_C = \dfrac{900}{1.125} = 800$ W

The remainder of the power $(900 - 800) = 100$ W is shared equally between the upper and lower side frequencies.

(b) For $m = 1.0$

$$900 = P_C \left(1 + \frac{1^2}{2}\right)$$

So $P_C = \dfrac{900}{1.5} = 600$ W

For 100% modulation, 300W is contained in the side frequencies. This is only a third of the total power in the wave.

It is important to realise that *the carrier does not carry any information* even though it has to be there for the frequency translation process. The information is contained in the side frequencies. DSB amplitude modulation is not an efficient use of power.

Single sideband suppressed carrier (SSBSC)

The upper and lower sidebands in an amplitude modulated wave contain identical information and so it is *not* necessary to transmit both of the sidebands. Furthermore, since the carrier does not contain any information that does not have to be transmitted either. Thus, in practice the transmission of only one sideband is sufficient. The technique is called single sideband suppressed carrier (SSBSC) transmission.

A method of producing a **single sideband signal** is shown in Figure 10.5. It uses a special amplitude modulator called a **balanced modulator** and a **band-pass filter**.

The band of frequencies containing the information together with the carrier f_c are applied to the balanced modulator. The modulator produces the normal

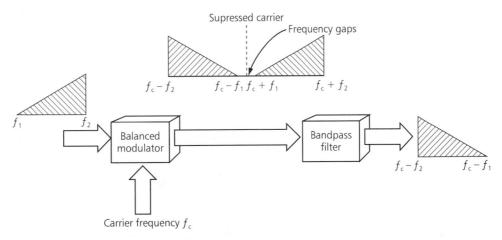

Figure 10.5 SSBSC generation

DSB signal but, because of its design, eliminates (or suppresses) the carrier f_c from its output. The bandpass filter allows the wanted sideband to pass through it but eliminates the unwanted sideband. In Figure 10.5, the frequency range of the bandpass filter only allows the lower sideband to be transmitted.

Some of the advantages of SSBSC compared with normal DSB are:

- SSBSC required half the bandwidth of a DSB signal that has been modulated with the same signal. This means that the available frequency spectrum for transmission can be more efficiently utilised since more channels can be accommodated.
- Since the bandwidth required per channel is halved then the signal-to-noise ratio will be improved when compared to the DSB signal.

The main disadvantage of SSBSC is that the carrier must be reinserted at the receiver with the correct frequency and phase before demodulation can occur. This means that radio receivers would be more expensive. This is one reason why SSBSC is not used for medium wave broadcasting.

SSBSC is used extensively in telephone systems when frequency division multiplexing is used, (see below).

Frequency division multiplexing (FDM)

The basic principles of FDM were outlined at the beginning of this element. With FDM the available bandwidth is divided into a number of non-overlapping frequency 'slots' with each channel occupying a particular slot. The way the channels are assembled and the frequency ranges used are laid down by international agreement. The multiplexing process starts by assembling collections of 12 channels into what are known as 'basic groups'.

Basic 12-channel group

The technique used to form a basic group is illustrated in Figure 10.6(a). Each of the 12 telephone channels of bandwidth 0–4 kHz is applied to its own balanced modulator fed with a carrier frequency. The carrier frequencies are spaced by

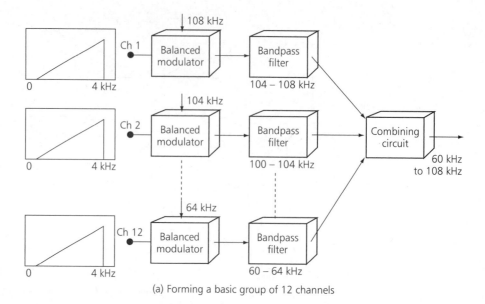

(a) Forming a basic group of 12 channels

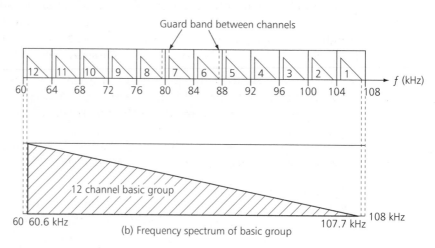

(b) Frequency spectrum of basic group

Figure 10.6 Technique used to form a 'basic group'

exactly 4 kHz. For example, channel 1 amplitude modulates a 108 kHz carrier and produces upper and lower sidebands from 108–112 kHz and 104–108 kHz. The bandpass filter removes the upper sideband and allows the lower sideband to pass to the **combining circuit**. Thus channel 1 has been translated to the band 104 kHz–108 kHz. A similar process is followed for the 11 other channels. For instance, channel 12 has a carrier of 64 kHz. After modulation the lower sideband is selected so channel 12 now occupies the frequency slot of 60–64 kHz. All of the translated channels are combined together in the combining circuit.

Remember that even though a bandwidth of 0–4 kHz is allowed per channel, the actual voice information only occupies 300 Hz–3.4 kHz. After translation, there is a bandwidth of 900 Hz between each 'voice slot'. This is called a **guard band** and is necessary because the bandpass filter cannot be absolutely perfect. The guard band prevents one channel interfering with another. The 12-channel basic group occupies a precise frequency range of 60.6 kHz to 107.7 kHz, or approximately 60 kHz to 108 kHz. The frequency spectrum is drawn in Figure 10.6(b).

**Progress
check 10.3**

What does a 1 kHz frequency in channel 2 and channel 12 become in the basic group frequency spectrum?

The supergroup and hypergroup

Further multiplexing is achieved by taking five separate groups of 12 channels and translating them in frequency to form a 60-channel supergroup. A supergroup occupies the frequency range 312 kHz to 552 kHz.

The final baseband for transmission is assembled by translating supergroups using similar techniques. For example, 15 supergroups can be combined to form a **hypergroup**. This contains 900 channels (15 × 60 channels) and occupies the frequency range 312 kHz to 4028 kHz.

Case study

When a colour television signal is transmitted from a TV transmission station a form of amplitude modulation known as **vestigial sideband modulation** (VSB) is used. VSB produces a saving in bandwidth and the basic principle is as follows.

A television signal contains frequencies from 0 Hz (d.c.) up to 5.5 MHz. If double sideband amplitude modulation were used, a bandwidth of 2 × 5.5 MHz = 11 MHz would be required. Trying to transmit using single sideband would also be extremely difficult because the filter which selects the single sideband would have its lowest frequency exactly at the carrier frequency. As shown in Figure 10.5, this is not a problem when the lowest frequency in the signal is *not* d.c. (e.g. 300 Hz for speech) because there is a frequency gap between the sideband and the carrier. Any filter needs a frequency gap between the wanted and unwanted frequencies so that the filter's attenuation can increase sufficiently. For this reason, SSB is not possible for signals containing frequencies down to d.c.

A vestigial sideband signal has a full upper sideband of bandwidth 5.5 MHz plus a 'little bit of' or 'vestige of' the lower sideband. The first 1.25 MHz of the lower sideband is transmitted normally at full amplitude and then the amplitude falls to zero over the next 1 MHz of bandwidth. Thus the total bandwidth occupied by the VSB signal is 5.5 + 1.5 + 1 = 8 MHz. This is a considerable saving in bandwidth when compared to the corresponding DSB bandwidth of 11 MHz.

Frequency modulation (FM)

With frequency modulation (FM), the amplitude of the carrier is kept constant but its frequency is varied in sympathy with the instantaneous value of the modulating signal as illustrated in Figure 10.7. Thus the carrier 'swings' in frequency above or below its unmodulated value f_c.

Notice that because the amplitude of the carrier wave is not changed when it is frequency modulated, the power contained in the wave stays constant.

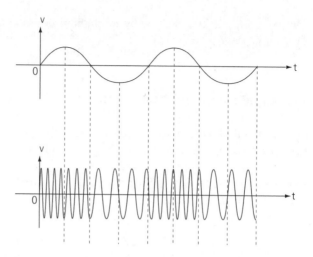

Figure 10.7 Frequency modulation of a carrier with an analogue signal

Frequency deviation

The amount by which the carrier frequency is changed from its unmodulated value is called the frequency deviation (FD) and is proportional to the instantaneous value of the modulating signal. The size of the frequency deviation allowed is decided on by the designer of the frequency modulator circuit. For example, one FM modulator might be designed to produce a deviation of, say, 10 kHz for every volt of modulating signal. If a modulating sinewave had a peak value of, say, 3 V then the carrier would deviate by ± 30 kHz around its unmodulated value, f_c.

The maximum value of deviation that is allowed to take place in an FM system is called the **rated system deviation** (FD_{max}) and occurs when the modulating signal reaches its peak value.

Bandwidth of an FM wave

The bandwidth of an FM signal is, in general, wider than that of an amplitude modulated wave.

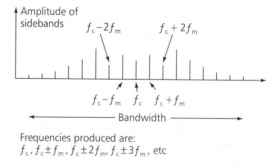

Figure 10.8 Spectrum of an FM wave

As shown in Figure 10.8, a whole range of side frequencies is produced when the carrier f_c is frequency modulated with the signal f_m.

Although not easy to calculate mathematically, it can be shown that the bandwidth is given by:

$$\text{Bandwidth} = 2(\text{FD}_{max} + f_m) \text{ (hertz)}$$

The next example shows that the bandwidth needed for FM music transmissions is much wider than that required to transmit the same music signal in an AM system.

Activity 10.2

Examine the frequency scales on your radio receiver once again. Write down the frequency range used for the carriers in the FM band.

WORKED EXAMPLE 10.3

The maximum audio frequency transmitted by an FM radio station is 15 kHz and the rated system deviation is 75 kHz. Calculate the bandwidth required and compare it with an AM DSB system transmitting the same range of frequencies.

Solution
Required bandwidth is:

$$\text{Bandwidth} = 2 \times (\text{FD}_{max} + f_m) = 2 \times (75 \text{ kHz} + 15 \text{ kHz}) = 180 \text{ kHz}$$

For an amplitude modulated DSB transmission the bandwidth required is:

$$\text{Bandwidth} = 2 \times 15 \text{ kHz} = 30 \text{ kHz}$$

This is a sixth of that required for an FM transmission. Thus, FM uses far more of the available frequency spectrum than AM does.

FM vs AM
The question is, why bother to use FM anyway since bandwidth is such a precious commodity? There are many reasons why FM is generally superior to AM. The two major advantages are:

- The signal-to-noise ratio at the output of an FM receiver is usually greater than that produced by a DSB receiver.
- An FM receiver possesses an effect called the capture effect. What this means is that if two signals are received at or near the same frequency, the FM receiver suppresses the weaker signal so there is less interference. AM receivers do not have this ability.

Case study

If you completed Activity 10.3, you would have discovered that the frequency range for FM radio broadcasts is 88 MHz to 108 MHz. In this case study we will investigate the signal flow through a **typical FM receiver**. A simplified block diagram is shown in Figure 10.9.

Notice that some of the blocks are the same as that in an AM receiver except that in this case they operate at much higher frequencies. Other blocks are different.

The r.f. amplifier, mixer and local oscillator are gang tuned as in the AM receiver. The r.f. amplifier can amplify any frequency in the 88 MHz to 108 MHz range, but its tuned circuit will only select the carrier frequency f_c selected by the tuning dial and the bandwidth of the FM transmission centred on f_c. The local oscillator produces a steady sinusoidal wave which is 10.7 MHz above f_c. The mixer operates in the same way as in the AM receiver except that its output is a 10.7 MHz frequency modulated signal. This 10.7 MHz i.f. frequency is amplified by the i.f. amplifier.

Noise and interference picked up in transmission causes amplitude variations in the FM signals. A circuit called a **limiter** removes these unwanted amplitude changes and produces a constant output FM signal at 10.7 MHz.

The discriminator is equivalent to the detector in an AM receiver. It demodulates the FM signal to recover the original modulating audio signal.

The **de-emphasis network** is not used in an AM system. For special reasons in FM systems, the higher frequencies in the audio signal are amplified more than the lower frequencies before they are transmitted. The circuit used at the transmitter is called a **pre-emphasis** circuit. The function of the de-emphasis block is to return the higher audio frequencies back to their correct amplitude relationship to the lower audio frequencies.

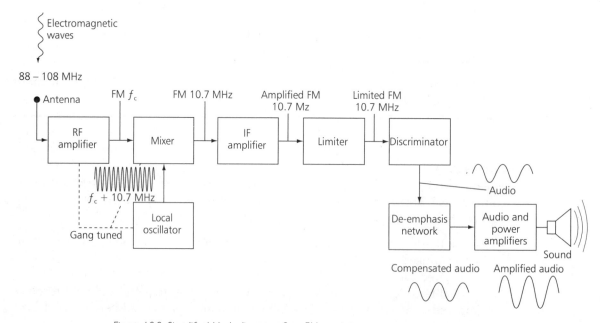

Figure 10.9 Simplified block diagram of an FM receiver

The audio amplifiers 'boost' the signals as necessary to drive the loudspeaker. Usually, in an AM/FM receiver the same amplifiers would be used whether the receiver was used for AM reception or FM reception.

Activity 10.3

Use your radio receiver at home and 'scan' all the way across the FM band. Make a list of the stations you receive and note their carrier frequencies.

Case study

The operation of a **satellite communications system** has been outlined already in Element 9.2, Chapter 9. In this case study we will consider how amplitude and frequency modulation are used to send telephone traffic via satellite. The system is illustrated in Figure 10.10.

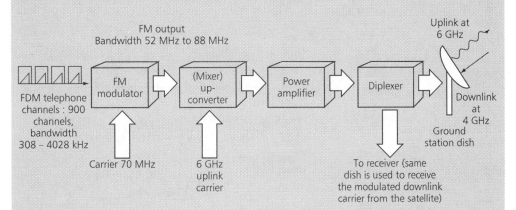

Figure 10.10 Telephone traffic via satellite

Using frequency division multiplexing, (utilising SSBSC amplitude modulation), the individual telephone channels are assembled into the baseband signal. In Figure 10.10, 15 × 60 channel supergroups occupying the frequency range 308 kHz to 4028 kHz form the 900-channel baseband signal. This signal frequency modulates a 70 MHz carrier and produces an FM signal of bandwidth 52 MHz to 88 MHz. This signal, together with a 6 GHz local oscillator signal (called the uplink carrier), is applied to a mixer which is now called an up-converter. In this type of mixer the *sum* of the two frequencies is selected. Thus the output of the up-converter is a frequency modulated 6 GHz uplink carrier which is now amplified until it has a power of several kilowatts and passed to the earth station parabolic antenna. This antenna is large (typically 30 m diameter) and so the microwave power is focused into a very narrow beam and pointed directly at the satellite.

The circuit called the diplexer allows the same antenna to be used for transmission and reception. The received modulated downlink frequency (typically 4 GHz) is passed to the receiver system for demodulation and demultiplexing.

Frequency shift keying (FSK)

When digital data is sent from one location to another, the **Public Switched Telephone Network** (PSTN) is often used to carry the data. Although nowadays the majority of PSTN operates digitally internally (PCM etc.), the connection to the exchange is still mostly analogue. Thus, the digital data must be converted into a suitable analogue form for transmission over the PSTN. One method is to use a frequency modulation technique called frequency shift keying (FSK). With FSK, '0' bits are transmitted at one frequency and '1' bits at another. The net carrier frequency is the average of the two frequencies and the data waveform continually switches between the '0' and '1' frequencies according to the data stream that is transmitted. The device that changes the data bits into the two audio frequency signals is called a **modem**. The principle of FSK is illustrated in Figure 10.11.

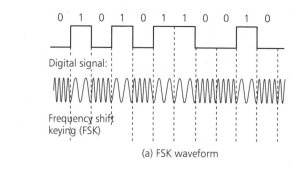

(a) FSK waveform

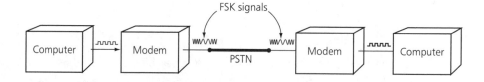

(b) Modems are needed at each end

Figure 10.11 Frequency shift keying (FSK)

Note in Figure 10.11(b) that another modem is required at the other end of the PSTN to convert the FSK tones back into the data waveform. Typically, the frequencies used could be, '0' = 2100 Hz and '1' = 1300 Hz.

Progress check 10.4 What is the net carrier frequency for the FSK signal above?

Time division multiplexing (TDM)

Time division multiplexing (TDM) is now the favourite method of multiplexing different information signals so that they can share a common transmission medium. It is ideally suited for the multiplexing of digital signals – whether from digital sources like computers etc. or digital versions of analogue signals such as PCM signals. Although the basic principles of TDM were outlined at the beginning of this element, in this section a more detailed investigation is given with reference to how TDM is used in PCM telephone systems.

As illustrated in Figure 10.1(b), each channel is sampled in sequence so that only during the time 'slot' that a channel is being sampled is it connected to the common transmission medium. Each full set of samples is known as a **frame** and the frames travel down the transmission system one after another.

In a TDM/PCM telephone system each sample is coded into an 8-bit digital code so that each frame is a 'string' of 1's and 0's propagating down the common transmission medium. Of course, the sampling theorem still applies to this system so each channel must be sampled at a frequency of at least twice the maximum frequency in the signal. A commercial telephone channel is assumed to have a highest frequency of 4 kHz (even though the voice signal is from 300 Hz to 3.4 kHz), so each channel is sampled at 2×4 kHz $= 8$ kHz $= 8000$ times per second. Thus, the time for one complete frame will be, $\frac{1}{8000} = 125\ \mu$s.

In the UK, a frame is divided into 32 time slots with only 30 of the time slots being used for speech signals. The other two time slots are used for synchronisation purposes (called **frame alignment**) and signalling. Frame alignment is necessary of course because the multiplexer and demultiplexer have to be synchronised with each other. The frame structure of a 30-channel TDM/PCM system is shown in Figure 10.12. Study the diagram carefully and note the following:

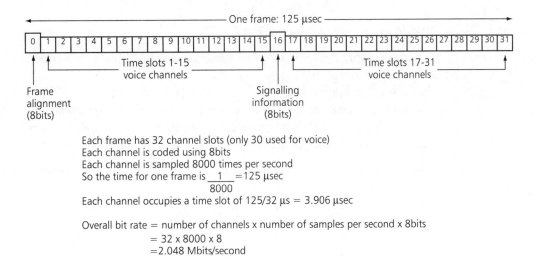

Each frame has 32 channel slots (only 30 used for voice)
Each channel is coded using 8bits
Each channel is sampled 8000 times per second
So the time for one frame is $\frac{1}{8000} = 125\ \mu$sec
Each channel occupies a time slot of 125/32 μs = 3.906 μsec

Overall bit rate = number of channels x number of samples per second x 8bits
= 32 x 8000 x 8
=2.048 Mbits/second

Figure 10.12 Frame structure of a 30-channel TDM/PCM system used in the UK

- The bit rate per channel is 64 Kbits/sec (8 × 8000).
- The overall bit rate is 32 × 8000 × 8 = 2.048 Mbits/sec

This is usually 'rounded' to a whole number and is described as being a 2 Mbits/second system.

Higher-order TDM systems

The basic 30-channel TDM/PCM system operating at 2.048 Mbits/sec is called a primary or first-order system. This can be used as a basic 'building block' to give higher-order TDM systems in a similar way in which the basic 12-channel groups are multiplexed to form supergroups etc. in a frequency division multiplex system.

Assignment 10

This assignment provides evidence for:
Element 9.3: Investigate communication techniques
and the following key skills:
Communication 3.2, 3.3
Information Technology 3.1, 3.2, 3.3
Application of Number 3.3

The evidence indicators for this element require you to investigate various communication techniques and to perform modulation and demodulation with simple circuits provided at your Centre.

Report

Your report should describe:

- the baseband transmission of audio signals; i.e. telephone signals and data telephone signals
- the principles of frequency translation; i.e. AM and FM
- how the available frequency spectrum is shared; i.e. frequencies and bandwidths for radio, TV, satellite systems and multichannel telephone systems
- multiplexing and demultiplexing techniques; i.e. FDM and TDM.

Additionally, you should submit a log of the practical work carried out on modulation and demodulation.

PART FOUR: COMPUTER SYSTEMS AND APPLICATIONS

Chapter 11: Computer hardware for engineering applications
Chapter 12: Software packages for engineering applications
Chapter 13: Configuration and use of computer systems

- Computers are now common place; wherever we go we are confronted by them. Sometimes they are obvious, as in the case of a desktop computer; sometimes they are less obvious, such as in the case of an ATM machine which we all use to obtain cash from our bank accounts. There is an incredible range of applications for computers and software from word processing, computer-aided design to complex monitoring, control and simulation. To not be able to understand how to use and apply computer systems is like not being able to read and write. It is therefore important for engineers and technicians to understand the basic principles associated with the configuration and application of both computer hardware and software.
- Part four covers the basics of computer systems and investigates different types of software packages and their applications. A method for the selection of both hardware and software for a particular application is also developed.

Chapter 11: Computer hardware for engineering applications

> **This chapter covers:**
> Element 10.1: Evaluate computer system hardware for engineering applications.
>
> **... and is divided into the following sections:**
> - Characteristics of computer hardware
> - Interfacing devices
> - Peripheral devices
> - Communications between computers
> - Selection and evaluation criteria
> - Engineering applications.

Computers are now used for many different and varied applications which range from simple word processing to computer-aided design and monitoring and control. It is therefore important to gain an understanding of the basic characteristics of computer systems and associated hardware to enable the most appropriate system to be selected for a given application.

When you complete this element you should be able to:

- describe the characteristics of computer hardware
- describe the functions of computer system components
- identify and show examples of interfacing and peripheral devices
- select and evaluate hardware for specific engineering applications.

Characteristics of computer hardware

The heart of any computer system is the microprocessor; the internal architecture and operation of this is covered in chapter 20. In this chapter we will consider the use of the microprocessor in the personal computer or PC and briefly look at its historical development in terms of speed, capacity and word length and then consider in detail the various component parts which go to make up the average personal computer.

Central processing unit

The main part of a computer is the microprocessor or CPU as it is often called. This particular device has developed at an alarming rate to meet the demands of higher speed, more complex software applications and graphical user interfaces or GUI's. When the now famous PC hit the market place in the early 1980's it used an 8086 16-bit microprocessor typically running at a speed of 4 or 8 MHz. This was a significant step forward on the previous 8-bit microprocessor-based computer systems of the day. As the PC gained popularity so the demand for even more processing power spawned the development of faster and more complex processors; this led to the development of the 80186 and then the 80286 and an extra processor called a maths co-processor the 80287. This co-processor took over the heavy burden of performing complex mathematical tasks as demanded by spreadsheets and computer-aided design software packages.

Very rapidly software designers started to demand the presence of a maths co-processor in the computer systems running their particular applications. Up to this point in time the word length used was still 16 bits and typical clock speeds were now up to 12 MHz.

The next major development came with the introduction of the 80386 which was a true 32-bit processor cable of operating at clock speeds of up to 33 MHz. This particular processor was also available with a built-in maths co-processor and in this case was called a 386DX, the DX indicating the presence of a co-processor. Improvements in operating systems and the introduction of higher resolution graphics and the new Windows graphical operating environment pushed processor development so that a new generation of 80486 processors began to emerge with ever higher clock speeds, in fact the last in the 486 series operated at clocks speeds of 100 MHz. Then came another major development that of the 80586 or Pentium processor which brought high-speed 64-bit capability to the PC with clock speeds of 200 MHz. To match this new processor yet another operating environment was developed, that of Windows 95. With the changes in processing power has come the ability to use greater and greater amounts of memory from the meagre 256 Kbytes of RAM in the early 1980's to the commonplace 16/32 Mbytes of RAM in the mid 1990's.

Storage devices

Various types of memory storage devices are required in modern computers: temporary and permanent, portable and built-in.

Floppy disk

The floppy disk, also called the diskette, is an extremely popular storage medium. The first floppy disks introduced were 8 inch, of low storage capacity and not particularly reliable. The 8 inch disk was superseded by the 5¼ inch disk which in turn has been superseded by the 3½ inch micro floppy. The floppy disk is inserted into the computer's disk drive where it is rotated at high speed. A **read/ write head** is then brought very close to the surface of the disk.

The read/write head is moved to and fro across the disk in order either to record data on the disk surface (write data), or to take back data (read data) which has previously been recorded or stored. The typical **storage capacity** for a Double Density (DD) disk is 720 Kb, and for a High Density (HD) disk is 1.44 Mb. The 3½ inch floppy disk drive is now the standard disk drive found on every type of personal computer and is shown in Figure 11.1.

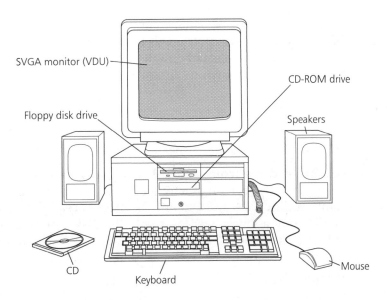

SVGA monitor (VDU)

CD-ROM drive

Floppy disk drive

Speakers

CD

Keyboard

Mouse

Figure 11.1 Typical permanent computer setup

Hard disk

A hard disk drive unit is found in most personal computers. The hard disk unit can store considerably more information than a floppy disk. The hard disk itself is quite small with diameters of approximately 3 or less inches with a trend towards the smaller sizes becoming more common.

The method of reading and writing data to the disk is very similar to the floppy disk unit but the hard disk is usually not removed from the computer unless it requires maintenance or becomes damaged. The typical storage capacity of a hard disk drive is 500 Mb up to 2.5 Gb.

Tape streamer (magnetic tape storage device)

The main type of magnetic tape used for computer storage is ½ inch tape in reel-to-reel form. However, cartridge forms have become popular alternatives because they are easier to use. The magnetic tape storage device operates in much the same way as a conventional audio tape recorder. It has a read head for 'reading' the information stored on the tape and a separate write head for recording data onto the tape. Magnetic tape storage is generally used for backup in the form of

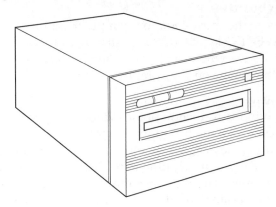

Figure 11.2 Tape streamer

a tape streamer for personal computers. The typical storage capacity ranges from 300 Mb to many Giga bytes. This type of device is useful for backing up or storing sensitive data which could otherwise be lost if the computer broke down. A typical tape streamer is shown in Figure 11.2.

Optical disk storage

There are three basic types of optical disk and all of them use lasers to read or write data and provide a means of storing very large volumes of data.

CD-ROM drive (compact disk read only memory)

These devices work on exactly the same principle as that used for domestic audio compact disk (CD) and can only be used to read data. Data is read from the disk by directing a low power laser onto the disk surface and detecting the pattern of light reflected back. CD-ROM is used to store reference works, catalogues, encyclopaedias, software as well as sound, typical storage capacity is 650 Mb.

WORM drive (write once read many)

Data is written to the disk by burning a permanent pattern into the surface of the disk by means of a high precision laser. Data is read back using the same method as that for reading a CD-ROM disk.

EO (erasable optical) or MOD (magneto optical disks)

These are the latest technology re-writeable optical discs. The disc is similar to the WORM disc but works on a slightly different principle. Instead of burning a pattern on the disk surface the laser heats the disk which is made of a special alloy. Magnetic molecules in the alloy surface can be aligned by a magnetic field when warmed by the laser but cool again to leave a semi-permanent magnetic pattern. The data can therefore be deleted and rewritten when required.

Progress check 11.1

What is the main component part of a computer system?
What is the advantage of a 32-bit word length over a 16-bit word length?
List three of the main storage devices found in a typical computer system.

Input devices

Keyboard

This is the most common computer input device and in general has a standard 'QWERTY' layout which is similar to that of a typewriter. The keyboard allows entry of a whole range of different character types and controls the movement of the cursor on the computer display screen.

Mouse

With the introduction of the Windows operating environment this device has now become a standard input device on all personal computers. It is essentially a pointing device which moves the cursor around on the monitor screen of a computer but in addition to this it can also perform simple control functions using its buttons. The most common type of mouse is the roller ball type which runs on the desktop and it usually has two or three buttons which can be programmed to perform various control functions.

Typically the outer buttons would be programmed to emulate the 'Return and Esc' keys of the keyboard. The mouse is often programmed differently when used with say a computer-aided draughting system; in this case the left button would be used to pick a point on the monitor screen, the middle button would duplicate the 'Return' key on the keyboard and the right button to activate special menus.

Digitising tablet

The digitising tablet consists of two main component parts, the tablet and the puck as shown in Figure 11.3.

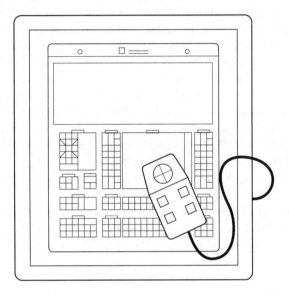

Figure 11.3 Digitising tablet

The puck is similar to the mouse except that it can position a point with a higher degree of accuracy, in addition to performing the same functions as a mouse in conjunction with the tablet it can be used to digitise existing drawings. This technique is particularly useful for digitising maps and can save a great deal of time. The process of digitising is quite straightforward and involves fixing the drawing down onto the tablet and then the various features on the drawing are picked by the puck until all of the drawings features have been entered. This process is still fairly time consuming but is still quicker than redrawing say a complex map. It is often the preferred input device for professional computer-aided draughting systems.

Joystick

A joystick is an input device and allows the computer user to manipulate items on the computer screen with ease. It is particularly useful when manipulating 3D images on the screen, as a joystick often has up to 6 axes of movement. Although it is often associated with computer games joysticks are now used for 3D applications such as **virtual reality**.

**Progress
check 11.2**

What are the two main input devices found on a typical modern day computer?
What is the difference between a mouse and a digitising tablet?

Display devices

VDU or monitor

The **visual display unit** (VDU) is the screen used by the computer to provide a visual output of data to the operator or user.

The pictures or characters on a VDU screen are made up of a number of discrete dots. High quality VDUs have more addressable dots available and therefore are able to display pictures and characters more clearly. To achieve a display the whole screen is organised into rows and columns of individual 'dot' positions called **pixels**. On a black-and-white screen each pixel can simply either be ON (white) or OFF (black). On a colour screen each pixel may comprise a cluster or red, green and blue primary-coloured dots which can be ON and OFF in different combinations or partially ON or OFF to allow for different colour intensity.

A VDU screen is normally matched to a particular type of graphics card. For instance a 15-inch VDU with a **scan rate** of 60 Hz would be matched to a **versatile graphics adapter** card (VGA) with a pixel resolution of 640 × 480. On the other hand a higher resolution 17-inch VDU with a scan rate of 75 Hz would be matched to a Super VGA (SGVA) card with 1 Mbyte of video RAM and a resolution of 800 × 600 or 1024 × 768 pixels. The best resolution at present requires the VDU to have a scan rate of 87 Hz matched to an SGVA card with 2 Mbytes of video RAM giving 1280 × 1024 pixels.

Interfacing devices

One of the main advantages of the PC is that it can support a very wide range of plug-in interface boards which can be used to monitor and control engineering systems, complex communication systems such as networks, high resolution graphics, hard disk and floppy disk drives, CD-ROM drives and sound systems.

Converters

Interface boards used for monitoring and control make use of converters such as **analogue-to-digital** and **digital-to-analogue converters**. The characteristics and operation of these devices will be covered in Chapter 17 and will not be covered in any more detail except to say that a typical interface board of this type may contain a number of ADCs and DACs and when used with the appropriate software can be used to monitor and control various engineering systems.

Interface board/card

Typical interface boards found in a modern day PC will include a graphics board, a hard disk and floppy disk controller board, a serial/parallel interface board and a sound card with CD-ROM controller. The most important thing to understand about these interface boards or cards is not so much how they work in detail but how they are installed in a typical PC. This particular subject will be covered in greater depth in Chapter 13, but for now we will briefly mention each type of card typically found in a modern day PC.

Graphics board/card

A graphics board is a device which is normally installed inside a computer and usually plugs into one of the spare slots found in the average personal computer. These cards are available with many different specifications and are used to control the size, resolution and colour attributes of the computer display screen.

Sound board/card

These days many personal computers have multimedia capability which means that they have to be able to process sound. Like the graphics card, the sound card also plugs into a spare slot inside the computer. Most modern sound cards have both sound recording and playback facilities often with stereo capability. When a computer has a sound card fitted it opens up many application possibilities including music, sound recognition, signal analysis and so on.

Hard disk/floppy disk controller card

This interface card has the necessary circuits for controlling up to two hard disk drives and two floppy disk drives. Later generation cards have the facilities to control up to four hard disk drives and two floppy disk drives. This type of interface card also has additional facilities such as two serial ports and a parallel port. Later generation cards also have a games port to which a joystick can be connected.

Serial/parallel interface card

The standard hard disk/floppy disk controller card will give the standard computer two serial ports and one parallel port. However, additional interface cards can be purchased which have additional serial and parallel ports. This would be useful when two peripheral devices such as a printer and a plotter which both have a parallel connection are required to be connected to the same computer. These cards are available with different combinations of serial and parallel ports to meet the requirements of most applications.

Peripheral devices

Printers

There are two basic types of printer:

- **The impact printer** Impact printers have print heads that hit inked ribbons against paper. The print head may be in the shape of the character to be printed, or a series of pins which hit the ribbon in a certain order depending on the character required.
- **The non-impact printer** Non-impact printers use a variety of methods for producing characters. **Thermal printers** burn an image onto special paper; **ink jet printers** fire tiny drops of ink onto paper controlled by an electrostatic field; **laser printers** transfer their image to paper from an electrostatically charged drum coated with ink toner.

Impact printers

Dot matrix

The dot matrix printer mimics the action of a typewriter by printing single characters at a time in lines across the paper. The print is produced by a small print head that moves backwards and forwards across the page stopping momentarily in each character position to strike a print ribbon against the paper. The characters are made up of a number of dots. According to the quality of the printer the characters may be made up of a matrix of either $7 \times 5, 7 \times 7, 9 \times 7, 9 \times 9$ or 24×24 dots. The more dots the better the image. The print speed for dot matrix printers ranges from 30 to 600 characters per second, and the print quality ranges from poor to near letter quality (NLQ).

Daisywheel

This is another popular type of low speed printer and is used when high quality print is required. These printers are fitted with exchangeable print heads called daisywheels. To print each character the wheel is rotated and the appropriate spoke is struck against an inked ribbon. The print speed for daisy wheel printers ranges from 15 to 100 characters per second, and the print quality is near letter quality (NLQ).

Non-impact printers

Ink jet printer

These printers operate by firing very tiny droplets onto the paper by using an electrostatic field. The shape of the character is controlled by electrical fields acting on the electrically charged ink droplets. They are very quiet but of low speed. Some models print colour images, by means of multiple print heads each firing droplets of a different colour. The print speed for ink jet printers ranges from 25 to 250 characters per second, and the print quality is fair to NLQ. A typical ink jet printer is shown in Figure 11.4.

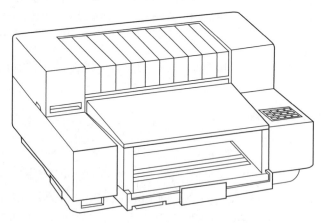

Figure 11.4 Ink jet printer

Thermal printer

This type of printer impresses a print head onto aluminium coated paper. A column of pins form characters and an electrical charge burns the characters onto the paper. The print speed is approximately 5 sheets per minute and print quality is poor. The main advantage of thermal printers is that they are extremely quiet.

Laser printer

An electronically controlled laser beam marks out an electrostatic image on the rotating surface of a photo-conductive drum. Ink toner is attracted onto the electrostatic pattern on the surface of the drum. The ink is transferred onto the paper as it comes in contact with the drum. A typical desktop laser printer looks similar to a photocopier, and is able to print both text and pictures. The print speed for laser printers is approximately 10 sheets per minute with excellent print quality. Figure 11.5 shows a typical laser printer.

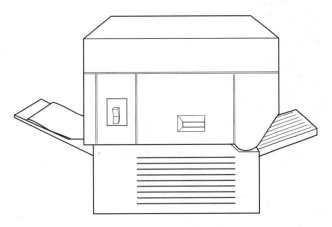

Figure 11.5 Laser printer

Paper feed mechanisms

Various methods are used to feed paper into printers. The three most common paper feeding devices are friction, traction and form feed.

Friction feed

Paper is gripped between two rollers, as in a typewriter. This type of feed is suitable for single sheets. However, some printers are equipped with a form feed or

Figure 11.6 Friction feed

cut sheet feeder. This is a device which automatically feeds sheets of paper into a friction feed printer and is a popular method of feeding paper into a printer and is found on most low cost printers. A modern low cost printer with cut sheet feeder is shown in Figure 11.6.

Traction feed

Specially designed paper with holes along the edges fits over wheels with corresponding teeth for holes. The wheels revolve, drawing the paper through the printer as shown in Figure 11.7.

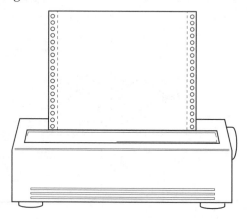

Figure 11.7 Traction feed

This type of feed is mainly used for draft letters and computer program listing. The paper is supplied in extremely long continuous lengths and printed documents are torn off at the nearest perforations. At one time this was popular with low cost printers but has been superceeded by the cut sheet feeder. However, it will be found on large printers used by organisations such as local council offices for printing cheques, demands and invoices etc.

Activity 11.1

Take two different types of printer, carefully remove their casings and make detailed sketches showing the type of paper feed mechanism and printhead types used.

From the printer manuals produce a specification for each printer detailing all of the different facilities available in each.

Plotters

Graph plotters are used for scientific and engineering purposes. One special application is CAD (**computer-aided design**) in which engineering or architectural designs are created and then drawn by graph plotters. There are two basic types.

Flatbed

The pen moves up, down, across and side-to-side, and the paper is held in position by mechanical or electrostatic means. This type of plotter is quite popular

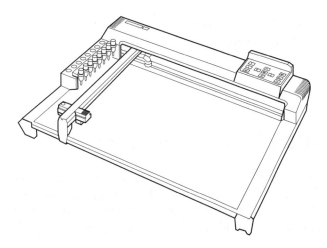

Figure 11.8 Flatbed plotter

for producing circuit diagrams and small mechanical drawings up to a maximum of A3 size. Many plotters of this type have up to six different coloured pens and can produce drawings with different details highlighted in different colours. The basic construction of a typical flat bed plotter is shown in Figure 11.8.

Drum plotter
The pen moves up, down and across and the paper rolls around the drum in either direction to produce the sideways movement as shown in Figure 11.9.

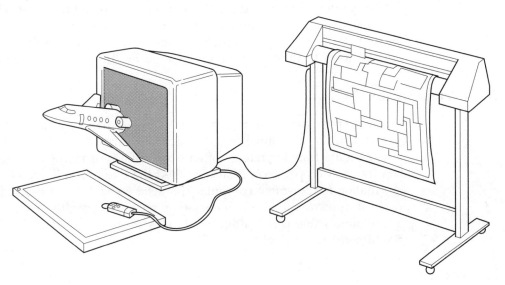

Figure 11.9 Draughting system

This type of plotter is more popular for producing large architectural drawings typically up to A0 size. As with the flat bed plotter a number of coloured pens can be accessed to highlight various details within the drawing.

**Progress
check 11.3**

For a particular application how would you choose between having a printer or a plotter as the main output device?

Communications between computers

One computer can communicate with another computer at a distance using a device called a modem which in turn is connected to the public telephone network. With special software one computer can automatically dial up the other and establish a communication link between the two.

Modems

Modem is an abbreviation for modulator/demodulator and is a device for sending and receiving digital data down standard telephone lines.

Computer systems operate by transferring and processing data in a digital form. This means that all data used by the system is formed by a series of 1's or 0's as in a digital signal. The 1's and 0's represent data or **logic levels** permitting calculations and control functions to be executed. Computer data can be sent down a standard telephone line but the telephone system expects to be sent analogue signals (analogue signals are those which vary continuously, not in two logic levels). In order to transmit computer data along a telephone line it must first be converted from a digital signal into an analogue signal. This is done by a modem which plugs directly into the telephone wall socket in the home or office. At the receiving end another modem converts the analogue signal back to a digital form so that it can be understood by the receiving computer.

Data transmission

A data transmission link is normally called a channel and one very important characteristic of a channel is its capacity to carry data. The actual rate of data flow, i.e. **data transfer rate**, is often expressed in terms of the number of bits of data transmitted per second and is called the **baud rate**. It usually takes 8 bits of information to transmit a computer character so a 500 baud transmission link will transmit approximately 40 characters a second. Typically, telephone lines, and optical fibre cable can handle baud rates of 28,800 baud and 500,000,000 baud (500 Mbaud) respectively.

Networks

Large numbers of computers can now be connected together to form what is called a network. A network allows any computer in the network to communicate with any other computer on that network. Networks can also be interconnected to form even larger networks. Ultimately networks can be connected together across the globe. The **Internet** is really an example of a truly global network where any computer in the world can communicate with any other computer.

A computer that is not connected to any other computer is often called a standalone system while an interconnected set of two or more computers is called a computer network. There are two types of computer network, the Local Area Network (LAN) and the wide area network (WAN).

Local area networks (LAN)

Networks used to interconnect computers in a single room, rooms within a building or buildings on one site are normally called local area networks (LANs). LANs transmit data between computers in digital form using wires or optical cables. Sometimes multiple LANs are used on the same site; for example one LAN per floor on a multi-storey building.

Wide area networks (WAN)

The networks which interconnect computers on separate sites, separate cities or even separate countries are called wide area networks (WANs) or long haul networks. WANs often use optical fibre cables, satellite communications links and microwave links as well as conventional telephone lines.

Server

A server is a special computer used as part of a communications network of other computers. It is normally the controller of the network and stores all of the software which can be used by the other computers on the network.

Activity 11.2

Discuss the relative merits of using a computer network as opposed to standalone computer systems.

Selection and evaluation criteria

Fitness for purpose

When choosing hardware for a particular application it is important to ensure that it is fit for the purpose required. For instance, when selecting hardware for an engineering application such as draughting, selecting a computer system with a small low-resolution monitor screen and with only a keyboard as an input device would be completely unsatisfactory. Careful thought must be given to the selection of the computer and peripheral devices to ensure that they meet the requirements of the particular application.

Reliability

A computer system can be the most highly specified system imaginable but if it is unreliable it is worse than useless. When selecting a computer system be sure to choose a supplier with a good track record for reliability and ideally choose one who will offer a warranty and supply a replacement system if problems occur. This is one way of minimising system downtime and ensuring that a computer system will always be available. Many computer suppliers will now offer this service and generally speaking a warranty period of at least two years is common.

Cost

When a customer specifies the type of computer system required for a given application a maximum amount of money will often be allocated for the purchase of the equipment. Unfortunately in many cases the money available does not necessarily match with the specification of the system. It is therefore quite a difficult job to choose a system which both meets the customer's requirements and stays within the set budget. This will often involve some compromises in relation to the specification and therefore close liaison with the customer until an agreement is reached in terms of the overall budget or specification.

Availability

When choosing hardware it is important to select a hardware supplier who keeps a stock of equipment and can therefore supply it ex-stock and deliver within a few days. It is quite easy to be caught out when ordering state-of-the-art hardware as delivery times can sometimes be very long, especially for new product lines. This could be embarrassing for instance if a customer was waiting for a computer system that was almost complete and ready for delivery but was missing a graphics board which was on six weeks delivery. Always check delivery times with suppliers before placing an order for equipment.

Ease of maintenance

Most modern computer systems need very little maintenance but some of the peripheral devices such as printers may need cleaning, replacement print heads and so on. Choosing a device which has simple access to routine maintenance items is therefore important. If the peripheral device is difficult to maintain because it has to be completely stripped down to replace a print head then system down times are going to be significant. Therefore before specifying a particular peripheral device check on the maintenance aspects of the device and avoid if possible those devices with complex maintenance requirements or consider duplication for back-up purposes.

Engineering applications

In the first instance an idea of the basic system requirements for given engineering applications is important. Once this has been established the system can then be configured to match the exact specification for the application and software used. Choosing software and configuring computer systems is covered in Chapters 12 and 13. In this chapter we will establish the basic system requirements for the following engineering applications.

Draughting

Figure 11.10 shows the computer and peripheral device requirements for a typical draughting or **computer-aided design** (CAD) system. Notice that the most important aspects of the system are the size and resolution of the monitor screen, input devices and output devices. Typically a digitising tablet would be used as the input device or possibly a mouse depending on the accuracy required and a flatbed **plotter** would be a suitable output device.

Design

A system used for the design of electronic circuits would need a system similar to that specified for the draughting system except that the preferred input device

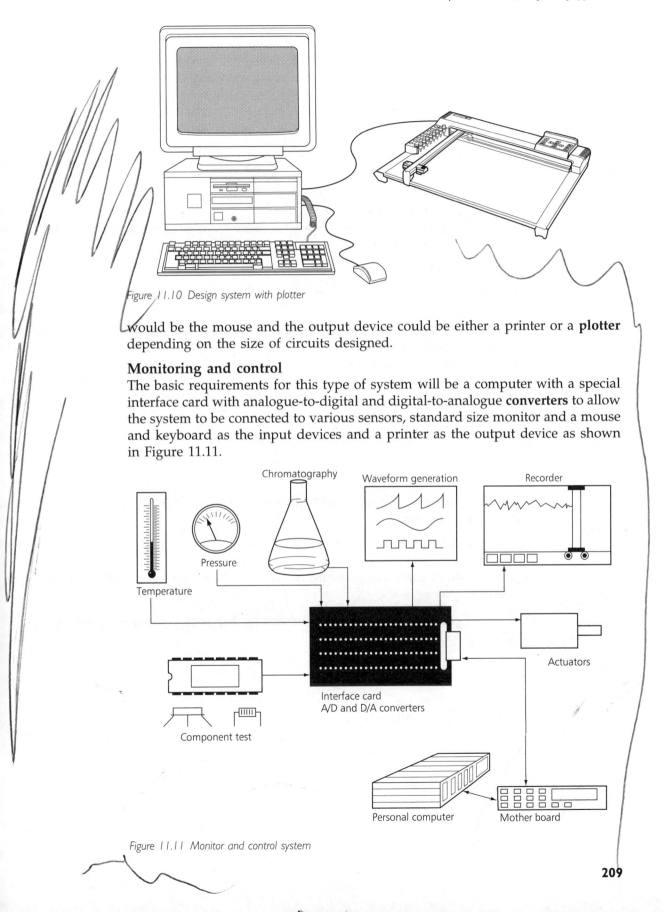

Figure 11.10 Design system with plotter

would be the mouse and the output device could be either a printer or a **plotter** depending on the size of circuits designed.

Monitoring and control

The basic requirements for this type of system will be a computer with a special interface card with analogue-to-digital and digital-to-analogue **converters** to allow the system to be connected to various sensors, standard size monitor and a mouse and keyboard as the input devices and a printer as the output device as shown in Figure 11.11.

Figure 11.11 Monitor and control system

Data storage

In all of the above cases a **storage device** will be required and typically this will be a hard disk drive and a floppy disk drive. The size requirements of the hard disk drive will vary depending on the application and software used.

Assignment 11

This assignment provides evidence for:
Element 10.1: Evaluate computer system hardware for engineering applications
and the following key skills:
Communication 3.2, 3.3
Information Technology 3.1, 3.2, 3.3, 3.4

You have been given the task of defining the basic component parts of a computer system to design simple electronic circuits; the largest size of drawing to be produced will not exceed A3. Write a simple report listing the various component parts required with details of price and suitable suppliers. Your choices of actual devices should be based on the selection criteria described in this chapter.

Report

The evidence indicators for this element suggests that a report be produced covering the following:

● The evaluation and selection of computer hardware for each of two engineering applications.
● Briefly identify computer hardware required by the remaining applications in the range.

There are opportunities to generate evidence for key skills in Communications and IT within this element.

Chapter 12:   Software packages for engineering applications

This chapter covers:
Element 10.2:   Select and evaluate software packages for engineering applications.

. . . and is divided into the following sections:
● Software
● Engineering applications
● Selection and evaluation criteria.

For a computer system to undertake a particular task it requires special programming software to control it so that it will perform the required task. The purpose of this section is to investigate various software packages for given engineering applications.

When you complete this element you should be able to:

● describe the requirements of software packages for given engineering applications
● describe software packages meeting the requirements of given engineering applications
● select a software package for given engineering applications
● evaluate selected software packages for given engineering applications.

Software

The term 'hardware' is used to describe all the physical devices found in a computer system (disk drives, keyboard, processor etc.) whereas 'software' is the general term used to describe all the various programs that may be used on a computer system. The are two types of software: systems software and applications software.

Systems software

These are programs which control the way the computer operates or provide facilities that extend the general capabilities of the system. The systems software includes a program or collection of programs called the **operating system**. The operating system controls the performance of the computer to ensure the efficient use of hardware by applications programs. Most applications programs can only work when used in conjunction with the operating system.

Applications software

This is software which is designed to be put to specific practical use. Word processors, databases, spreadsheets and drawing packages are all examples of applications software. In this chapter the main consideration will be applications software. Typical examples include:

- *Word processing*   Software that assists in the preparation of textural documents.
- *Spreadsheets*   A collection of related numerical and textural data in the form of text, numbers and formulae. A change in the value of one piece of data automatically changes the value wherever that particular piece of data has been used.
- *Databases*   A structured set of records comprising a large amount of data or information which is usually interrelated.
- *Desktop publishing (DTP)*   Text and graphical images can be mixed and manipulated by this type of software to create newspaper pages, magazines and posters etc.
- *Computer numerical control (CNC)*   Listings of X, Y and Z co-ordinates, together with other information such as tool speed and feed rates, are fed to a machine tool to enable the repetitive production of artifacts.
- *Computer-aided design (CAD)*   Complex drawings can be produced in both 2 dimensions (2D) and 3 dimensions (3D) and may include options such as surface shading/rendering and different view points.
- *Computer-aided manufacture (CAM)*   The integration of CAD and CNC manufacturing techniques, where encoded graphical images produced using a CAD system are transferred electronically to the machine tool.
- *Electronic design aid (EDA)*   Engineers use this type of software to assist them to design electronic products, circuits and systems. The software often allows the actual operational characteristics of the design to be simulated on the computer. This technique considerably reduces amount of development time required to produce a working design.
- *Supervisory control and data acquisition (SCADA)*   Software which allows a computer system to monitor parameters such as temperature, pressure etc. and to save this information on disk for later analysis. Often the software has facilities for controlling external devices such as PLC systems, see Chapter 23.

There are many different types of software packages available, one of the major problems is deciding on the correct type of software for a given application. This requires an understanding of the application and the particular requirements of that application which must be met by the software. To start with we need to consider some examples of engineering applications and then investigate some typical software packages which could be used with those applications.

Progress check 12.1

Explain the difference between system software and applications software. Give one example of system software and two examples of applications software.

Engineering applications

Draughting

This particular application relates to the requirement of being able to produce a detailed drawing of an object on the screen of a computer. The object may be a complex mechanical piece-part, electronic circuit diagram, a complete system drawing of say a chemical plant or an architectural drawing of a single house or a complete housing estate. Figure 12.1 gives some idea of the complexity of detail required for a typical piece-part drawing.

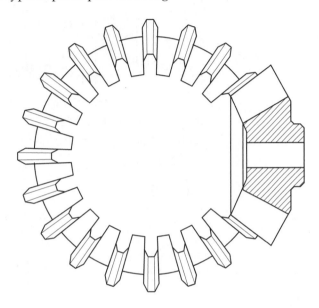

Figure 12.1 Piece-part drawing

Design

Modern day engineers not only use the computer to draw their designs they also use it to help them test their designs. In the case of a mechanical design this may involve simulating the effects of stress or temperature and in the case of an electronic circuit it may involve analysing the frequency response of the circuit as shown in Figure 12.2.

Monitoring and control

Most industrial processes require careful monitoring to ensure that all aspects of the process are operating correctly. Many also require each part of the process to be controlled and supervised centrally. A typical application is shown in Figure 12.3 and involves the constant monitoring of a chemical process. Various parameters are displayed on the screen and are also stored on computer disk at regular intervals. The storing of data at regular intervals is known as **data logging**. Another popular requirement is to have a **mimic diagram** of the process being monitored or controlled, displayed on the screen.

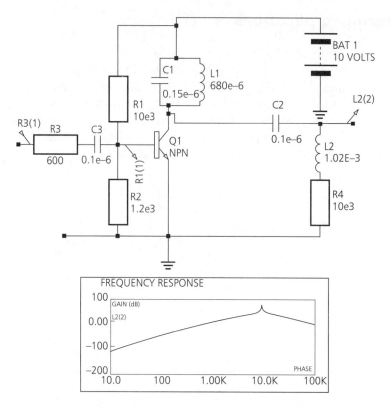

Figure 12.2 Circuit design and simulation

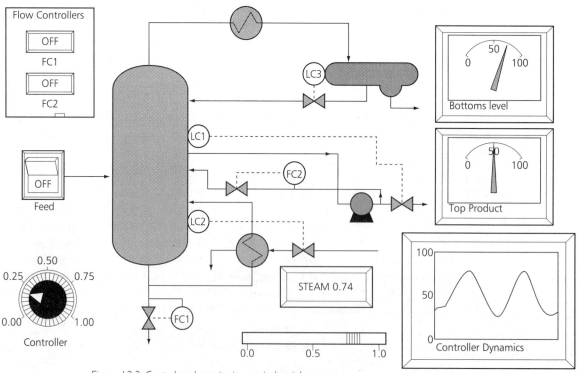

Figure 12.3 Control and monitoring an industrial process

Selection and evaluation criteria

Fitness for purpose

In the first instance the software must have all of the facilities required by the application. For example, a draughting application may call for both 2D and 3D facilities and therefore a software package supporting only 2D draughting would be unacceptable in this case. Having found a software package which has facilities matching those called for by the application it is then important to establish that the software package meets a number of basic requirements.

Speed

When using a software package it is very important to the user that the software operates quickly. It can be very frustrating if it takes many seconds for the computer to renew a drawing on the screen every time a simple modification is made to the drawing. This is particularly important when drawings become very complex. In some low-cost packages redrawing the screen can take many minutes and in some cases this makes the package totally unusable. (The same can happen with low-cost hardware.) The best test is to use a known drawing of reasonable complexity and to actually try out the facilities of the software package with this drawing. This is possible as many software suppliers with give you the software on 30 days approval. If it does not meet with your requirements within this time scale you can return it free of charge.

Ease of use

A software package may have all of the facilities required for a given application and may operate very quickly but may actually be very difficult to use. In this case consideration has to be given to the frequency of use and the implications of training.

If a software package is going to be used on say a daily basis, then one that is difficult to use may not be a problem as long as the people using it are properly trained. On the other hand if the software package is only going to be used occasionally it becomes very important that it is easy to use and requires only a small amount of learning time or set-up time.

Reliability

This is important for any product including software: the package should operate reliably and be error free (bug free). Often the best test for this is to find out from the software supplier how many people are actually using the software. It is then possible to contact any of the individuals or organisations concerned and get direct feedback regarding persistent problems etc.

Fault tolerance

With complex draughting packages which rely on mathematical calculations, it is important that the software correctly deals with mistakes which can cause errors in the mathematical calculations, such as division by zero. The software should simply flag up the mistake as an error and allow the user to take corrective action. The software should not respond in an undefined way such as crashing out before the user has time to save his/her work. This particular problem can waste a great deal of time and hence cost a company a great deal of money in the long term. Again, the best test is to talk to people who actually regularly use the software and finally to actually test the software yourself by deliberately trying to make the software misbehave.

Backup features

This particular subject is often forgotten but is very important when much time and effort is used to create something on the computer screen. There is nothing worse than spending two or three hours producing a drawing only to find that when a power cut occurs all of your work has been lost.

Most modern software packages have an automatic back-up or file saving facility which means that if power is lost, you will at worst have only lost that bit of work which was completed after the last automatic back-up. It is therefore important that this facility is included in your chosen software package; if it is not, then people using the package must be trained to make regular back-ups of their work.

Technical support

With complicated software packages with many hundreds of different facilities it is now very important that technical support is to hand. Many software suppliers have what is called a Help Desk or support hot line. The number is often clearly printed on the inside cover of the software manual. When the number is rung the help given should be of good quality and if the identified problem is associated with a bug in the software then a software update should be sent to you within 24 hrs at no charge.

The best test for this is to choose a particularly difficult-to-use feature of the software and then to ring the Help Desk for advice and assistance and then judge the quality of support for yourself.

Availability

Ideally the software for a particular engineering application should be **off-the-shelf software**, in other words it is a standard product which can be used for many different applications without modification. This is often the cheapest solution but real consideration has to be given to whether or not the software truly meets all of your requirements.

If you have a particularly difficult application the requirements of which cannot be meet with a standard software product then **custom design software** may be worth considering. However, it has to be said that custom designed software is very expensive and is usually only considered by large organisations with an application requiring widespread use of the software throughout the organisation. Examples include Banks, Building Societies and other commercial organisations. Custom software has more limited use in the engineering environment.

Cost

Cost is always an important consideration when choosing software and must be weighed against the facilities offered by a particular software package. For instance a software package may meet the requirements of a particular application and cost £500 whereas another package may meet the requirements of the present application but have some additional facilities for future applications and cost £600. The question is: Do you take the cheaper package and then find that as your application grows the software shows its limitations or do you pay the extra knowing that the software has some spare capacity to deal with your growing application? This can sometimes be a difficult decision to make but at the end of the day will be ultimately governed by the available budget.

Documentation
This should be clear, concise, well structured, easy to use and above all should clearly state how to install the software. Good documentation can bring hidden cost benefits in training and should include tutorials and examples of how to use the software rather than over-complicated explanations of how to use particular facilities. Of even more use is the inclusion of example files which can be used immediately to allow you to get a good feel for the software.

Compatibility
This is often where many software packages fall down: the suppliers will often state a minimum hardware requirement but when the software is actually run on such equipment its performance is often unacceptable leading to hidden cost penalties. In addition to this there is often a requirement to use the information produced by one software package in another package. This can sometimes cause problems if the file format used by one package is not supported by another. It is therefore important to establish the file formats supported by a particular software package and to test the package on the minimum hardware configuration stated by the suppliers. This will ensure that under worst case conditions the performance of the software still meets your requirements.

Assignment 12

This assignment provides evidence for:
Element 10.2: Select and evaluate software packages for engineering applications
and the following key skills:
Communication 3.2, 3.3, 3.4
Information Technology 3.1, 3.2, 3.3, 3.4

A company called Design Consultants has asked you to select a piece of software for them to meet their requirements to produce 2D drawings of their products and to monitor and control their manufacturing processes. Using the selection criteria described in this chapter you are required to produce a report analysing the facilities offered by at least two different pieces of software, for each application, then come to a conclusion with regards to which one the company should purchase.

Report

The evidence indicators for this element suggest that a report should be produced on the selection of software packages for two engineering applications. Include:

- a description of the software requirements of each engineering application
- a description of at least two software packages meeting the requirements for each engineering application
- notes on the selection of one software package for each engineering application
- notes on the evaluation of the selected packages.

Chapter 13:   Configuration and use of computer systems

Deciding on the correct configuration of computer hardware for a particular engineering application can be difficult. It is the purpose of this chapter to give some guidelines to allow the correct configuration of hardware to be identified and to indicate how the engineering application is supported.

When you complete this chapter you should be able to:

- describe the configuration requirements of computer systems for given engineering applications
- identify the parameters required by a computer system for an engineering application
- configure the computer system for an engineering application
- use the computer system to carry out an engineering application.

System specification

For a particular engineering application a complete computer system will need to be specified and configured. This will involve the correct choice of both hardware and software.

Hardware
The main component parts relating to computer hardware have been described in Element 10.1, Chapter 11.

Software
The requirements of software packages for engineering applications have been covered in Element 10.2, Chapter 12.

Engineering applications
To apply the knowledge gained in Elements 10.1 and 10.2 we shall consider in detail the configuration requirements of a computer system for two engineering applications: draughting/design and monitoring/control.

Computer system requirements

Draughting and design applications are highly demanding on system resources such as memory, graphics and processing power. The need to store complex graphical images implies large amounts of storage capacity. To help with the problem of identifying system components we start by drawing up a specification for the required system, this is best done by considering a case study.

Case study

A company called System Designs Ltd wants to introduce **computerised draughting systems** into their design office. Initially they want to introduce two systems one for producing simple 2D drawings and another for producing more complex 3D drawings. The company has little or no experience of using computers for this particular type of application. They need advice in terms of defining precisely what they need to purchase by way of hardware, software and peripheral devices to enable them to meet their requirements.

General system requirements

Start by thinking about the overall requirements for both types of system; for instance both systems will need a computer, video monitor, input device, output device and storage system which we represent by means of a **block diagram**.

   Having established the general form of the system we now need to look more specifically at the requirements of each of the 'blocks' detailed.

Specific requirements

Computer

Let us assume that standard PC's will be used and that the main internal component parts will be defined mainly by the software. The company has indicated a preferrence for the Windows operating system and would like to use Turbocad 2D and Autocad for the 3D work. This decision helps us to define the requirements for each PC as the software manufacturer always states the minimum requirements for correct software operation.

2D system

This system will run Turbocad 2D in Windows and the software requires a 486DX66 processor as a minimum, 4 Mbytes of memory minimum and 20 Mbytes of hard disk space for installation, a mouse or graphics tablet and an SGVA graphics card with 1 Mbyte of video RAM.

   Most of the internal requirements of the computers have now been defined, and at this point we need to consider **storage** requirements such as hard disk size and floppy disk drives and whether a back-up drive will be required.

Hard disk size

To allow sufficient space for Windows and the Turbocad software 100 Mbytes of storage will be required. To allow space for storing working drawings etc. allow another 150 Mbytes, hence the minimum disk size required is 250 Mbytes.

Floppy disk drive

Smaller drawings may be supplied to the company for modification etc. and these would normally be sent on 3.5 inch floppy disks and therefore it would be essential to have a 3.5 inch drive in the main computer.

Backup

A detailed drawing may take many man hours of work to produce and therefore it is important that a copy of the the drawing is kept on a medium which can be removed and stored in a safe place. It would therefore be sensible to have a magnetic tape backup system installed in the main computer.

Input/output devices

The specification for the computers is just about complete, so we now turn our attention to peripheral devices.

Input devices

A keyboard and mouse will be required for general inputting but for accurate placement a **graphics tablet** with a puck will be required. As both the mouse

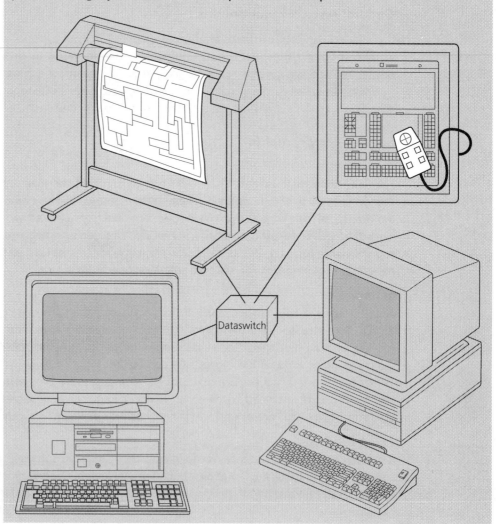

Figure 13.1 Plotter and graphics tablet with dataswitch shared between two PCs

and graphics tablet will require serial inputs the computer will need an interface card which will support at least two serial inputs.

Output devices

As both the 2D and 3D systems will need to produce hardcopies of drawings up to A0 size it would be a good idea to purchase a parallel port A0 plotter with a **data switch** which will allow either computer system to have access to the plotter. This arrangement is shown in Figure 13.1.

As we have chosen a parallel port plotter the computer interface card specified must also support a **parallel output port**.

Finally the video monitor will need to be large enough to show the maximum amount of drawing detail and will need to have SGVA resolution as defined by the software. A 17-inch **SGVA monitor** would therefore be suitable for both the 2D and 3D systems.

2D system specification

The final specification for the 2D system will be as follows: .

- Processor: 486DX66
- Memory: 4 Mbytes
- Hard disk: Size: 250 Mbytes
- Floppy disk: Type: 3½-inch
- Back-up: Type: Internal magnetic tape
- Input/output ports: 2 serial and 1 parallel
- Keyboard: Standard Qwerty
- Monitor: 17-inch SGVA
- Mouse: 2 button
- Graphics tablet: 12 × 12 inch summergraphics tablet
- Dataswitch
- A0 plotter (shared resource)
- Parallel cables

This completes the specification for the 2D draughting system which permits the job of **costing** the various components so that a price for the whole system can be calculated.

Activity 13.1

Using information from the case study you are required to undertake the following activities:

- Complete the costings for the 2D draughting system and come up with an overall price for the system. You will need to contact computer suppliers or use catalogues and magazines to get up-to-date prices.
- Produce a specification for the 3D draughting system using the same format as for the 2D system.

Activity 13.2

To extend your understanding of defining system requirements and to produce evidence for your portfolio you are required to produce a specification with costings for a computer-based monitoring system. The general system requirements are as follows:

- A company called SCADA systems Ltd requires a PC-based monitoring system running under Windows which can monitor up to 8 analogue channels with voltages in the range ±5 V and data log to disk and produce hardcopy of the monitored data. The highest input frequency is 20 kHz and the smallest input signal is ±10 mV.
- Your task is to define the system requirements in terms of computer hardware (including details of any special interface cards) and the choice of suitable software.
- The documentation produced should have sufficient detail to enable the customer to order the component parts of the system.

System configurations

So far we have only considered how to specify the system for a given application; this of course is only part of the story. Once the system has been specified and purchased it will need to be configured. Configuring the system may involve both hardware and software – for instance to configure the system to use a mouse will require the mouse to be physically plugged into a suitable serial port and then the mouse driver software is installed.

The software part of the configuration is usually quite simple. A disk is normally supplied with the mouse and then a program called INSTALL.EXE is usually run which will automatically install the required mouse driver. This will also automatically update the two configuration files called CONFIG.SYS and AUTOEXEC.BAT. Once the installation is complete the computer has to be reset to enable the modified configuration files to take effect. The mouse is now ready to use with any of the application software on the computer.

More complex configurations

Most computer systems purchased these days will already be configured to operate correctly. However, there comes a time when a computer may need to be ugraded with a new processor, more memory, an additional interface card or new software. In this situation configuration is slightly more complex and requires careful reading of the manufactures' installation manuals. To highlight the link between hardware and software when it comes to configuring a system let us consider the following case study.

WORKED EXAMPLE 13.1

A computer is purchased with the following specification:

- Processor: 486SX25
- Memory: 4 Mbytes
- Input/output: 1 serial 1 parallel port.

It is required that the system is upgraded to meet the following specification:

- Processor: 486DX66
- Memory: 8 Mbytes
- Input/output 2 serial 2 parallel ports.

This upgrade will require the purchase of a 486DX66 processor, 4 Mbytes of memory and an interface card with one serial and one parallel port.

Solution

This job will require the computer case cover to be removed and then the processor will need to be located on the motherboard. Before the cover is removed the computer must be disconnected from the mains supply. The motherboard manual will normally have a diagram indicating the positions of the main components as shown in Figure 13.2.

Installing a new processor

Having located the processor it will need to be carefully removed from its socket and stored safely away. The new processor can now be removed from its protective box and pushed into the socket on the motherboard. However, before doing this the position of pin 1 must be established on the motherboard socket and on the processor. Figure 13.3 shows the position of pin 1 on the motherboard socket and Figure 13.4 shows the position of pin 1 on the processor. Notice that pin1 is normally indicated by the square at the bottom of the pin on the processor and the shaped cut-out on the socket.

The processor can now be inserted into the motherboard socket ensuring that all of the pins are correctly engaged. The first stage of the installation is now complete, the next task is to find a spare power supply connector into which the processor fan is connected; the fan is required to supply extra cooling.

The physical installation of the processor is now complete. The next task is to change the various jumpers on the motherboard to change the clock frequency, enable the co-processor etc. The various jumper settings are defined in the motherboard manual. The number of changes required to the jumpers depends on the type of motherboard; some require only one or two changes, other motherboards require a significant number of changes as shown in the tables of Figure 13.5. To enable the jumper connectors to be changed their position will need to be established on the motherboard as shown.

Each jumper must now be checked and adjusted according to the tables shown in Figure 13.5. This is best done by changing one jumper at a time and then ticking the table until each jumper has been checked to be in the correct position. Having completed this task the computer cover can then be replaced and secured and the computer connected to the mains supply. Ensuring that all other cables are correctly connected turn on the computer; the system will now automatically detect the new processor and clock speed and display this information during the startup sequence.

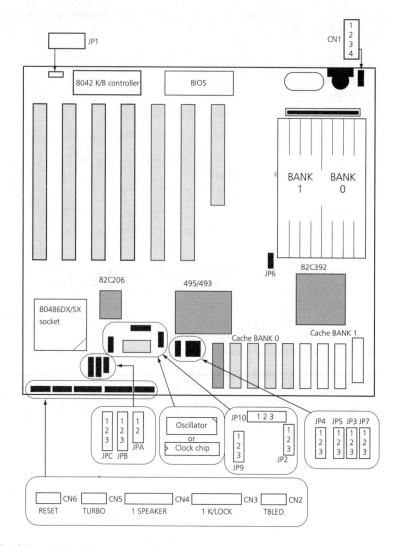

Figure 13.2 486 motherboard layout

The computer has now been successfully upgraded with a 486DX66 processor, to ensure that the upgrade is working correctly each piece of software installed on the system should be checked for correct operation.

Installing more memory

As with the processor upgrade the computer case cover will need to be removed and as before ensure that the mains supply is disconnected from the machine. Having removed the cover the sockets which hold the memory cards need to be located. This is done using the diagram supplied in the motherboard manual a typical example of which is shown in Figure 13.6 in which the memory board socket positions are highlighted.

Each motherboard is different and may have differing types and numbers of sockets. Typically an older style motherboard may have eight 30-pin sockets whereas later motherboards may have four 30-pin and four 72-pin sockets or there may be just eight 72-pin sockets. The memory cards which plug into these sockets

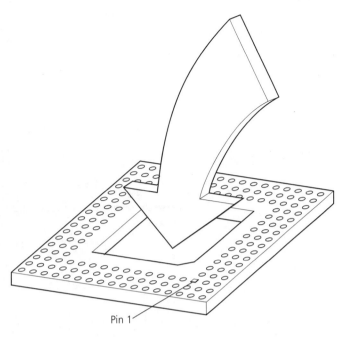

Figure 13.3 Pin 1 socket

Figure 13.4 Pin 1 processor

are called SIMM's or single-in-line memory modules. Care must be taken to ensure that the correct type of SIMM is purchased and that it matches the number of pins in the motherboard sockets. On motherboards with 30-pin SIMM's it must be remembered that this type of SIMM must be installed in groups of four, whereas the later 72-pin SIMM's can be installed singly.

Therefore, if your motherboard has 30-pin sockets, four would already be occupied with 1 Mbyte SIMMs; to increase the memory to 8 Mbyte would require the insertion of four 1 Mbyte SIMMs. If the motherboard had 72-pin sockets only one would be occupied with a single 4 Mbyte SIMM and to upgrade to 8 Mbyte would require the insertion of another single 4 Mbyte SIMM. To help with the identification of the different types of SIMM, typical examples are shown in Figure 13.7.

Having established the correct type of SIMM and inserted it into the correct socket or sockets also ensuring that it is correctly orientated, the physical part of

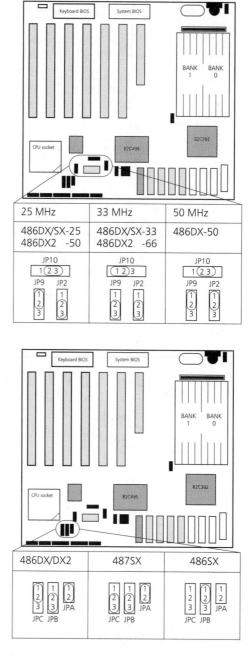

Figure 13.5 Motherboard jumper tables/positions

the memory upgrade is now complete. Figure 13.8 shows an example of a four-SIMM upgrade with the upgraded memory positions highlighted. The computer case cover can now be replaced and the mains supply reconnected.

Having completed the physical part of the upgrade, the computer can now be switched on. During the start-up sequence a memory check takes place which involves a count of the actual amount of memory in the computer. If all is well the on-screen count should indicate that the computer now has a total of 8 Mbytes of memory.

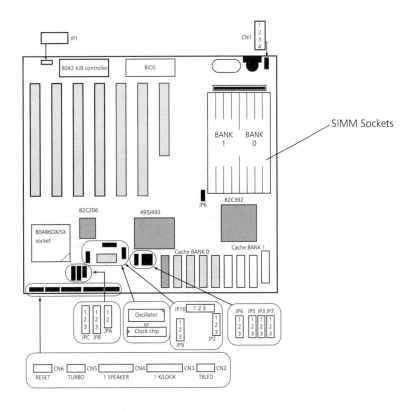

Figure 13.6 Motherboard diagram showing SIMM sockets

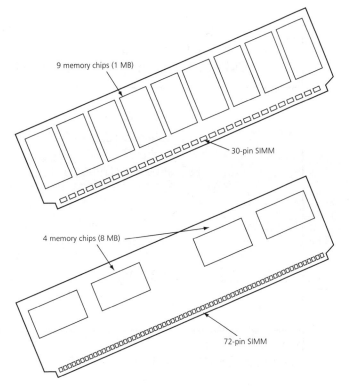

Figure 13.7 Examples of different SIMMs

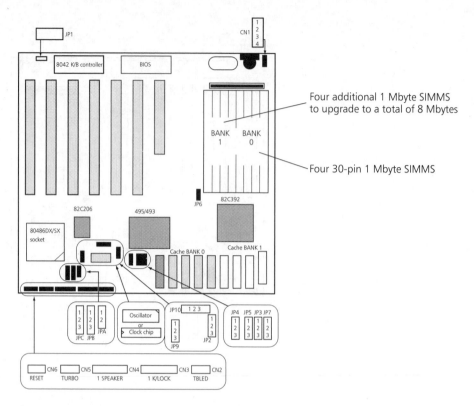

Figure 13.8 Memory upgrade

BIOS SETUP PROGRAM - AMI BIOS SETUP UTILITIES
(C) 1992 American Megatrends Inc., All Rights Reserved

STANDARD CMOS SETUP
ADVANCED CMOS SETUP
ADVANCED CHIP SETUP
BIOS SETUP DEF. AUTO CONFIGUARTION OPTION
POWER-ON DEF. AUTO CONFIGURATION OPTION
CHANGE PASSWORD
AUTO DETECT HARD DISK
HARD DISK UTILITY
WRITE TO CMOS AND EXIT
DO NOT WRITE TO CMOS AND EXIT

Advanced CHIPSET Setup for Configuring the CHIPSET Registers

ESC:EXIT ⇓⇒⇑⇐ :Sel F2/F3:Color F10:Save & Exit

Figure 13.9 Setup screen

This indicates that the memory has been correctly installed and is operating correctly, however, the amount of memory available is stored in the CMOS RAM of the computer and on startup this figure is checked against the actual amount of memory installed. In one case there will be a difference and this will cause an error message to be displayed on the screen: RAM CHECK SUM ERROR.

To change the information stored in CMOS RAM requires the computer setup to be modified, this is done by running the SETUP utility by pressing either the 'DEL' or 'F1' key on the keyboard during the startup sequence. We are then presented with a screen similar to the one shown in Figure 13.9.

To update the setup to register the increase in memory is simply a case of choosing the 'write to CMOS and exit' option from the screen and selecting 'YES' to save the new setup.

The computer now knows exactly how much memory it has and will now operate correctly with the new memory. Having completed the memory upgrade, the final job is to install the new interface card.

Installing a new interface card

Each interface card sits in an expansion slot located on the motherboard. The processor communicates with each interface card by using an **interrupt** (see Chapter 22) or IRQ line and an address. Each standard type of interface such as video, hard disk controllers, serial/parallel cards, are preallocated certain addresses and interrupt lines; for instance the address 3F8H and interrupt line IRQ3 may be allocated to the serial interface. The major problem with adding additional interface cards is to ensure that the same address and IRQ line is not used twice otherwise there will be conflict and the system will not operate correctly. Therefore, before physically installing the new interface card we need to known what address and IRQ lines are already in use.

Activity 13.3

If you have access to a PC running Windows use the control panel option to discover the following:

- How many serial and how many parallel ports are being used.
- What addresses have been allocated to the serial and parallel ports.
- What IRQ lines have been allocated to the ports.

Following the same procedure as for the previous upgrade examples, once the case cover has been safely removed the expansion slots need to be identified on the motherboard as indicated in Figure 13.10.

Having established the position of the expansion slots, find a spare slot, remove the screw holding the small blanking plate to the back of the computer casing, remove the blanking plate and store safely. The expansion slot is now ready to accept the new interface card. Prior to inserting the new card into the spare slot the correct address and IRQ lines must be selected. This is done by changing the position of various jumpers on the **interface board**; the available address and IRQ numbers are normally printed on the back of the card with the position of the various jumpers. Ensure that the jumpers are positioned so that the selected address and IRQ number does not clash with any of those already in use. Once

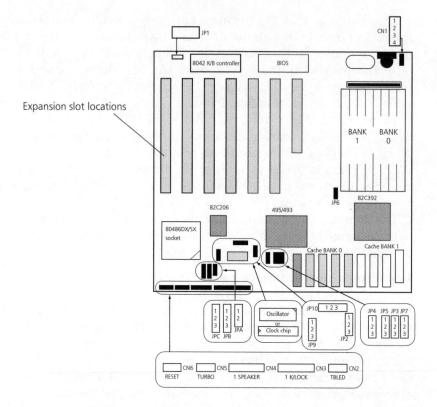

Figure 13.10 Expansion slot locations

this has been done the card can be physically inserted into the spare slot and secured to the back panel by a small retaining screw. The physical installation is now complete and the casing can now be replaced and the computer connected to the mains supply.

Testing a newly installed card

To test the installed card turn on the computer and carefully monitor the screen during the startup sequence. If all is well during startup the screen should display information relating to the number and addresses of the serial and parallel ports.

The screen will actually show that the serial ports have been assigned as COM1 and COM2 with addresses 3F8H and 2F8H and that the parallel ports have been assigned as LPT1 and LPT2 with addresses 378H and 380H. The interface card is now fully functional and can be used when required. This completes the case study which has shown how a PC can be upgraded in terms of a new processor, more memory and a new interface card.

Assignment 13

This assignment provides evidence for:
Element 10.3: Configure and use computer systems for engineering applications
and the following key skills:
Communication 3.2, 3.3
Information Technology 3.1, 3.2, 3.3, 3.4

To fully understand the problems associated with configuring computer systems it is necessary to undertake actual upgrades, re-configurations and if possible the complete configuration of a new computer. It is suggested therefore that this particular assignment be modified to suit the available resources and is therefore only presented here in skeletal form.

You are required to carry out the following activities:

(a) On a fully functional computer you are required to install an operating system such as DOS or Windows and to demonstrate its correct operation.
(b) On a fully functional computer with an installed operating system you are required to install appropriate engineering software and demonstrate its correct operation.
(c) You are required to install a 3½-inch floppy disk drive into a functioning computer system and test for correct operation by successfully formatting a disk.
(d) A hard disk drive has failed in a computer, you are required to remove the defective drive and replace with a new one. You are then required to correctly install the new drive, make it into a system drive and reload any approriate software. The correct operation of the disk drive is to be demonstrated by the computer correctly starting up and by running an appropriate piece of software.
(e) You are required to upgrade a functioning computer and this will involve changing the processor, increasing the size of memory and installing a sound card and CD ROM drive.

Report

The evidence indicators for this element suggest that a log is required of the configuration and use of computer systems for two engineering applications. The record should be supported by:

- a description of the configuration requirements of computer systems for each of two engineering applications.
- identification of the parameters required by a computer system for each of the two engineering applications.

Key skills in IT might be covered throughout this element.

PART FIVE: ELECTRONICS

Chapter 14: D.C. power supplies
Chapter 15: Small-signal amplifiers
Chapter 16: Logic elements and digital circuits

- Since the beginning of the twentieth century there have been enormous advancements in science which in turn have led to a fantastic growth in technology. There is no-one who has not shared in the consequences of this, but it is the growth in electronics/microelectronics and its associated technologies which has had the greatest impact on the way we live. The availability of low cost and powerful PCs, networking of PCs and the access to the Internet, etc. means that our everyday lives are very different to those of our parents and grandparents.
- In Part Five of this book, we explore some aspects of electronics, both analogue and digital. A number of practical activities are included to extend and reinforce basic electronic principles. Case studies look at the step-by-step design of some important circuits.

Chapter 14: D.C. power supplies

This chapter covers:
Element 13.1: Investigate d.c. power supply units.

. . . and is divided into the following sections:
- Diagram of a basic power supply
- Basic atomic theory
- Rectifying circuits
- Smoothing the waveform
- Voltage regulation of a power supply.

The majority of electronic circuits require a supply of d.c. power before they can operate. If the equipment has low power consumption (e.g. digital watches, calculators, etc.), the d.c. can be provided easily from batteries. It would not be practical, however, to run complex computer systems, TVs, and hi-fi's etc. from battery supplies because you would be forever replacing the batteries.

In practice, the d.c. supply for most equipment is derived from the a.c. mains by using a circuit whose job it is to produce a constant d.c. output voltage from the a.c. mains supply input. Such circuits are called **power supplies** and are an important class of electronic circuit in their own right.

When you complete this chapter you should be able to:

- describe the properties of pn junction diodes.
- describe the characteristics of d.c. power supply units.
- identify and give examples of functions of constituent circuits within d.c. power supply units.
- explain the operation of the constituent circuits within the power supply units.
- measure voltage and connect within d.c. power supply units.

Diagram of a basic power supply

The block diagram of a basic power supply is shown in Figure 14.1. The basic power supply itself is made up of the first three blocks. The last block, the load, is the circuit or component to which the power supply is giving direct current. We will consider each block in turn.

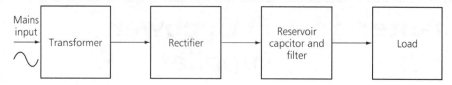

Figure 14.1 Basic power supply unit

The transformer

In most practical applications, the magnitude of the mains supply voltage is different to the d.c. voltage level required by electronic circuits. For example, the d.c. voltage needed by integrated circuits have low values such as 5 V, 9 V, etc. whereas the peak value of the 230 V_{RMS}, 50 Hz UK mains supply is about 325 V (1.414 × 230 V) and therefore has to be reduced in magnitude. The mains voltage can be changed to almost any other level by using a transformer.

For low voltage power supply applications, the transformer will be **step down**, meaning that the a.c. secondary voltage (V_{SRMS}) will be lower than the mains supply.

The rectifier

The rectifier is the subcircuit of the power supply which actually converts the alternating voltage at the transformer secondary into pulses of direct voltage. It is made up of one, two or four semiconductor components called **pn junction diodes**.

All diodes are devices which have two terminals. One is the **anode** of the diode and the other is the **cathode**.

The **diode symbol** when used in electronic circuits and its typical structure are shown in Figure 14.2. The end marked with a ring identifies the cathode and, in many cases, the type number of the device is also printed on the body.

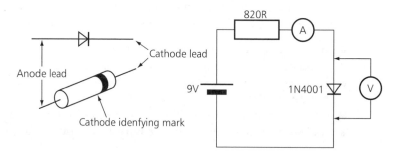

Figure 14.2 Diode symbol and single diode circuit

The diode is formed by creating a junction between two types of **semiconductor** material – called **n-type** and **p-type**. As a result, current will flow easily in one direction (from anode to cathode) with almost no current flow in the opposite direction (from cathode to anode).

The characteristics of a **pn junction** are fundamental to most semiconductor component operation. For example, a transistor is effectively two pn junctions in the same semiconductor material, and a good understanding of its properties is very important. Before we can continue with our investigation of power supplies the characteristics of pn junctions need to be examined in greater detail.

You can confirm a diode's more important properties with the first practical activity.

Do this 14.1

Simple diode investigations

Equipment and components required:

- IN4001 silicon power diode
- IN4148 silicon signal diode
- 820R (¼ watt) resistor
- laboratory power supply
- ammeter and voltmeter

Procedure:

1 Construct the circuit in Figure 14.2. Here the diode is connected so that its anode is connected to the positive side of the 9 V supply. This is called **forward biased** connection and will be examined in detail later. Note the current through the diode and measure the voltage across it.
2 Replace the IN4001 with the IN4148 and repeat. In each case you should see a current of a few mA's with a voltage drop across the diode of about 0.7 V.
3 Reverse both diodes and repeat. Now the anode is negative with respect to the cathode and is said to be **reverse biased**. For both diodes you should measure zero (or almost) current with 9 V across the diode.

In summary:

- When forward biased a diode allows current to flow through it with a small voltage drop (about 0.7 V) – it therefore has a low resistance when connected this way in a circuit.
- When reverse biased very little current flows and the voltage drop is high. The resistance is therefore very high when connected this way round in a circuit.

To understand why diodes have these dual properties you need a deeper knowledge of semiconductors. Firstly, some basic atomic theory.

Basic atomic theory

The basic building block of all material is the **atom**. The atom is the smallest part of an element that retains the characteristics of that element. Different elements have different atomic structures. Any atom has a central **nucleus** around which revolve minute negative electric charges called **electrons**. The nucleus itself is composed of two types of particles (except hydrogen – see Figure 14.3(a)): **neutrons** which have no charge and **protons** which have a positive charge equal to that of the electron. Isolated atoms are electrically neutral which means that the number of protons and electrons is equal.

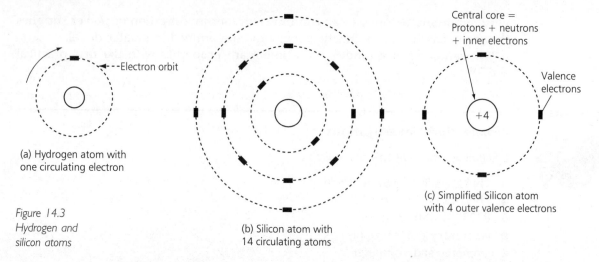

*Figure 14.3
Hydrogen and
silicon atoms*

(a) Hydrogen atom with
one circulating electron

(b) Silicon atom with
14 circulating atoms

(c) Simplified Silicon atom
with 4 outer valence electrons

Figure 14.3(a) shows the structure of the hydrogen atom which is the simplest of all the elements.

Silicon atoms

The most widely used semiconductor material is **silicon**. It is more complicated than hydrogen but note that the outermost orbit has four electrons, Figure 14.3(b). These electrons furthest from the nucleus are called **valence electrons** and, compared to the electrons in the inner orbits, experience only a relatively weak attractive force from the positive protons in the nucleus. When silicon atoms bind together to form solids it is these valence electrons which mainly determine the electrical properties of the semiconductor material. Because of their importance, the simplified picture of a silicon atom shown in Figure 14.3(c) is often used. Remember why the core has a positive charge of +4 – it contains the

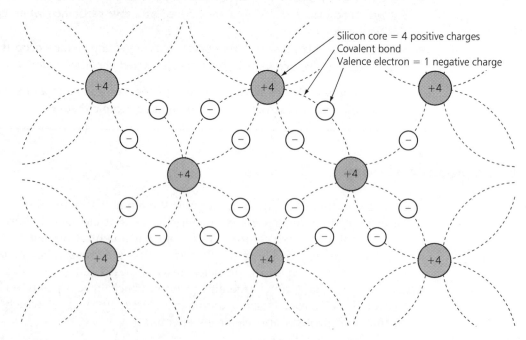

*Figure 14.4
Pure silicon
crystal structure*

positive protons in the nucleus and all of the other electrons *except* the four valence electrons.

Pure crystalline silicon

When silicon atoms combine to form solid silicon, they arrange themselves in a very precise structure called a **crystal lattice**. The valence electrons are relatively free to interact with each other, and each silicon atom shares its valence electrons with four neighbouring atoms to form stable **covalent bonds**. A simplified picture is shown in Figure 14.4.

The valence electrons, shown in circles, are firmly held in their bonds and can only move along the dotted lines, which represent these bonds between adjacent silicon cores. You can think of these covalent bonds as a sort of a scaffolding system that binds the atoms together to make the solid crystal.

Doped silicon

The electrical properties of pure silicon can be dramatically altered if certain selected 'impurities' are introduced into the otherwise pure silicon crystal. Such a process is known as **doping**. Two types of semiconductor material can be reproduced – n-type and p-type. We will consider each in turn.

N-type semiconductor

N-type silicon is produced by introducing impurity atoms which have *five valence electrons*. These impurities are called **pentavalent atoms** and the effect of artificial introduction of a pentavalent atom into pure silicon is shown in Figure 14.5(a).

Only four of the valence electrons are used to form **covalent bonds** so the fifth electron becomes free. The pentavalent atom is called a **donor impurity** since it donates a free electron to the crystal. Furthermore, before the impurity atom was introduced into the silicon crystal its central core was surrounded by five valence electrons – thus the atom was neutrally charged. When introduced into the silicon crystal each donor atom receives a fixed **positive** charge equivalent to the last

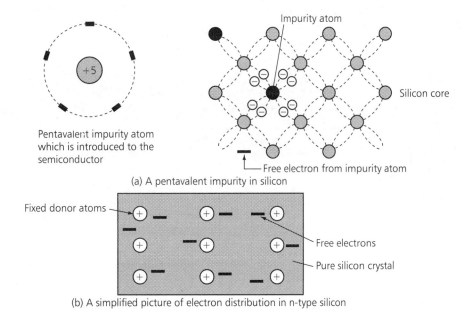

(a) A pentavalent impurity in silicon

(b) A simplified picture of electron distribution in n-type silicon

Figure 14.5 The effect of doping pure silicon to create n-type semiconductor material

negative charge of the fifth valance electron which has nothing to bond to or with. A simplified picture of n-type silicon is shown in Figure 14.5(b). This depicts free electrons (originating from the donor atoms) moving through the once-pure silicon crystal lattice which contains fixed donor atoms in various places, each one holding a tiny positive charge of electricity. Phosphorous and arsenic atoms are examples of donor impurities.

P-type semiconductor

P-type silicon is produced by doping with impurities which have only three valence electrons. These impurities are called **trivalent atoms** and the introduction of a trivalent atom into pure silicon has the opposite effect to n-type silicon: there is a shortage of valence electrons which create **electron 'holes'** or gaps in the lattice. An electron from a nearby covalent bond can move in to fill a hole leaving another hole in the bond from which that electron came. Since it accepts an electron from a nearby silicon bond the impurity is called an **acceptor** impurity.

Electron holes (absence of negatively-charged electrons) are said to be positively charged. Before the trivalent atom was introduced into the crystal, its central core was surrounded by three valence electrons – the atom was neutral. When an electron fills the hole, the +3 core is surrounded by four electrons and so at the position of each acceptor atom is a negative charge of -1. Acceptor atoms in p-type silicon therefore carry a negative charge. The hole is described as being 'free' because when an electron fills the hole, another hole is created at the position from which the electron came. Another electron could fill this and leave another hole at its original position and so on. Thus you can imagine a hole 'hopping' through a crystal from covalent bond to covalent bond moving in the **opposite direction** to free electrons in n-type silicon. They move in the opposite direction to electrons, as if they were positive charges. Aluminium and boron atoms are examples of acceptor impurities.

The pn junction

Figure 14.6(a) shows the simplified pictures of p and n-type silicon with their fixed impurity atoms and free holes and electrons.

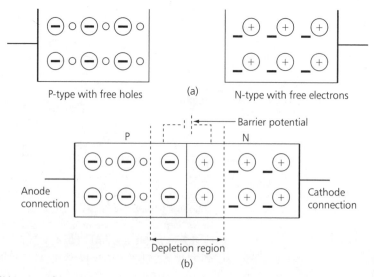

Figure 14.6 PN junction formation

Suppose that a continuous crystal is made by combining the p and n-type silicon. When this occurs, the electrons and holes near the junction combine with each other and effectively disappear. (Remember, all that has happened is that an electron has filled an incomplete covalent bond which was the hole.) The recombination results in Figure 14.6(b). Notice that there is now a barrier of positive and negative charges at the junction which stops any more electron and hole movement across the junction. It is as if there were an imaginary battery across the junction stopping any more hole and electron movement. This imaginary battery is called the **barrier potential**, and since the region around the junction is devoid of free holes and electrons it is called the **depletion region**. In practice, it is not possible to produce a useful pn junction simply by 'sticking' pieces of p and n silicon together. Although several manufacturing techniques are used, basically a piece of pure silicon is taken and each half is *doped with acceptor and donor impurities* from opposite ends. The **anode connection** is made to the p-type silicon and the **cathode connection** to the n-type silicon. The resulting device is called a **pn junction diode**.

The reversed-biased pn junction

If an external voltage is connected across a pn junction (or diode) so that it is of the same polarity as the barrier potential then the junction is said to be reverse biased. As shown in Figure 14.7(a), the positive terminal is to the cathode and the negative to the anode.

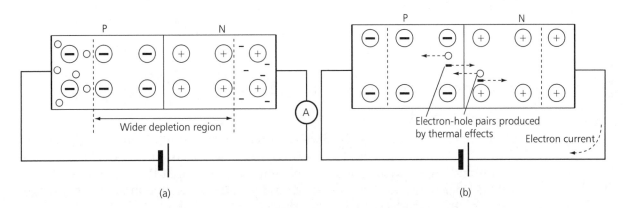

Figure 14.7 A reverse biased junction diode

Now, the electrons in the n material, and holes in the p material will be attracted away from the junction so increasing the width of the depletion region. At first sight it appears that there is no possibility of current flowing in this circuit because of the increased barrier potential. However, as you may have found in Do This 14.1, a very tiny current does flow called the **reverse current**. At normal temperature (room temperature), some electrons in covalent bonds can gain enough energy to break free and leave a hole behind. (As temperature increases further, more and more bonds will be broken.) As shown in Figure 14.7(b), it is these electron-hole pairs thermally produced near the junction that are attracted across the junction and are responsible for the minute reverse current flowing across the junction and forming a measurable current flow in the circuit.

(a) Do you think the size of the reverse voltage will have much effect on the reverse current?
(b) In Do This 14.1 the reverse voltage was 9 V and you probably measured no current. What would happen if you heated the diode artificially – say with a soldering iron?

Reverse breakdown

Providing the temperature remains fairly constant, the reverse current remains relatively unchanged over a wide range of reverse voltages. For any junction however, there will be a value of reverse voltage at which the current will suddenly increase. Unless adequate steps are taken the junction will burn out and destroy itself. This effect is called reverse breakdown. The **reverse characteristic** of a pn junction showing breakdown is shown in Figure 14.8.

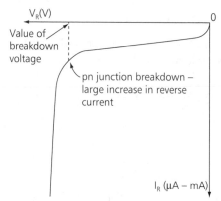

Figure 14.8 Reverse breakdown of a pn junction (temperature constant)

Obviously, the fact that breakdown occurs when V_R is high enough can be restrictive in many practical circuits. The diodes you used in Do This 14.1 are **power** and **signal diodes** and you have to be careful that the maximum reverse voltage that they could be subjected to does not exceed the breakdown voltage given by the manufacturer for that diode.

Types of diodes called **zener diodes** are deliberately designed to operate at breakdown. Providing the maximum current through the zener diode at breakdown does not exceed a safe value given by the manufacturer the diode will not destroy itself. Zener diodes are mainly used to provide reference voltages (the breakdown voltage) and are available with a wide range of values. The breakdown voltage is controlled by the level of doping during manufacture. We will investigate zener diodes later in this element.

The forward biased pn junction

A pn junction (or diode) is forward biased by connecting a voltage in such a way to reduce the barrier potential. Thus the anode is made positive with respect to the cathode as shown in Figure 14.9(a).

As the applied voltage V_F is increased from zero, little current is measured until a value of voltage called the **turn-on voltage** is reached when the barrier

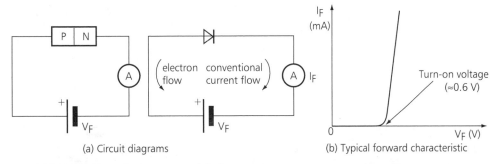

(a) Circuit diagrams
(b) Typical forward characteristic

Figure 14.9 Forward biased pn junction and diode

self-potential is overcome. When this happens, electrons and holes move easily across the junction and forward current I_F begins to flow. We say that the diode is **conducting**. Further increase in voltage causes I_F to increase rapidly with the voltage across the diode remaining essentially fixed or locked at the turn-on value. A graph of I_F against V_F is called the **forward characteristic** and a typical characteristic for a silicon diode is shown in Figure 14.9(b). For the range of silicon diodes available the turn-on voltage lies between about 0.6 V to about 1 V.

You should study Figure 14.9(b) carefully and note the following points:-

● As you would have discovered in Do This 14.1, forward currents, I_F, are much larger than reverse currents I_R – the diode has a low 'forward' resistance, and high 'reverse' resistance.
● I_F does not begin to flow until the turn-on voltage is reached.
● Once conducting in the forward direction, the voltage dropped across the diode is locked at about 0.6–0.7 V is this case.

Activity 14.2

The original semiconductor material used before silicon was germanium. Produce a short report comparing the properties and characteristics of silicon and germanium diodes.

Do this 14.2

Measuring your own diode characteristics

Equipment and components required:

● 1N4148 silicon signal diode
● 1N4001 silicon power diode
● Resistors (8K2, 1K8, 1K, 820R, 270R) ¼ W (5%)
      (220R, 180R) ½ W (5%)
      (1R, 82R) 1 W (5%)
● Multimeter
● Power supply (0–25 V)

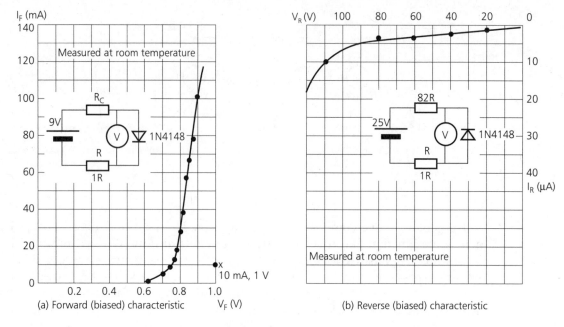

Figure 14.10 Characteristic of 1N4148 signal diode

Procedure (forward characteristic):

1 Construct the circuit in Figure 14.10(a) with R_C = 8K2
2 Measure the voltage across the diode V_F.
3 Measure the forward current I_F by *measuring the voltage across the 1R resistor.* (Because of Ohm's law, I_F is equal to the voltage reading since voltage reading = $I_F \times R$ and the value of R = 1.)
4 Measure I_F and V_F for the range of R_C resistor values and complete the table.

R_c	I_F(mA)	V_F(V)
8K2		
1K8		
1K		
820R		
270R		
220R		
180R		
82R		

5 The graph in figure 14.10(a) is a forward characteristic measured and plotted by the authors. Plot your own results in a similar way.

6 Closely compare your graph with the one given – they will almost certainly be slightly different. This is not only due to the different equipment used in the measurement (accuracy, etc.) but also because it is impossible to manufacture diodes (and in fact any semiconductor devices) that are identical in all respects. There will always be a **production spread** of results. This is highlighted by the way manufacturers' give data on their components. For example, the diode used in Figure 14.10(a) was manufactured by ITT Semiconductors. ITT guarantees that V_F at $I_F = 10$ mA will not exceed 1 V. (see point X on graph). The forward characteristics of all ITT 1N4148 diodes will lie to the left of this point.

Procedure (reverse characteristic):

1 Switch off the power supply, reverse the diode and replace R_C with R = 82R. (The inclusion of resistance in series is to protect the power supply if a short circuit occurs).

2 Switch on the supply and for 5 V steps from 0 V to 25 V measure the diode voltage and the reverse current, (if any), as before and complete the table.

Voltage	I_R (μA)	V_R (V)
5		
10		
15		
20		
25		

3 Figure 14.10(b) is a reverse characteristic measured by the authors with a wider range power supply. Plot your own results. Note carefully how small the reverse current is (if it is indeed measurable with your equipment).

Signal and power diodes

The 1N4148 is classed as a **signal diode** whilst the 1N4001 is described as being a **rectifier** or **power diode**. As you will have seen, they differ in construction but they also differ in their electrical characteristics as well. The 1N4148 is designed for fast switching applications in high frequency circuits involving relatively small currents (up to about 150 mA). The reverse voltage rating is 75 V. The 1N4001, is a member of the 4001 to 4007 series of power diodes which can handle much larger forward currents (up to 1 A) with a reverse voltage rating of 1300 V for the 1N4007 from ITT Semiconductors. They are designed for use as low frequency diodes (typically 50 Hz) in power supply circuits. Diodes from this range will be used in the power supply circuits which follow.

Activity 14.3

Obtain data sheets for the IN4148 signal diode and the IN4001 to IN4007 power diode series. Produce a brief report comparing their forward current and reverse voltage ratings.

WORKED EXAMPLE 14.1

An IN4148 diode is carrying a d.c. current of I_F = 60 mA. Determine the voltage drop across it, V_F, its d.c. resistance and the power dissipated.

Solution

The values I_F = 60 mA and the corresponding value V_F are called the **operating point** values. Using the forward characteristic in Figure 14.10(a) the voltage drop corresponding to I_F = 60 mA is about 0.85 V.

The forward d.c. resistance of the diode R_{dc} is

$$R_{dc} = \frac{V_F}{I_F} = \frac{0.85}{60 \times 10^{-3}} = 14.2 \ \Omega$$

Whenever a diode is carrying current it will dissipate power – it will get warm! The power dissipated P_{dc} will be:

$$P_{dc} = V_F \times I_F = 0.85 \times 60 \times 10^{-3} = 51 \ mW$$

Progress check 14.2

For the IN4148 diode calculate the d.c. resistance and power dissipated at an operating point of:

(a) I_F = 10mA
(b) I_F = 50mA

Rectifying circuits

You now have enough knowledge of diodes to return to the main topic of this unit which is the investigation of power supplies. The **rectifier block** in the diagram of Figure 14.1 will utilise a power diode (or diodes) to convert the alternating voltage at the secondary into pulses of direct current and voltage. This is because, as we have seen, the diode offers a low resistance in one direction but blocks flow of current in the reverse direction because of its high 'reverse resistance'. This characteristc of a diode enables alternating current to be changed (rectified) into direct current.

The halfwave rectifier

A halfwave rectifier which uses a single power diode is the simplest form of rectifying circuit and is shown in Figure 14.11.

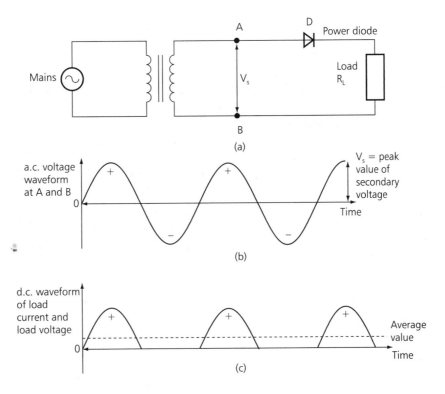

Figure 14.11 The halfwave rectifier

The diode will only conduct when its anode is positive with respect to the cathode, i.e. when the secondary voltage is such that A *is positive with respect to* B. Current will only flow through the diode and the load for each half cycle of the input. On the negative halves of the cycle, the diode becomes reverse biased by the negative value of the secondary voltage. You can see that the load current is definitely direct (flows in one direction only).

Average current rating I_{FAV}

The diode in Figure 14.11 must of course be capable of handling the average forward current that flows. The *maximum average forward current rating*, I_{FAV}, is an important rating for a power diode and is effectively a measure of the power that the diode can dissipate.

Repetitive peak reverse voltage V_{RRM}

Every time the alternating voltage goes negative in Figure 14.11, the diode will be repeatedly reverse biased to the peak value of this voltage. Obviously, the diode must be rated so that it will not breakdown under these conditions. Manufacturers describe a power diode's ability to withstand a repetitive reverse voltage in terms of its *repetitive peak reverse voltage rating* V_{RRM}. I_{FAV} and V_{RRM} are the most important ratings for a power diode.

Progress check 14.3

In Activity 14.2 you were asked to obtain data sheets for the 1N4001–1N4007 series of power diodes. What values for I_{FAV} and V_{RRM} are given for this series?

Fullwave bridge rectifier

The halfwave rectifier is not particularly efficient because only one half of the secondary alternating waveform is utilised. The circuit that is now almost universally employed is called a fullwave bridge rectifier and is shown in Figure 14.12. Four diodes are used.

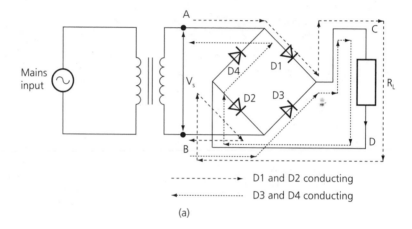

- - - - - - ▸ D1 and D2 conducting
◂ · · · · · · · · · · · · · D3 and D4 conducting

(a)

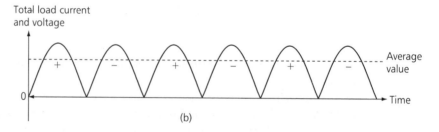

(b)

Figure 14.12 Fullwave rectifier and load waveform.

It operates like this. For the positive half of the secondary alternating voltage, A is positive with respect to B and so diodes D1 and D2 will be forward biased (D3 and D4 are reverse biased). Current flow is shown by the dashed lines in Figure 14.12(a). For the negative halves of the input voltage, D3 and D4 are forward biased and the current flow is indicated by the dotted lines in the diagram. The net result is that *current always flows in the same direction through the load*. The waveform is shown in Figure 14.12(b). In a practical circuit, there would be a voltage drop of about 1.4 V (2 × 0.7 V) across each conducting pair of diodes so the average load current and voltage would be approximately twice that obtained for the halfwave circuit. The frequency of the half sinewave pulses is now twice the frequency of the mains supply, i.e. 100 Hz.

The V_{RRM} rating for each diode in the bridge still has to be at least V_S. To understand why this is look at Figure 14.12(a) again and assume for example that A is positive with respect to B. D_1 and D_2 are conducting and D_3 and D_4 are reverse biased. When the secondary voltage reaches its peak value Vs, Vs will be applied across D_4 and D_2 for example. The voltage across D_2 will only be about 0.7 V since it is forward biased and so $(V_s - 0.7)$ will be dropped across the reverse biased diode D_4. (A similar argument could be applied to D_1 and D_3 under these conditions). Thus, the V_{RRM} rating for each diode has to be at least V_s.

Progress check 14.4

A halfwave rectifier circuit uses a transformer with a secondary voltage of 170 V_{RMS}. The average load current to be provided is 0.5 A. Use your data sheets to select a power diode from the IN4001–IN4007 range.

Smoothing the waveform

Both the half- and fullwave circuits produce currents that are direct (d.c.) but are far too rough to be used immediately; the pulses of current have to be smoothed by the addition of further components (the third block in Figure 14.1). Smoothing of the waveform also raises the average value of the load current. The first step

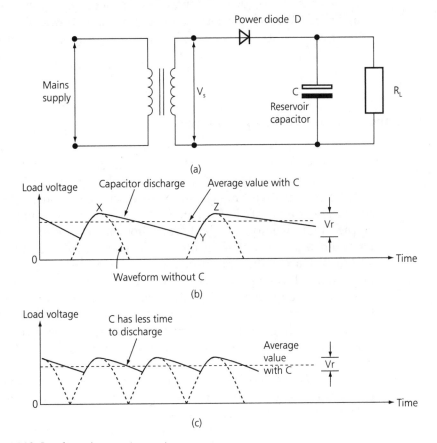

Figure 14.13 Rectifier with reservoir capacitor

in smoothing the rectifier output is to connect a capacitor is parallel with the load. This capacitor is often called the **reservoir capacitor** since it acts as a reservoir (or store) of electric charge.

Consider the simple halfwave rectifier in Figure 14.13(a) with the smoothing capacitor C across the load.

When the secondary voltage first goes positive at switch-on, D conducts and allows current to flow through R_L and C. C will charge to (almost) the peak value of the secondary voltage, which is also the voltage across the load. As the secondary voltage falls below the peak value, D becomes reverse biased and current flow through D ceases until the secondary voltage again rises above the capacitor voltage, nearly a cycle of the input later. During the period of time that D is cut off, C will discharge through R_L and so maintain a current flow to the load. As C discharges, the voltage across it and R_L (because they are in parallel) will also fall. The resulting load voltage waveform is drawn in Figure 14.13(b). A much higher average load voltage is produced but it is still not perfectly smooth. The small variation that still remains is called the *ripple voltage*, V_r.

For a bridge circuit (fullwave rectifier), the load voltage is as shown in Figure 14.19(c). For the same value of C and R_L, V_r will be smaller for the bridge because C would have less time to discharge before being topped up again, half a cycle later.

Progress check 14.5

(a) What is the frequency of the ripple produced by the bridge rectifying circuit?
(b) To be effective, C has to be large and will be **electrolytic**. What two important precautions would you take when using an electrolytic capacitor as a reservoir capacitor?

Further effects of the reservoir capacitor

It is not so much the value of C that is important but the product C × R_L which determines the size of the ripple. The product C × R_L is the **time constant** of the circuit and is a measure of how quickly the capacitor discharges. (A capacitor will take about five time constants to discharge). You might say, 'Why not use the largest value of C you can find and so get the smallest possible value of ripple'. Well, we do use a large capacitor but in choosing C other factors have to be taken into account. Consider the waveform of the halfwave circuit (Figure 14.13(b)). During the period of time between the points X and Y, C discharges and the load voltage falls. The value of voltage at Y is also the voltage at the cathode. Since the diode will only conduct when its anode is positive with respect to the cathode, diode current will only flow to charge C for the time it takes the secondary voltage to increase from the voltage value just above Y to that of Z. The time interval can be a very small part of the cycle and the peak current can be high.

Thus *without C*, current flows through the diode for very nearly the whole half cycle, but *with C* diode current is a short duration pulse.

If you refer to the data sheet for the 1N4001 (Do this 14.2) you will see a *repetitive peak forward current rating* I_{FRM}. In practice you must ensure that the peak value of the current is below this rating. (I_{FRM} is quoted as 10 A for a 1N4001).

The inclusion of C also means that we have to carefully consider the V_{RRM} rating of the diodes. Consider the halfwave circuit again. If the ripple is not too large, the capacitor voltage remains approximately equal to the peak value of the secondary voltage. When the secondary voltage goes negative on every other half cycle, the diode will be repeatedly reverse biased to about *twice the peak value of the secondary voltage*. Thus, the diode V_{RRM} rating needs to be at least twice V_S. Similar arguments could be applied to the diodes in a bridge circuit.

> **You should only attempt the following practical activity if you have access to a boxed and correctly fused mains transformer giving a nominal 12 V_{RMS} across the secondary.**

Do this 14.3

Measurements within a basic power supply

Equipment and components required:
- 4 × IN4001 diodes
- 1 × 270R resistor (¼ W)
- 1 × IR resistor (¼ W)
- 1 × 15R resistor (¼ W)
- 1 × 4700µF capacitor (25 V)
- 1 × 100µF capacitor (25 V)
- oscilloscope.

Procedure Part 1: Basic fullwave circuit

1 Construct the circuit shown in Figure 14.14(a). (The IR resistor is included so that you can measure and monitor the current into the bridge by connecting an oscilloscope across it.)
2 Use the oscilloscope to observe the 12 V a.c. waveform at the secondary and note its peak value.
3 Use the oscilloscope to observe and note the waveform across the 270R load.
4 Use the a.c./d.c. button on the oscilloscope to measure the average value (d.c. value) of the load voltage.
5 Use the oscilloscope to observe and note the current waveform across the IR resistor.

Procedure Part 2: Effect of reservoir capacitor

1 Switch off and connect a 4700 µF electrolytic capacitor across the load. (Check that you have the polarity correct).
2 Switch on and observe the new load voltage waveform and measure the peak-to-peak ripple.
3 Use the a.c./d.c. button to measure the d.c. level of the load voltage.
4 Observe and note the current waveform across the IR resistor.

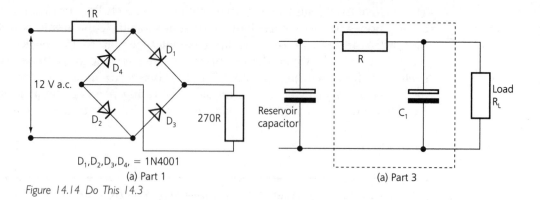

D₁,D₂,D₃,D₄, = 1N4001

(a) Part 1

(a) Part 3

Figure 14.14 Do This 14.3

Procedure Part 3: Additional filtering

To reduce the voltage ripple further, additional components can be included between the reservoir capacitor and the load. One simple method is to add a resistor and capacitor connected as a **low pass filter** as shown in Figure 14.14(b). In practice C_1 is chosen so that its reactance:

$$X_C = \frac{1}{2\pi fC}$$

at the ripple frequency is about a tenth of R.

1 Connect the filter between the reservoir capacitor and the load.
2 Monitor and observe the new load voltage waveform, the peak-to-peak ripple and the d.c. level.

You should see that although the ripple is reduced, so is the d.c. output voltage at the load. Often a ripple filter is not used in low voltage, high current supplies because of the voltage drop across R. Instead only a reservoir capacitor is used together with a **voltage regulator circuit**. Regulators will be investigated later in this chapter.

Activity 14.3

Produce a short report detailing your findings for Do This 14.3. Include waveform sketches and measured values in your report and relate the theory given to the results.

Voltage regulation of a power supply

If you have carried out the practical activities, then you should be convinced that your power supply is reasonably good. With the inclusion of the filter, you have probably measured a d.c. output of about 10 to 11 V or so with a ripple of a few mV. An important point to note, however, is that the measurements have been taken with a **fixed load** (i.e. 270R).

Practical power supplies will have a range of specifications but one of the most important aspects is the voltage regulation. Voltage regulation is a measure of a power supply's ability to maintain a *constant output voltage even if the current taken from it varies*, i.e. different loads are connected. Voltage regulation (VR) is normally expressed as a percentage and is given by:

$$VR = \left(\frac{\text{no-load voltage} - \text{full-load voltage}}{\text{no-load voltage}}\right) \times 100\%$$

Here, **no-load voltage** is the output voltage on open circuit, and **full-load voltage** is the output voltage when the unit is giving the maximum current for which it was designed. For example, suppose the output voltage drops by 100 mV from a no load voltage of 10 V, then

$$VR = \frac{(10 - 9.9)}{10} \times 100\% = 1\%$$

The lower the VR the better the power supply will be at maintaining a constant output voltage.

Why should the d.c. voltage change at all as the current supplied to the load varies? Firstly, (if included) the load current must flow through the ripple filter resistance R. As the current increases, the voltage drop across R will increase giving a lower output voltage. Secondly, since the secondary winding of the transformer has resistance, a varying load will produce a varying voltage drop. Thirdly, when the load current increases, the capacitor voltage will fall further from its peak value. This causes the average value of the voltage to fall. (The a.c. ripple will also increase.) Finally, we know that *the forward characteristic of a diode is not vertical but has a finite slope.* As the forward current through it changes, so will the forward voltage drop, albeit by a small amount.

To obtain detailed information about a power supply's performance a **voltage regulation curve** is plotted using a circuit similar to that in Figure 14.15(a). I_L is varied by changing R_L for a range of values from no load to full load. For each value of I_L, V_L would be noted. Finally, a graph of V_L against I_L is drawn and would be similar to that in Figure 14.15(b).

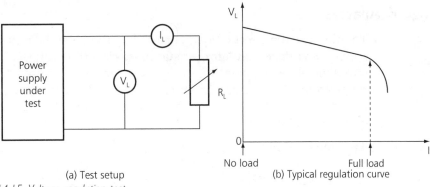

(a) Test setup (b) Typical regulation curve

Figure 14.15 Voltage regulation test

Do this 14.4

Voltage regulation measurement of the basic power supply

Equipment and components required:

- power supply from Do This 14.3
- suitable ammeter and voltmeter
- decade resistance box.

Procedure:

1 The 1R resistor in the circuit for Do This 14.3 is not needed so remove it and connect the rectifier directly to the secondary.
2 Measure V_L and I_L for no-load conditions (open-circuit).
3 We will assume that full-load is, say, $I_L = 40$ mA, so obtain a range of values of V_L and I_L up to this maximum value by changing R_L.
4 A regulation curve measured for the basic power supply is shown in Figure 14.16. Plot your own points on this graph. They will almost certainly not be the same but you should see V_L falling as I_L increases.

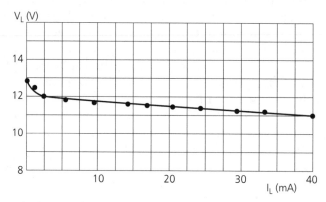

Figure 14.16 Measured voltage regulation curve for the basic power supply

Voltage regulators

To minimise the effect of load current variations on the output voltage of a power supply, a **voltage regulator** (or **stabiliser circuit**) is connected just before the load. (Another block is added to the block diagram in Figure 14.1.) The regulator can range from a very simple circuit using a single **zener diode** up to fairly complex integrated circuit device. We will investigate how a simple zener diode regulator operates first of all.

The zener diode

You have seen that any diode which is reverse biased to a high enough voltage will breakdown. This would mean disaster for a signal or power diode because the diode will burn out and destroy itself. A zener diode, however, is designed so that it can operate at reverse breakdown without destruction. Providing the reverse current flowing is not allowed to exceed a certain value the zener diode will not be damaged. Zener diodes are available which will breakdown at a reverse voltage of as low as about 2 V with a very wide range of other voltages easily obtainable. Although the forward characteristic is similar to a signal or power diode, zeners are almost invariably used and connected in reverse bias mode and so they can be used to provide a **standard or reference voltage** within a circuit. For this reason zener diodes are often called **reference diodes**.

The symbol for a zener together with a typical reverse characteristic is drawn in Figure 14.17.

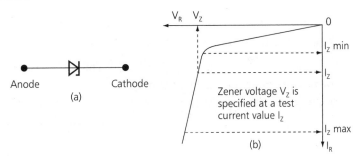

Figure 14.17 Zener diode and reverse characteristic

Ideally we would like the breakdown characteristic to be vertical but this is never the case. In view of this, manufacturers specify the breakdown voltage V_Z of their diodes at a chosen value of reverse current I_Z. This value of I_Z is often called the **test current**.

Besides V_Z, the other important parameter of a zener diode is its **maximum power dissipation rating**, P_{TOT}. If V_Z and P_{TOT} are known then it is a simple matter to calculate the maximum current that can be allowed to flow through the diode. This is shown in the following worked example.

WORKED EXAMPLE 14.2
Calculation of maximum current for a zener diode

'A 5V6 zener diode has a maximum power dissipation of 500 mW. What is the maximum current that the diode can handle?'

Solution
The maximum power dissipation P_{TOT} is equal to the maximum zener current I_{ZMAX} multiplied by the breakdown voltage V_Z, i.e.

$$P_{TOT} = V_Z \times I_{ZMAX}$$

$$I_{ZMAX} = \frac{P_{TOT}}{V_Z} = \frac{500 \times 10^{-3}}{5.6} = 89.3 \text{ mA}$$

Note that since a zener characteristic is not vertical, the actual voltage across the diode when carrying I_{ZMAX} will be greater than the zener voltage V_Z. Therefore, the calculation is only a fair approximation. Usually, manufacturers provide details of the **admissible zener current** that the diode can withstand in their data sheets. (For the 5V6 ITT give 70 mA at 25°C). Since it is good practice to make sure that the diode *operates well within its I_{ZMAX} value*, the calculation above is used as a good guide.

Activity 14.4

Locate data sheets for a range of 500 mW zener diodes (the ZPD1–ZPD51 series from ITT Semiconductors is very useful).

Do this 14.5

Reverse characteristic determination

Using a similar technique to that in Do This 14.2, design and perform a measurement to obtain the reverse characteristic of a 500 mW, 5V6 diode. Report on your results and compare with those given by the manufacturer.

The zener diode regulator

To illustrate how the zener can be used to provide regulation, consider the circuit in Figure 14.18.

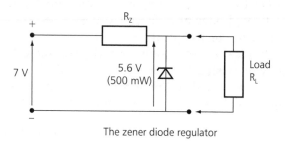

The zener diode regulator

Figure 14.18 The zener diode regulator

Here, it is necessary to provide a fixed voltage of 5.6 V across the load from an unregulated 7 V supply. On no load, maximum current will flow through the diode. We have already calculated this to be 89.3 mA in Worked Example 14.2. The voltage dropped across the resistor R_Z, will be $(7 - 5.6) = 1.4$ V so the minimum value required for R_Z is:

$$R_Z = \frac{1.4}{89.3 \times 10^{-3}} = 15.7 \ \Omega$$

When the load is connected, current will flow in it and less will flow through the diode. As the load changes, provided the zener current lies above I_{ZMIN}, the zener remains in the breakdown condition and the load voltage remains stabilised at 5.6 V. If the load is taken too low so that the current is below I_{ZMIN}, the zener will not be in its breakdown region and there will be no regulation.

Integrated circuit voltage regulators

There is a range of off-the-shelf integrated circuit (IC) voltage regulators available which are inexpensive and make the design of stabilised power supplies quite straightforward. The 78 series for positive voltages and the 79 series for negative voltages are IC regulators with three terminals which can be used to produce a range of fixed voltages from 5 V to 24 V. We will use a 7805 device which will produce a 5 V supply in the following worked example.

Activity 14.5

Obtain a data sheet for the **78/79** series of 3-terminal regulators.

WORKED EXAMPLE 14.3

Design a power supply with the following specification for use in a mains powered computer console. The power supply should provide +5 V for a full load of 300 mA. A 12 V_{RMS} supply is available from a 6 VA transformer. For documentation purposes, a regulation curve should be provided.

Solution

1   Choice of circuit
    Use a bridge rectifier and a 7805 3-terminal regulator. The final circuit is shown in Figure 14.19.

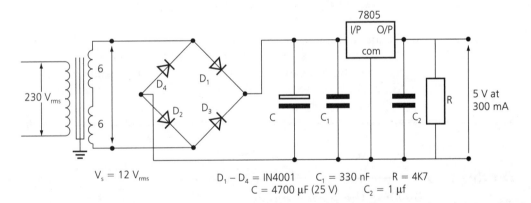

Figure 14.19 5 V regulated power supply

2   Choosing the diodes
    The peak value of the secondary voltage is:

$$Vs = 12 \times 1.414 = 16.97\,V = 17\,V$$

Because of the reservoir capacitor, the diode V_{RRM} rating has to be at least twice this, i.e. 34 V.

    Use 1N4001 diodes whose I_{FAV} and V_{RRM} ratings are sufficient for this specification, i.e. 1 A and 100 V.

3 Reservoir capacitor and regulator
When each pair of diodes is conducting about 1.4 V will be dropped across them. Therefore, the peak voltage into the regulator is (17 − 1.4) = 15.6 V. The data sheet for the 7805 gives that the minimum input should be 7 V and the maximum 25 V. Thus, the maximum allowed ripple is 15.6 − 7 = 8.6 V

It can be proved mathematically that the minimum value of reservoir capacitor is given by:

$$C_{MIN} = \frac{I_L}{2 \times f_r \times V_r}$$

where I_L = load current, f_r = ripple frequency (100 Hz), and V_r = maximum ripple. Therefore:

$$C_{MIN} = \frac{300 \times 10^{-3}}{2 \times 100 \times 8.6} = 173 \ \mu F$$

This is a minimum value, so any reasonable larger value can be used. The 4700 μF capacitor should have a working voltage of 25 V.

4 C1, C2 and R
C1 and C2 are included to prevent the regulator producing high frequency oscillations. They should be *mounted as close to the regulator as possible* and their values are given in the manufacturer's data sheet as C_1 = 330 nF; and C_2 = 1 μF. R is included to prevent damage to the regulator if it is operated with no load connected. (If there is always a load connected R can be left out of the circuit). A few K is required – 4K7 is used in Figure 14.19.

5 Heatsinking and regulation
When the power supply delivers current to the load it dissipates power and gets warm. Manufacturers give details of the heatsinking requirements in the data sheet. Since the 7805 is rated at 1 A, but is being used at 300 mA maximum, only a small heatsink (if any) needs to be used.

Do this 14.6

Building the power supply

Construct the power supply from Worked Example 14.3 and measure its voltage regulation from 0 mA to 300 mA. Calculate the percentage regulation. This power supply can be used in Element 14.3 (Chapter 19) to power the digital logic circuits if required.

Assignment 14

This assignment provides evidence for:
Element 13.1: Investigate d.c. power supply units
and the following key skills:
Communication 3.2
Application of Number 3.1

The evidence indicators for this element require you to produce a report of an investigation into d.c. power supply units.

Report

Your report should include:

● a description of the properties of pn junctions, i.e. characteristics, power dissipation and ratings
● a description of the characteristics of d.c. power supply units, i.e. output voltage, load current, regulation, ripple voltage etc.
● an identification of the functions of, and operation of, the constituent circuits within power supplies, i.e. rectification, smoothing, filtering and stabilisation
● details of measurements taken within power supply units, i.e. no-load and full-load voltage, ripple voltage and load current.

Chapter 15:   Small-signal amplifiers

One of the most common operations in electronics is **amplification**. In many cases we are dealing with voltages and currents so small that they must be increased in amplitude before they can be used. For example, the signal received by the antenna of radio is only a few tens of microvolts. This is too small to drive a loudspeaker and it is therefore necessary to have circuits which will increase or **amplify** the signal to the desired level. In this element we will investigate how the **bipolar junction transistor** (BJT) and the **field effect transistor** (FET) may be used with other components to *amplify a.c. signals*. Practical activities are included to emphasise the main learning points.

When you complete this chapter you should be able to:

- describe direct current operation of amplifying devices
- describe amplifying device characteristics
- describe small signal amplifier characteristics
- describe amplifiers using small signal amplifier models
- determine the operational features of small-signal amplifiers.

Specifying amplifiers

The typical representation of an amplifier is shown in Figure 15.1(a).

Amplifier gain

The function of the amplifier is to make the output signal larger than the input signal. If this operation is successful then we say that the amplifier has **gain**. For

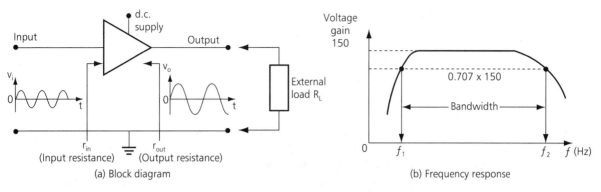

Figure 15.1 Typical amplifier representation and frequency response

the amplifier in Figure 15.1(a), the output voltage signal v_o is larger than the input signal v_i so this device has **voltage gain**. The voltage gain A_v is defined as:

$$A_v = \frac{v_o}{v_i} \text{ (no units)}$$

If the output signal current i_o is larger than the input signal current i_i then the amplifier would also have **current gain**, A_i. A_i is defined as:

$$A_i = \frac{i_o}{i_i} \text{ (no units)}$$

The power gain of an amplifier, A_p, is the product of A_v and A_i, i.e.

$$A_p = A_v \times A_i \text{ (no units)}$$

The power to provide gain is derived from the power supply (see Figure 15.1a).

Frequency response

An amplifier is designed to amplify a specific band of frequencies which will depend upon its intended application. An audio frequency amplifier, for example, will be designed to amplify all frequencies in the range 20 Hz up to 20 kHz. A graph of gain versus frequency is called the **frequency response** of the amplifier. A typical plot is illustrated in Figure 15.1(b). This amplifier has a gain of 150 for most of the frequency range over which it operates.

The gain of an amplifier falls off at high and low frequencies. The **bandwidth** is defined as the frequency difference between those frequencies at which the gain is 0.707 of the maximum gain, i.e.

$$\text{Bandwidth} = (f_2 - f_1) \text{ (Hz)}$$

D.C. and a.c. operation

It is very important to realise that all amplifiers have to be considered from two different points of view. Notice in Figure 15.1 that a d.c. supply must be connected to the amplifier circuit before it can operate. The d.c. sets the **operating point** of the device(s) used in the circuit. We say 'The d.c. sets the necessary bias for correct operation.' Thus, *the operation of the circuit under d.c. conditions* must be understood.

When the a.c. signal is applied, it causes small variations in the operating point values of the amplifying device previously set by the d.c. bias. Therefore, the operation of *the circuit under signal conditions* must also be investigated. D.C. and a.c. analysis is always required in amplifier design and understanding. We will investigate this in more detail later.

Input and output resistance

The input resistance is the resistance 'seen' by the applied signal at the input terminals of the amplifier. At the output the signal will experience a resistance which is usually different to that at the input. The resistance at the output terminals of the amplifier is called the output resistance. Input and output resistance are indicated in Figure 15.1.

The bipolar junction transistor

A bipolar junction transistor (BJT) is a single crystal of semiconductor which is 'doped' to produce a three-layer device or semiconductor sandwich of n-type and p-type material. Since a semiconductor, like silicon, can be doped either n- or p-type, then two possible BJT structures can be manufactured, i.e. both npn and pnp devices are available. The two possible constructions are shown in Figure 15.2(a).

Three separate regions are formed called the **emitter** (E), **base** (B) and **collector** (C). Metal connections are made to each.

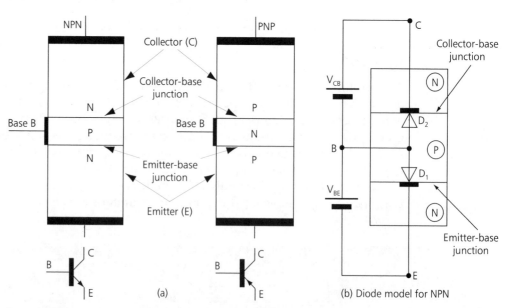

Figure 15.2 *The two types of BJT with circuit symbols and diode model*

In view of the sandwich construction, there are two **pn junctions** involved. In real BJTs the collector region is larger than the emitter and bonded to the transistor case. This is because most of the heat generated within the device occurs at the collector-base junction. Bonding the collector to the case allows the heat to be dissipated to the surrounding air.

Biasing the BJT

Before a BJT can be of any operational use d.c. supply voltages must be connected to the emitter, base and collector. This procedure is called biasing and we say 'the d.c. bias sets the operating point of the BJT'. For normal operation two conditions must be satisfied:

- The base-emitter junction must be forward biased.
- The collector-base junction must be reverse biased.

Bias for the npn BJT

From the point of view of biasing, it is useful to think of a BJT as two diodes connected in series. This is shown in Figure 15.2(b) for an NPN transistor. The base-emitter diode is labelled D_1, and the base-collector diode D_2.

To **forward bias** the base-emitter junction, the anode voltage of D_1 must be positive with respect to its cathode, i.e. *The base must be positive with respect to the emitter.*

To **reverse bias** the collector-base junction, the cathode of D_2 must be positive with respect to its anode i.e. *The collector must be positive with respect to the base.*

The correct bias for the NPN using batteries V_{BE} and V_{CB} is shown in Figure 15.2(b).

What can we say about the size of V_{BE} and V_{CB}? You will remember that when a diode is forward biased the voltage across it becomes 'locked' at around 0.7 V. Since the base-emitter junction is essentially a diode we should expect similar behaviour. For BJTs V_{BE} *locks at a value between about 0.6 V up to about 1 V* or so depending on the type of transistor. This is a most important point for you to remember! We cannot put a definite value on V_{CB} because this depends on the circuit that the transistor is in and what value of V_{CB} was decided upon by the circuit designer. Manufacturers give a *maximum rating for V_{CB} which should not be exceeded.*

Progress check 15.1 Sketch a diagram similar to Figure 15.2(b) to show the correct bias for a pnp transistor.

Current flow in the npn BJT

When a BJT is correctly biased, current will flow through it and in the external circuit. Understanding what happens will not be as easy as it was for a single diode because transistors have two pn junctions. Nevertheless, a step-by-step approach will make the operation easily understood. The *final analysis* is summarised in figure 15.3. V_{BE} has been taken to be about 0.7 V.

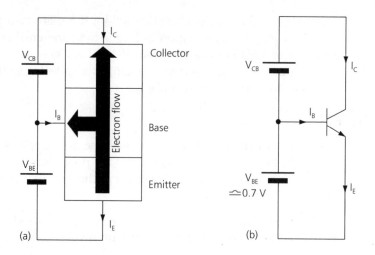

Figure 15.3 Current flow in an npn transistor

Notice that **conventional current flow** has been shown in the external circuit but **electron flow** has been shown inside the device. Base and collector currents flow into the BJT and emitter current flows out (in the direction of the arrowhead). The rule 'what goes in equals what comes out' (Kirchoff's current law) applies for the BJT so:

$$I_E = I_B + I_C$$

This is a most important equation. We will now describe how the device works. Consider Figure 15.4(a). This shows what happens when the base-emitter junction is forward biased but the base-collector junction is unbiased (open circuit). The depletion region at the base-emitter junction is decreased, while the collector-base depletion region stays the same. The device acts like a forward biased diode and current will flow in the base-emitter circuit. Figure 15.4(b) shows what happens when both junctions are biased at the same time. Because of the simultaneous biasing, most of the electrons leaving the emitter will be attracted by the donor atoms in the collector and the positive terminal of the collector battery. These electrons will move rapidly through the base, are 'swept' into the collector region and flow out to the external circuit. A few electrons will recombine with holes in the base and to make up for this 'loss' a small base current I_B flows.
Thus,

- a current I_E flows out of the emitter
- a slightly smaller current I_c flows into the collector
- a much smaller current I_B flows into the base .

Mathematically,

$$I_E = I_B + I_C$$

For low power BJTs, I_E and I_C will be a few mA's and I_B will be a few μA's.

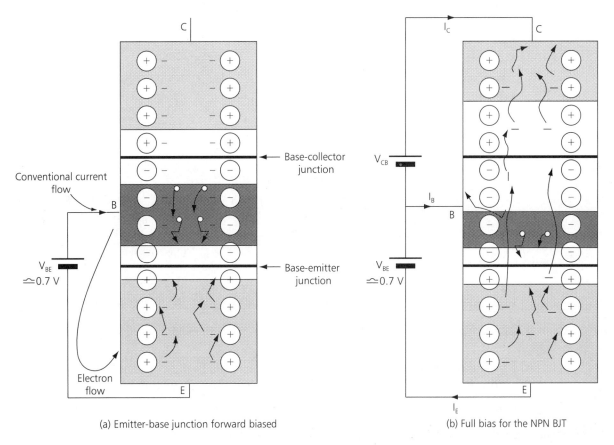

Figure 15.4 *NPN transistor operation*

A useful approximation

The above current equation is EXACT, but in many cases we can make a very important approximation. Since I_B is always much smaller than I_E and I_C, we can make the approximation:

$$I_E \approx I_C$$

This approximation is very useful when transistor circuits are being designed and analysed.

Progress check 15.2
For a particular BJT, the base current is 1% of the emitter current. If the emitter current is 40 mA, what are the values of the base and collector currents?

Connection of transistors in circuits

Since the BJT has three terminals, there are three possible ways in which it can be connected in circuits. In each case, one terminal is common to both the input and output parts of the circuit. Consequently, the three modes of connection are referred to as:

- Common base
- Common emitter
- Common collector

Each is shown in Figure 15.5.

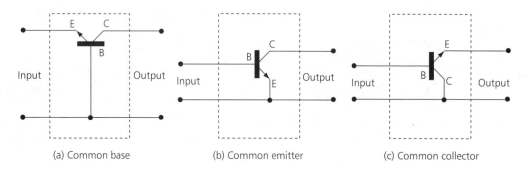

(a) Common base   (b) Common emitter   (c) Common collector

Figure 15.5 BJT modes of connection

The one most widely used in practical circuits is the common emitter. We will only investigate this mode of connection for the rest of the element.

Common emitter input and output voltages and currents

When correctly biased, current will flow in the transistor and the external circuit. A correctly biased BJT in common emitter is shown in Figure 15.6(a). Notice the largest battery is connected between collector and emitter to ensure that the collector base junction is reverse biased.

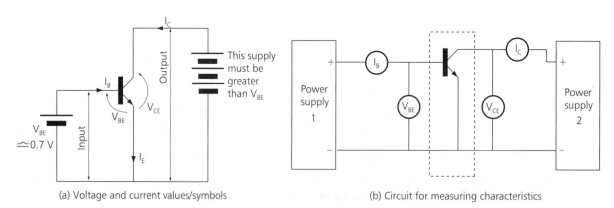

(a) Voltage and current values/symbols   (b) Circuit for measuring characteristics

Figure 15.6 BJT in common emitter connection

Study Figure 15.6(a) carefully and note the following points:

- The input current is I_B and the input voltage is V_{BE}
- The current at the output is I_C and the output voltage is V_{CE}

Since I_C is much larger than I_B, there is more current flowing in the output than the input. Thus, in common emitter mode, the transistor has **current gain** from input to output. The ratio of I_C to I_B is called the d.c. **forward current gain** and is represented by the symbol h_{FE}:

$$\text{Current gain, } h_{FE} = \frac{I_C}{I_B} \quad \text{(no units)}$$

h_{FE} is one of the most important transistor parameters and is quoted in manufacturers' data sheets. Usually, a typical value is given as well as the maximum and minimum values you can expect for that transistor type.

Progress check 15.3

When a transistor is connected in a certain circuit $I_C = 20$ mA and $I_B = 60$ μA. Determine: (a) I_E and (b) h_{FE} at this operating condition.

Other common emitter characteristics

In our investigation of the diode it was a fairly simple matter to measure its V/I characteristic and use the graph obtained to understand how diodes behave in circuits. Similar measurements can be made on transistors and provide most of the details of the device that we need to know. The measurements are carried out using d.c. voltages and currents and the resulting graphs show how the input and output voltages and currents are related to one another. A circuit suitable for measuring the common emitter characteristics of an npn transistor is illustrated in Figure 15.6(b).

Generally, manufacturers provide a mass of information about their devices and several sets of characteristics are given in their data sheets. The two most important are the **input characteristic** and **output characteristic**.

Progress check 15.4

What changes would you make to Figure 15.6(b) if the BJT was a pnp device?

The input characteristic

In graphical form, this shows how I_B changes for various values of V_{BE} when V_{CE} is held at a fixed value. In Figure 15.6(b), V_{CE} would be adjusted to some chosen value using supply 2. Power supply 1 would then be varied so that V_{BE} provides a range of fixed values. For each value of V_{BE}, I_B is noted and finally a graph I_B versus V_{BE} is drawn. The input characteristics for two silicon npn transistors (BC108 and BFY51) are shown in Figure 15.7. Their shape looks rather familiar doesn't it!

As you may have expected, since the base-emitter junction is effectively a forward-biased diode, the characteristic should be similar to that of a diode. Notice

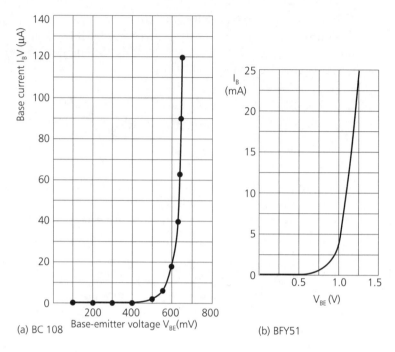

Figure 15.7 Input characteristic for BC108 and BFY51 (BFY51 courtesy Mullard)

each BJT turns on at about 0.6 V and then the voltage is essentially locked at this value. The BFY51 will carry more base current than the BC108, i.e. it can handle more power.

The output characteristic

The output characteristic is a graph showing the variation in I_C as V_{CE} is changed for different fixed values of the base current I_B. (Sometimes V_{BE} rather than I_B is shown). Figure 15.8 is a set of output characteristics for the BC107/108/109 range of transistors manufactured by Mullard.

Study the characteristics carefully and note the following points:

● For each value of I_B (or V_{BE}), the curves rise sharply and then flatten out.
● For the flat part of each characteristic a large change in V_{CE} does not produce much of a change in I_C, although I_C increases as I_B is increased.

If operating at a certain value of V_{CE}, the only way I_C can be changed or controlled is by changing I_B. For this reason the BJT is normally called a **current controlled device**.

Transistors in circuits – the d.c. loadline

We are now in a position to investigate how a BJT will behave in a real circuit and principally how it will amplify signals. Consider the circuit in Figure 15.9(a). Notice that a single supply battery (called V_{CC}) is used which is, of course, an important practical advantage. R_C is called the **collector load** and R_B is the **base bias resistor**. I_B flows through R_B producing the forward bias for the base-emitter junction.

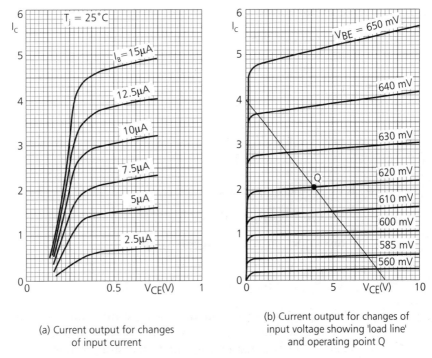

(a) Current output for changes
of input current

(b) Current output for changes of
input voltage showing 'load line'
and operating point Q

Figure 15.8 BC107 to BC109 output characteristics (courtesy Mullard).

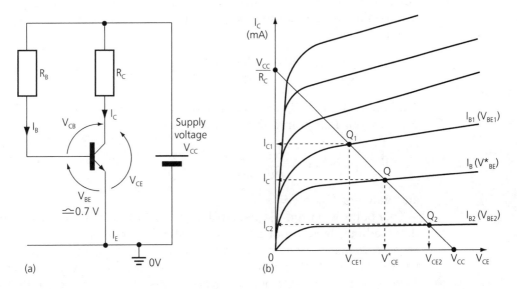

Figure 15.9 Common emitter circuit and load line determination

We can get some important information about this circuit:

- The collector-base junction must be reverse biased and so we must make sure that the voltage drop across R_B ($I_B \times R_B$) is larger than that across R_C ($I_C \times R_C$).
- Kirchoff's voltage law (KVL) to the input gives:

$$V_{BE} + I_B R_B = V_{CC}$$

i.e. $R_B = \dfrac{V_{CC} - V_{BE}}{I_B}$

- KVL to the output gives:

$$V_{CE} + I_C R_C = V_{CC}$$

If this equation is rearranged we get:

$$I_C = \frac{V_{CC} - V_{CE}}{R_C} = \frac{-1}{R_C} \times V_{CE} + \frac{V_{CC}}{R_C}$$

This equation is the equation of a straight line because it has the form $y = mx + c$, where $y = I_C$ and $x = V_{CE}$. I_C and V_{CE} *are used to plot the output characteristic*. Hence, using this equation a straight line can be drawn on the output characteristic. It is called the **d.c. loadline** for the circuit.

Drawing the d.c. loadline

Since only two points are needed to draw a line the easiest way of drawing the d.c. loadline on the output characteristic is as follows. If we put $V_{CE} = 0$ into the equation then:

$$I_C = \frac{-1}{R_C} \times 0 + \frac{V_{CC}}{R_C} = \frac{V_{CC}}{R_C}$$

This point is on the I_C axis, i.e. when $V_{CE} = 0$, and is shown in Figure 15.9(b). If we now put $I_C = 0$ into the equation then,

$$0 = \frac{-1}{R_C} \times V_{CE} + \frac{V_{CC}}{R_C}$$

and value of $V_{CE} = V_{CC}$.

This point is on the V_{CE} axis in Figure 15.9(b). Joining the two points gives the d.c. loadline. It is important to understand that *the loadline gives all the possible values of collector current and collector-emitter voltage* for the transistor in that circuit, i.e. for that particular load R_C and that particular supply V_{CC}. Thus in Figure 15.9(b), if the transistor operating point was Q*, then the operating point values would be I^*_B, V^*_{BE} I^*_C and V^*_{CE}. If the operating point were Q_1, then the corresponding values would be I_{B1}, V_{BE1}, I_{C1} and V_{CE1}, etc.

Signal input and output

If the circuit is to amplify a varying signal, the signal must be applied to the input and be taken from the output without upsetting the operating point of the

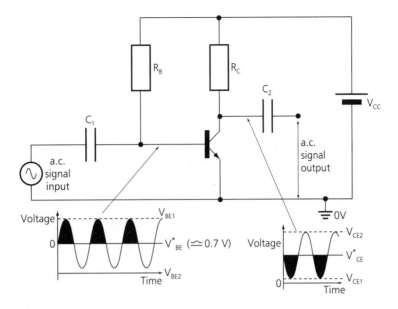

Figure 15.10 A common emitter amplifier

BJT. As shown in Figure 15.10, capacitor C_1 and C_2 can be used to pass a varying signal in and out of the circuit since a.c. signals can pass through the capacitor. C_1 and C_2 stop (or block) any d.c. current flow from the BJT circuit and so do not change the operating point values. C_1 and C_2 are chosen so that their reactance is negligible at the frequency of operation.

How the circuit amplifies

Suppose that the operating point Q* shown on the loadline in Figure 15.9(b) is chosen for the BJT and that the signal causes the base-emitter voltage to rise to V_{BE1} and fall to V_{BE2}, i.e. move up and down the loadline between Q_1 and Q_2. At Q_1, the base-emitter voltage has risen to V_{BE1} but the collector-emitter voltage has fallen to V_{CE1}. At Q_2, the base-emitter voltage has decreased to V_{BE2} and the collector-emitter voltage has increased to V_{CE1}. the change in V_{CE} is therefore an **inverted version** of the change in V_{BE}. Furthermore, the change in V_{BE} is in mV's, but the change in V_{CE} is in volts – so the circuit shows **voltage gain**. The input signal has been amplified! The value of the voltage gain is:

$$A_v = \frac{\text{peak-to-peak change in } V_{CE}}{\text{peak-to-peak change in } V_{BE}} = \frac{V_{CE2} - V_{CE1}}{V_{BE2} - V_{BE1}}$$

Because the output signal is an inverted version of the input we say *there is a 180° phase shift through the amplifier.*

The amplifier also has **current gain** because the swing in I_C (I_{C1} to I_{C2}) is in mA's, but the swing in I_B (I_{B1} to I_{B2}) is in μA's.

$$A_i = \frac{I_{C1} - I_{C2}}{I_{B1} - I_{B2}}$$

WORKED EXAMPLE 15.1

The amplifier in Figure 15.10 uses a BC108 whose output characteristics are shown in Figure 15.8.

(a) Plot the d.c. loadline using Figure 15.8(b) if $V_{CC} = 8\,V$ and $R_C = 2K$.
(b) What are the operating point values if V_{BE}^* is chosen to be 620 mV?
(c) If the operating point value base current is 60 μA calculate the value of the bias resistor R_B.
(d) Calculate the voltage gain if the input signal is 20 mV peak-to-peak.

Solution

(a) Using Figure 15.8(b), the loadline is drawn between the points:

$$V_{CE} = V_{CC} = 8\,V \quad \text{(horizontal axis)}$$

$$I_C = \frac{V_{CC}}{R_C} = \frac{8}{2000} = 4\,mA \quad \text{(vertical axis)}$$

(b) The operating point Q, will be at the intersection of the loadline and the 620 mV input characteristic. The operating point values are:

$$V_{BE}^* = 620\,mV$$

$$V_{CE}^* = 4\,V \quad \text{(approx.)}$$

$$I_C^* = 2\,mA \quad \text{(approx.)}$$

(c) The value of R_B is:

$$R_B = \frac{V_{CC} - V_{BE}^*}{I_B^*} = \frac{8 - 0.62}{60 \times 10^{-6}} = 123\,k\Omega$$

(d) The 20 mV peak-to-peak input causes movement up and down the loadline between the 630 mV and 610 mV characteristics. V_{CE} therefore varies between about 2.5V up to about 5.2V. Voltage gain A_v is:

$$A_v = \frac{\text{change in output voltage}}{\text{change in input voltage}} = \frac{5.2 - 2.5}{(630 - 610) \times 10^{-3}} = 135$$

The a.c. equivalent circuit of an amplifier

Any amplifier must be considered from two points of view. Both the d.c. and a.c. operating conditions must be understood. If we consider the amplifier in Figure 15.10, the d.c. operation is concerned with establishing the operating point Q*, choosing a suitable supply (V_{CC}) and calculating the necessary values of R_C and R_B to set that operating point. The capacitors C_1 and C_2 have no effect on the d.c.

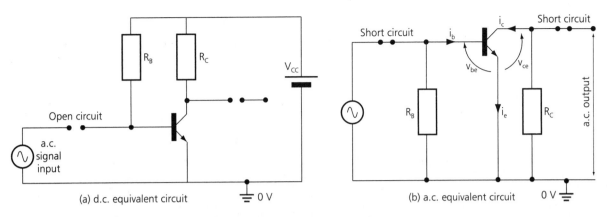

Figure 15.11 The d.c. and a.c. equivalent circuits

conditions because they behave like open circuits. The d.c. equivalent circuit is drawn in Figure 15.11(a).

C_1 and C_2 are chosen so that their reactance is negligible at the operating frequency – thus they behave as short circuits as far as the a.c. signal is concerned. *But there is another capacitor in the circuit!* The supply, V_{CC}, will probably be provided by a power supply similar to those investigated in Element 13.1 (Chapter 14), and you haven't forgotten the large reservoir capacitor have you? (Even if V_{CC} is derived from a battery then this also has a very large capacitance). To the a.c. signal this capacitor appears as a short circuit as well, so the top of R_B and R_C fold down and effectively connect to the 0 V line. Thus the circuit 'seen' by the a.c. signal is as shown in Figure 15.16(b).

Notice how lower case letters are used to show the signal currents and voltage. The a.c. output signal is the a.c. voltage across R_C. The bias resistor R_B is in parallel with the base-emitter of the transistor.

An improved bias circuit

The bias method using a single resistor R_B is simple but is rarely used in practice. This is because there is no stabilisation against temperature changes and the variations in h_{FE} values from one transistor to another of the same type. The circuit almost always used is shown in Figure 15.12. It is called *potential divider bias with emitter stabilisation*. Notice the way the power supply battery is not shown and two lines are drawn, one for V_{CC} and one for 0 V. These are called the **power supply rails**. This is the usual way of drawing amplifier circuit diagrams.

R_1 and R_2 form a **potential divider circuit** and their values are chosen so that the current through them is much larger than the base current I_B. Generally we choose about 10 × I_B through R_2 which gives 11 × I_B through R_1. Because the current through R_1 and R_2 is much larger than I_B, any changes in I_B produced by temperature variations or differing h_{FE}'s will have very little effect on the potential divider current. Consequently, the voltage at the base (point X) is essentially fixed.

The emitter current I_E produces a voltage across R_E ($I_E × R_E$) and so the forward bias voltage (V_{BE}) of the BJT is the difference between the voltage at X and the voltage at the emitter. Suppose that the temperature of the transistor increases for some reason. This causes a rise in I_C and a corresponding increase in I_E. The

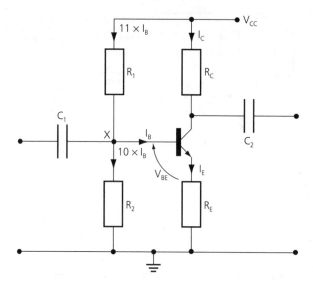

Figure 15.12 *Potential divider bias with emitter stabilisation*

voltage produced across R_E increases and V_{BE} decreases since the voltage at 'X' is fixed. This reduction in V_{BE} will reduce I_B and hence I_C and so tends to return I_C to its operating point value. A similar but opposite effect occurs if the temperature falls. This circuit configuration therefore provides a degree of **temperature compensation** or **stabilisation**.

The circuit stabilises in a similar manner to compensate for h_{FE} variations. For example, if a larger value h_{FE} transistor is substituted in the circuit I_C and I_E will increase causing a reduction in V_{BE}. This will reduce I_C and return the circuit to its normal operating condition.

In practice a well proven design rule is that R_E is chosen so that the voltage across it is about one tenth of the supply voltage V_{CC}.

Using a decoupling capacitor

The use of R_E will provide d.c. stability but will reduce the possible voltage gain available. To get good gain, *all* of the input signal at point X should reach the base-emitter junction to be amplified. However because of R_E, the a.c. input signal would cause an a.c. emitter current to flow in R_E and produce a voltage across it in phase with the applied signal. Thus, the net input signal to the base-emitter junction is the difference in the two signals. The final gain of the stage would be low since only a portion of the applied signal actually reaches the transistor to be amplified. This effect is called **negative feedback** and always results in a reduction in gain.

The negative feedback can be avoided, whilst still allowing R_E to stabilise by connecting a capacitor C_E across R_E. The value of C_E is chosen so that under a.c. conditions it shorts out R_E so that the emitter is connected to 0 V. This ensures that all of the input signal reaches the base-emitter junction of the BJT to be amplified.

When a capacitor is used in this way it is called a **decoupling capacitor**. To be effective, C_E is chosen so that at the lowest frequency of operation the reactance of C_E is no more than about one tenth the value of R_E.

There are several forms of negative feedback effect within amplifiers and most commercial circuits will have some sort of negative feedback deliberately designed into them. Although the gain is reduced, many desirable effects result. The most important is that the gain becomes very predictable and stable. This is important in the mass production of amplifier circuits. You can observe this in a practical exercise later in this chapter.

The inclusion of C_E also means that the circuit has two loadlines – a d.c. loadline and an a.c. loadline. This is best shown by a worked example.

WORKED EXAMPLE 15.2

The BJT used in the amplifier of Figure 15.13(a) has the output characteristic drawn in Figure 15.14.
(a) Draw the d.c. and a.c. loadlines and chose an operating point on the $I_B = 60\ \mu A$ characteristic
(b) If the input signal causes a base current swing of 40 µA peak-to-peak determine the current gain of the amplifier
(c) Describe how the voltage gain could be determined.

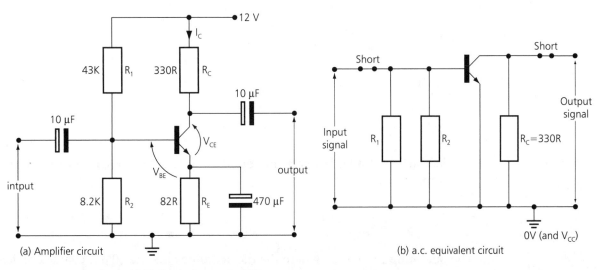

Figure 15.13 Worked example 15.2

Solution

(a) *Under d.c. conditions* all the capacitors appear as open circuits. Applying Kirchoff's voltage law to the output gives:

$$I_E \times R_E + V_{CE} + I_C R_C = V_{CC}$$

Since I_C is very nearly equal to I_E then:

$$I_C(R_E + R_C) + V_{CE} = V_{CC}$$

The d.c. loadline equation is:

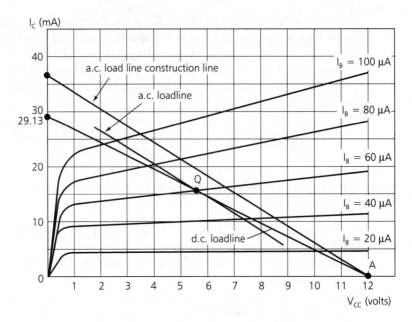

Figure 15.14 Output characteristic of transistor for Worked example 15.2

$$I_C = \left(\frac{-1}{R_E + R_C}\right) \times V_{CE} + \frac{V_{CC}}{R_E + R_C}$$

It has a slope of:

$$\frac{-1}{R_E + R_C} = \frac{-1}{330 + 82} = -\frac{1}{412}$$

The loadline is drawn between the points $V_{CE} = 12\,V$ (shown at point A on the graph) and:

$$I_C = \frac{12}{330 + 82} = 29.13\,mA$$

The operating point Q is where the d.c. loadline cuts the 60 μA characteristic. The operating point values are:

$$I_B^* = 60\,\mu A$$

$$I_C^* = 15.5\,mA \quad (approx.)$$

$$V_{CE}^* = 5.6\,V \quad (approx.)$$

Under a.c. conditions all the capacitors appear as short circuits and the V_{CC} rail folds down to the 0 V rail. (see Figure 15.13(b).)

The a.c. load is just $R_C = 330R$ because R_E has been 'taken out' by C_E. The a.c. loadline must pass through Q because if the a.c. signal is reduced to zero we must then be at the d.c. operating point. The a.c. loadline must have a slope of –1/330.

A simple way to draw the line is as follows. A point is located on the I_C axis of value $12/330 = 36.4$ mA. A line is drawn between point A (12 V) and this point. The line has a slope of $-1/330$. The actual a.c. loadline is parallel to this construction line and has to pass through Q. (See Figure 15.14).

(b) The input signal causes the operating point to move up and down the a.c. loadline between the 80 μA and 40 μA characteristics.

$$A_i = \frac{\text{peak-to-peak change in } I_C}{\text{peak-to-peak change in } I_B} = \frac{(20.5 - 10.5) \times 10^{-3}}{(80 - 40) \times 10^{-6}} = 250$$

(c) The peak-to-peak change V_{CE} can be found from the output characteristic in Figure 15.14. To find the corresponding change in V_{BE} as I_B swings between 80 μA and 40 μA, we would need the **input characteristic** of the transistor as well OR an output characteristic scaled with values of V_{BE} like the one in Figure 15.8(b). The voltage gain A_V would be:

$$A_V = \frac{\text{peak-to-peak change in } V_{CE}}{\text{peak-to-peak change in } V_{BE}}$$

WORKED EXAMPLE 15.3

A transducer which produces an output of 20 mV peak-to-peak requires a small-signal amplifier. Design a single-stage BJT circuit using a BC108A which will give a voltage gain of at least 80 from 100 Hz to 12 kHz. A 9 V power supply is available for the circuit.

Solution

1 **The transistor and amplifier configuration**
You are familiar with the typical BC108 characteristics because we have used it several times in this element. Data from Mullard gives that h_{FE} has a minimum value of 110, a maximum value of 220 and a typical value of 180 measured at $I_C = 2$ mA, $V_{CE} = 5$ V. We will use a potential divide bias circuit with emitter stabilisation for the amplifier.

2 **Choosing the operating point**
The typical output characteristics for the BC108 are reproduced again in Figure 15.15. (courtesy Mullard). The d.c. loadline will be drawn between the point $V_{CE} = 9$V (the supply rail) on the V_{CE} axis to a point on the I_C axis chosen by us. If we choose an intercept of $I_C = 3$ mA then the operating point Q is where this line intersects with the 610 mV characteristic. The operating point values are:

$$V^*_{BE} = 0.61 \text{ V (approx.)}; \quad V^*_{CE} = 4.75 \text{ V}; \quad I^*_C = 1.4 \text{ mA (approx.)}$$

Electronics

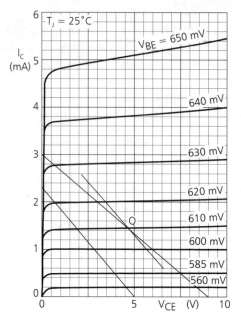

Figure 15.15 Typical output characteristics for BC108, Worked example 15.3 (courtesy Mullard)

3 Calculating R_E and R_c

The voltage across R_E should be about one tenth of the supply, i.e.

$$V^*_E = 0.1 \times 9 = 0.9 \text{ V}$$

Because I^*_E is almost equal to I^*_C, than $V^*_E = I^*_c \times R_E$, hence:

$$0.9 = (1.4 \times 10^{-3}) \times R_E$$

so $R_E = 642.8 \ \Omega$ (preferred value 680R). For the circuit, the intercept on the I_C axis (3 mA) is equal to:

$$I_C = \frac{V_{CC}}{R_E + R_C}$$

$$R_E + R_C = \frac{V_{CC}}{I_C} = \frac{9}{3 \times 10^{-3}} = 3000 \ \Omega$$

$$R_C = 3000 - 642.8 = 2357.2 \ \Omega \quad \text{(preferred value 2K2)}$$

4 Calculating the values of R_1 and R_2

Using the typical value of $h_{FE} = 180$, at the operating point we have:

$$h_{FE} = \frac{I^*_C}{I^*_B}$$

$$I^*_B = \frac{1.4 \times 10^{-3}}{180} = 7.8 \ \mu A$$

For the potential divider, 11 x I^*_B = 85.8 μA will flow through R_1 and 10 x I^*_B = 78 μA will flow through R_2. Using Kirchoff's voltage law:

$$V^*_E + V^*_{BE} + I_{R1} \times R_1 = 9$$

$$I_{R1} \times R_1 = 9 - 0.61 - 0.9 = 7.49$$

$$R_1 = \frac{7.49}{85.8 \times 10^{-6}} = 87.3 \text{ k}\Omega \quad \text{(preferred value 91K)}$$

Furthermore, I_{R2} x R_2 = $V^*_E + V^*_{BE}$

$$R_2 = \frac{1.51}{78 \times 10^{-2}} = 19.4 \text{ k}\Omega \quad \text{(preferred value 20K)}$$

5 Calculating C_E, C_1 and C_2

The lowest operating frequency is 100 Hz so the reactance of C_E must be no greater than about a tenth of R_E.

$$\frac{1}{2\pi \times 100 \times C_E} = \frac{1}{10} \times 680$$

$$C_E = \frac{1}{2\pi \times 100 \times 68} = 23.4 \text{ μF}$$

The nearest value is 22 μF but since the calculation is for a minimum value we'll use the next preferred value which is 47 μF.

If you look back at the a.c. equivalent circuit in Figure 15.13(b), the a.c. signal will 'see' an input resistance of R_1 in parallel with R_2 in parallel with the input resistance of the transistor. The input resistance of the BJT is given the symbol h_{ie}. (We'll investigate h_{ie} a little later).

The equivalent resistance of R_1 and R_2 in parallel is

$$\frac{91 \times 20}{91 + 20} = 16.4 \text{ k}\Omega$$

The input resistance of the amplifier is therefore 16.4 kΩ in parallel with h_{ie}. Mullard give the typical value of h_{ie} as 2.7 kΩ.

$$\text{Input resistance} = \frac{16.4 \times 2.7}{16.4 + 2.7} = 2.32 \text{ k}\Omega$$

The reactance of C_1 has to be no more than about one tenth of this at 100 Hz.

$$\frac{1}{2\pi f C_1} = \frac{1}{2\pi \times 100 \times C_1} = \frac{1}{10} \times 2320$$

$$C_1 = \frac{1}{2\pi \times 100 \times 232} = 6.86 \text{ μF}$$

The nearest preferred value is C_1 = 10 μF. The output resistance of the BJT is larger than its input resistance so choosing C_2 = 10 μF will be satisfactory here.

6 Drawing the a.c. loadline

Drawing the a.c. loadline allows us to determine the voltage gain of the amplifier. The loadline must have a slope of $-1/2200$ and pass through Q.

To draw the line we choose a convenient value of V_{CE} (say, 5 V), and find the point on the I_C axis equivalent to $I_C = 5/2200 = 2.27$ mA (say, 2.3 mA). This construction line has a slope of $-1/2200$.

As shown on Figure 15.15, the a.c. loadline is drawn parallel to this to pass through Q.

The 20 mV peak-to-peak signal moves the operating point up and down the a.c. loadline between the 600 mV and 620 mV characteristics.

$$A_V = \frac{\text{peak-to-peak change in } V_{CE}}{\text{peak-to-peak change in } V_{BE}} = \frac{5.75 - 3.6}{20 \times 10^{-3}} = 107.5 \text{ (approx.)}$$

Note: We have used typical values and typical characteristics so the actual gain value measured for the real amplifier is very unlikely to be exactly as calculated.

Use of software

Nowadays, most circuit design and analysis uses computer-aided design (CAD) software to speed up the design and analysis process. An example of the use of one package (called Microcap) to predict the voltage gain of the amplifier designed in Worked Example 15.3 is illustrated in Figure 15.16.

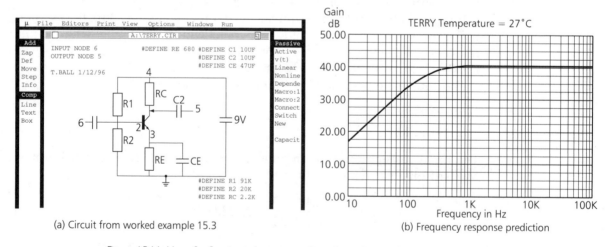

(a) Circuit from worked example 15.3

(b) Frequency response prediction

Figure 15.16 Use of software package to predict voltage gain

Firstly, the circuit is drawn on the screen by selecting the components required with a mouse and connecting them together. Figure 15.16(a) is a printout of what is shown on the screen. Finally, an a.c. analysis is performed by the software which predicts the **gain/frequency response** over the chosen frequency range. The prediction is that the gain will be about 40 dB $= 20 \log_{10} (V_{out}/V_{in})$ over most of the frequency range. In worked example 15.3 our prediction is $20 \log_{10} (107.5)$ $= 40.6$ dB.

Do this 15.1

Equipment and components required:

- BC108A BJT
- 680R resistor
- 2K2 resistor
- 20K resistor
- 91K resistor
- 2 × 10 μF electrolytic capacitors
- 47 μF electrolytic capacitor
- oscilloscope
- power supply

Procedure:

Build the amplifier circuit in the Worked Example 15.3. Measure its voltage gain from 100 Hz to 12 kHz and plot your results on the frequency response in Figure 15.16(b). Compare your value with the calculation and, if possible, other members of your learning group. It is most unlikely that you will get identical values. There could be some wide variations.

Adding negative feedback:

Remove the 47 μF decoupling capacitor so that you are deliberately adding negative feedback to the circuit. You should see the gain falling to a very low value and the new value of A_V for each member of your group should be very nearly the same. In fact it should be almost equal to

$$R_C/R_E = 3.23 \quad \text{(say, 3)}$$

This is the major advantage of negative feedback because it makes the gain very predictable and stable. The disadvantage of course is we would need more stages of amplification to satisfy our gain specification.

Positive and negative feedback

When negative feedback (NFB) is applied to an amplifier it always reduces the gain but makes the gain more stable and predictable. It also changes other properties of the amplifier as well. The most important effects are:

- the bandwidth increases
- the input and output resistances change.

Whether the input and output resistances increase or decrease depends on the method used to produce the NFB.

There are four ways in which NFB can be applied to circuits. They differ in how the feedback signal is produced and how it is introduced into the input of the circuit. Two examples are as follows.

Series current NFB

Here the voltage feedback is derived from the current in the output of the circuit and is applied in series with the input signal. This is the technique used in Do This 15.1 when the decoupling capacitor is removed from across R_E. The emitter current flows in R_E and produces a voltage which is in series with the input signal. As shown in Figure 15.17(a), sometimes a potentiometer is used for R_E or the emitter resistor is split to introduce series current NFB.

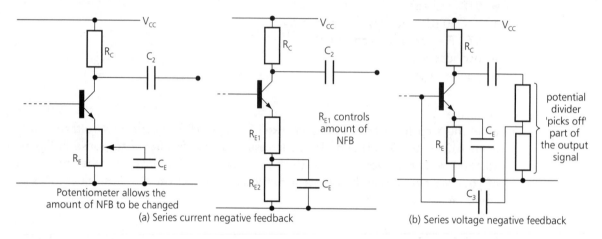

Potentiometer allows the amount of NFB to be changed

(a) Series current negative feedback

R_{E1} controls amount of NFB

potential divider 'picks off' part of the output signal

(b) Series voltage negative feedback

Figure 15.17 Two types of negative feedback

Series voltage NFB

As shown in Figure 15.17(b), a portion of the output signal voltage can be 'picked off' via a potential divider across the output and is applied in series with the input signal to provide voltage NFB.

Activity 15.1

Replace R_E in Do This 15.1 with a 1 KΩ potentiometer and connect C_E to the wiper. For several positions of the potentiometer measure the gain of the amplifier at a frequency of 10 kHz. Confirm that as more and more of R_E is not decoupled by C_E the gain falls because more negative feedback is being applied.

Positive feedback (PFB)

It is possible to feedback a portion of the output of an amplifier **in phase** with the applied input signal so that the effective input increases. This is called positive feedback. If PFB is used, the amplifier output will very quickly rise until it is limited by the power supply voltage. This is disastrous and so PFB is avoided in normal amplifier circuits.

Positive feedback does, however, have its uses and a class of electronic circuits called **oscillators** use PFB as their operating principle.

The transistor equivalent circuit

A transistor can be replaced by an equivalent circuit which models its behaviour in terms of simple components. An equivalent circuit which is easy to use at low frequencies is shown in Figure 15.18(a). It is called the **hybrid parameter equivalent circuit** and imitates the effect of small a.c. signals, i.e. signals which cause small changes in the operating point values.

- h_{ie} is the **input resistance** (i) in common emitter (e) measured with the collector-emitter voltage held fixed.
- h_{fe} is called the **small-signal current gain** measured with the collector-emitter voltage held fixed. (Remember h_{FE} was the d.c. current gain).
- h_{oe} is the **output conductance** (inverse of resistance) and is measured with the base current held fixed.

The value of h_{ie}, h_{fe} and h_{oe} can be obtained from the characteristics but it is easier to look them up in the manufacturer's data sheets. The table gives values for the BC108A from Mullard.

	Minimum	Typical	Maximum	Unit
h_{ie}	1.6	2.7	4.5	Ω
h_{fe}	125	220	260	none
h_{oe}	–	18	30	S

Using the equivalent circuit it is possible to work out equations for the voltage gain, current gain and input resistance of the common emitter amplifier.

AC equivalent circuit of common emitter amplifier

Figure 15.18(b) shows the h-parameter circuit for the transistor substituted in the a.c. equivalent circuit for the potential divider, emitter stabilised, common emitter amplifier. Since R_1 in parallel with R_2 is larger than h_{ie} then these parallel resistors can be ignored (remember when a large and small resistor are in parallel the equivalent resistance is close to the value of the smallest resistance).

Also, since h_{oe} is very small, $1/h_{oe}$ is very large and can be ignored compared to R_C. The approximate equivalent of the amplifier is shown in Figure 15.18(b).

The voltage gain equation:

$$A_v = \frac{\text{output signal voltage}}{\text{input signal voltage}} = \frac{v_{ce}}{v_{be}}$$

Since

$$v_{ce} = i_c \times R_c \quad \text{and} \quad v_{be} = i_b \times h_{ie}$$

$$A_v = \frac{i_c \times R_c}{i_b \times h_{ie}}$$

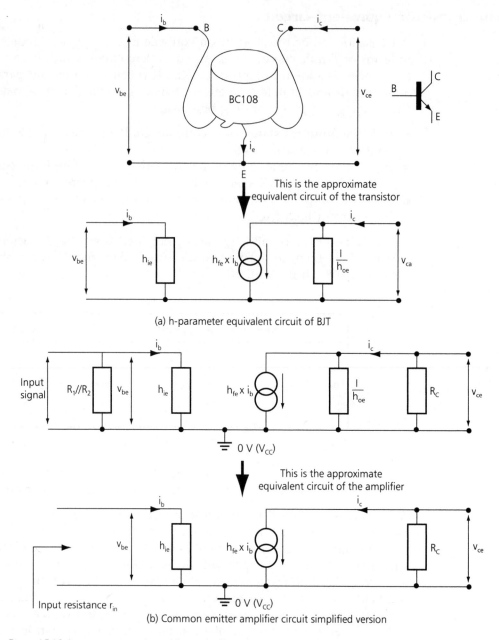

Figure 15.18 h parameters and amplifier equivalent circuit

but

$$\frac{i_c}{i_b} = h_{fe}$$

Hence

$$A_v = \frac{h_{fe} \times R_c}{h_{ie}} \quad \text{(no units)}$$

The current gain equation:

$$\text{Current gain } A_i = \frac{i_c}{i_b}$$

Hence

$$A_i = h_{fe} \text{ (no units)}$$

The input resistance equation:

$$R_{in} = h_{ie} \text{ (Ohms)}$$

Remember, these equations are approximate but will give a *guide* to the values expected.

Progress check 15.5

Use the voltage gain equation to estimate A_v for the amplifier in Worked Example 15.3. Compare the answer with the one obtained by measurement. Don't be surprised if they are very different!

The field effect transistor

The field effect transistor (FET) like the BJT is a three-terminal device made from semiconductor materials. Its principle of operation however is different to that of the BJT The BJT is called a **bipolar transistor** because both holes and electrons move through it to produce the current. In the FET, on the other hand, current flow is either by holes or electrons – not both. It is for this reason that the FET is often called a **unipolar device**. The term 'field' arises from the fact that the current flow is controlled by an electric field set up in the semiconductor material by an externally applied voltage.

There are two types of FET. The first is the **junction FET** (JFET) and the second is the **metal oxide semiconductor FET** (MOSFET). We will investigate the JFET first of all.

The junction field effect transistor

The basic structure of the JFET (n-channel) is shown in Figure 15.19 with circuit symbols for both p-type and n-type.

In the n-channel JFET, a bar of n-type material has p-type semiconductor doped into each side as shown. These P-type regions are joined by a common connection called the **gate terminal**. At either end of the bar there are two other terminals called the **drain** and the **source**. The space in the bar between the gate regions is called the **channel**. If a p-type bar is used with n-type gate regions then a p-channel JFET results.

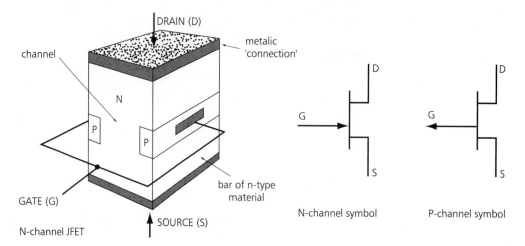

Figure 15.19 Basic structure of a JFET (n-channel) and circuit symbols

As you will now discover, a voltage applied to the gate will control the flow of current between the drain and the source in an analogous manner to the way in which base current controls the current between the collector and emitter in a BJT.

The operation of a JFET

To understand how the JFET operates we need to consider what happens when external voltages are applied between the terminals. The structure we will examine is the basic n-channel JFET. During manufacture, the n-type channel is more lightly doped than the p-type gate regions. Because of the unequal doping the depletion region at each pn junction extends further into the channel on the n side of the junction.

Consider what happens when the drain is made positive with respect to the source and the gate and source terminals are shorted together ($V_{GS} = 0$). As shown in Figure 15.20(a) as V_{DS} is increased slightly above zero volts, electrons in the n-type bar are attracted by the positive terminal of V_{DS} and **drain current**, I_D, flows. A slight increase in V_{DS} will cause I_D to increase in proportion to it. The channel is behaving at this stage like a fixed resistor and so Ohm's law is obeyed. As V_{DS} is increased further, the depletion regions penetrate further into the channel because the width of the depletion region in a pn junction depends on the size of the reverse voltage. The higher the voltage, the wider the depletion region.

As V_{DS} is increased still further, the channel becomes very narrow particularly at the drain end where the voltage across the junction is highest. This causes an increase in its resistance. Thus I_D is no longer directly proportional to the increase in V_{DS} and the graph begin to flatten out. **Pinch-off** occurs when the depletion regions meet at the middle of the channel. The value of V_{DS} at which this occurs is called the **pinch-off voltage** V_p, (see Figure 15.20(b)). I_D reaches its maximum value at pinch-off and this value is given the symbol I_{DSS} (the Drain to Source current with gate Shorted). Above pinch-off I_D stays about constant.

In normal operation, the gate-source junctions are reverse biased by an external voltage V_{GS}. Suppose $V_{GS} = -1$ V. As V_{DS} is increased from 0 V, pinch-off will occur earlier than in the $V_{GS} = 0$ V case. For each value of V_{GS} a different

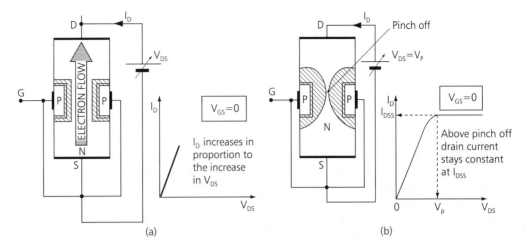

Figure 15.20 JFET operation for VGS = 0

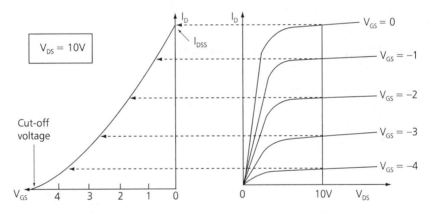

Figure 15.21 The transfer characteristic of a JFET derived from the drain characteristic

characteristic can be drawn as shown in the right-hand side of Figure 15.21, which gives the **drain characteristics** of the JFET.

- Note the similarity between the drain characteristics and the output characteristics of a BJT in common emitter operation.
- A value of V_{GS} is reached at which $I_D = 0$ no matter what V_{DS} is. Here V_{GS} is equal to the pinch-off voltage so the depletion regions meet all the way along the channel. This value of V_{GS} is called the **gate-source cut-off voltage**, $V_{(P)GS}$.
- For each characteristic (excluding $V_{GS} = 0$), the gate-source junction is reverse biased. The gate current is minute so the resistance between gate and source is extremely high – typically tens or hundred of megohms. (Remember the BJT has a forward biased junction between base and emitter giving a much lower resistance – a few kilohms).

Transfer characteristic of a JFET

The transfer characteristic of a JFET is a graph of I_D versus V_{GS} at a specified value of V_{DS}. As indicated in Figure 15.21 the transfer characteristic can be obtained from the drain characteristics by drawing a line at a chosen V_{DS} and then noting the value of I_D at the intersection of this line and the lines of constant V_{GS}.

Manufacturers usually provide typical characteristics in their data – we'll use a set later.

Using the characteristics

Two important parameters can be obtained from the JEFT transfer and drain characteristics by drawing tangents at the operating point. The parameters are needed for designing and understanding JFET amplifiers.

The **mutual conductance**, g_m, is a measure of the change in I_D produced by a change in V_{GS} at a fixed value of V_{DS}:

$$g_m = \frac{\delta I_D}{\delta V_{GS}} \quad \text{at } V_{DS} \text{ constant}$$

The symbol 'δ' (delta) means a 'small change in'. g_m has the units of Siemens.

The **drain-source resistance**, r_{ds}, is a measure of the change in V_{DS} produced by a change in I_D at constant V_{GS}:

$$r_{ds} = \frac{\delta V_{DS}}{I_D} \quad \text{at } V_{GS} \text{ constant}$$

Since the drain characteristics are fairly flat, r_{ds} is quite large and ranges from tens to hundreds of kilohms. Figure 15.22 shows the procedure for obtaining g_m and r_{ds} at an operating Q.

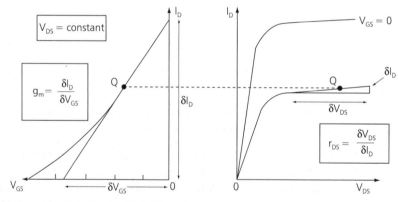

Figure 15.22 Procedure for determining g_m and r_{ds}

WORKED EXAMPLE 15.4

A set of typical characteristics for a BF245B N-channel JFET from Mullard are reproduced in Figure 15.23. Estimate the values of (a) g_m at an operating point of $V_{GS} =$ I V, $V_{DS} = $ 15 V and (b) r_{ds} at an operating point of $V_{GS} = -$I V, $V_{DS} = $ 10 V.

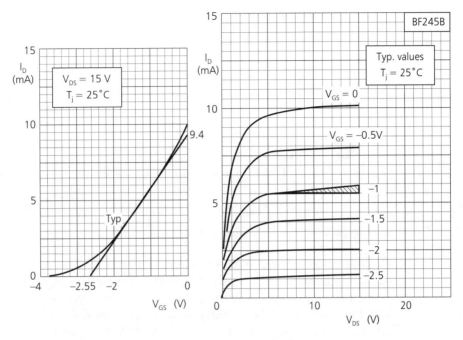

Figure 15.23 Typical characteristics for a BF245B (courtesy Mullard)

Solution

(a) A tangent is drawn to the transfer characteristic and $V_{GS} = -1$ V.

$$g_m = \frac{(9.4 - 0)}{(2.55 - 0)} \text{ mA (approx.)} = 3.7 \text{ mS}$$

(b) Since the $V_{GS} = -1$ characteristic is nearly straight, the tangent can be taken to be the characteristic itself. It is not easy to read accurately, but as V_{DS} changes from 5 V to 15 V, I_d changes from about 5.5 mA to about 5.8 mA.

$$r_{ds} = \frac{(15 - 5) \text{ V}}{(5.8 - 5.5) \text{ mA}} \text{ (approx.)} = \frac{10}{0.3 \times 10^{-3}} = 33.3 \text{ k}\Omega$$

Using the JFET as an amplifier

The FET like the BJT is a three-terminal component and may be connected in a circuit with either terminal common to the input and output part of the circuit. The most useful configuration for the JFET is the **common source mode**. The signal is applied between the gate and source and the amplified output signal is taken between the drain and source. Only small voltage gains are possible when compared to the common emitter amplifier, but a major advantage is the very much higher input resistance or impedance.

Like the BJT, the properties of individual FET's cannot be closely controlled and will vary from device to device of the same type number. Furthermore, device parameters will change because of temperature variations. The FET bias circuit

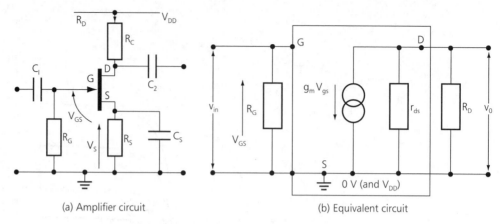

(a) Amplifier circuit   (b) Equivalent circuit

Figure 15.24 Common source amplifier using an n-channel JFET

must therefore provide some temperature stabilisation. A common source amplifier using a simple bias method, called **self-bias**, is shown in Figure 15.24(a).

For the n-channel JFET, the gate must be negative with respect to the source and the drain must be positive with respect to the source. The gate current is minute and can be taken as zero. R_G is chosen to have a large value (usually at least 1 MΩ). Because $I_G = 0$, there is no d.c. voltage drop across R_G so the gate is at ground potential. The drain current I_D flows through R_S and produces a voltage $V_S = I_D \times R_S$ across it. Thus, the source is positive with respect to the gate or (put another way) the *gate is negative with respect to the source*. This is the required bias condition for the n-channel JFET.

C_1 and C_2 couple the signal into and out of the amplifier and C_S 'takes out' R_S under a.c. conditions to prevent negative feedback. The design of a small-signal FET amplifier is very similar to that of a BJT amplifier, i.e. d.c. and a.c. loadlines can be drawn on the output characteristic.

Do this 15.2

Constructing a common source FET amplifier

Equipment and components required:
Construct the amplifier in Figure 15.24(a) using the following values:

FET = BF245B, R_S = 820R, R_D = 1K8, R_G = 1 MΩ,
C_1 = 22 nF, C_2 = 10 μF, C_S = 22 μF.

Operate the amplifier from a 15 V power supply and measure its gain from 100 Hz to 12 kHz. (The gain will be quite low – probably only about 4 or 5 over most of the frequency range.)

A.C. equivalent of JFET and common source amplifier

Under a.c. conditions C_1, C_2 and C_S are short circuits and the V_{DD} rail folds down to the 0 V rail. The equivalent circuit of the amplifier and JFET is shown in Figure 15.24(b). Notice that to model the very high input resistance of the JFET, an open circuit is shown between gate and source.

The input resistance of the amplifier is just R_G. Since r_{ds} is very much larger than R_D, the output resistance is R_D.

The output signal v_o is:

$$v_o = (g_m \times v_{gs}) \times R_D$$

and the input signal $v_{in} = v_{gs}$.

Therefore the voltage gain A_v is:

$$A_v = \frac{v_o}{v_{in}}$$

$$A_v = g_m \times R_D \quad \text{(no units)}$$

Activity 15.2

The FET amplifier in Do This 15.2 has d.c. operating point values of $V^*_{GS} = -2$ V, $V^*_{DS} = 8.5$ V, $I^*_D = 2.5$ mA. Use the characteristics in Figure 15.23 to estimate g_m at this operating point. Calculate A_v using the equation and compare with your measured value.

Metal oxide semiconductor FETs

If an extremely high input resistance (greater than 10^{12} Ω) is required than a type of FET called a **MOSFET** can be used. It is similar in many respects to the JFET in that it has three terminals (drain, gate and source) with current flow through the device being controlled by a voltage between the gate and the source. The major difference between them, however, is that the gate terminal of a MOSFET is insulated from its channel by a layer of insulating material called **silicon dioxide**. It is the construction that gives rise to such a high gate-to-source resistance.

Two types of MOSFETs are manufactured – called the **depletion type** and **enhancement type**. For each, n-channel and p-channel versions are available. Of the two, the enhancement type MOSFET has the simplest structure and as a result, an enormous number of these devices can be produced on a single silicon chip. Enhancement type MOSFETs are the fundamental building blocks of computer memories and microprocessors.

An n-channel enhancement type MOSFET is shown in Figure 15.25(a) with normal bias.

Notice that a channel is not deliberately 'doped' into the device during manufacture. Rather, a channel is induced below the oxide at the gate terminal when the correct external voltage is applied to the gate. The '+' sign indicates that the adjacent source and drain regions are heavily doped.

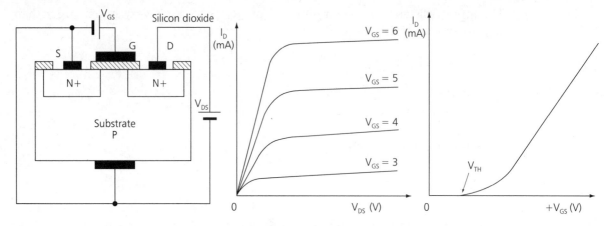

Figure 15.25 N-channel enhancement MOSFET with typical characteristics

Notice immediately that the gate is positive with respect to the source. The positive V_{GS} will repel holes away from beneath the gate terminal and attract minority electrons from the substrate. When V_{GS} is large enough, the region below the gate will have so many electrons that it will become n-type material. We say that *a channel has been induced in the device by V_{GS}*. This value of V_{GS} at which the channel is formed (so allowing current to flow between drain and source) is called the **threshold voltage**, V_{TH}. Typically V_{TH} is around 2 V or so. Increasing V_{GS} above threshold (V_{GS} = 3, 4, 5 V, etc.) allows a larger I_D for a particular value of V_{DS}. Typical characteristics are shown in Figure 15.25(b).

CMOS circuits

N and p-channel enhancement MOSFETs are usually called NMOS and PMOS devices respectively (pronounced N-MOS and P-MOS). As a result of their simple structure, it is fairly easy (and cheap) to produce NMOS and PMOS devices in integrated circuit form. High numbers of these devices can be fabricated on a silicon chip allowing powerful microprocessors and memory chips to be developed.

In the next Element, digital electronic circuits will be investigated. A type of integrated circuit called CMOS (pronounced C-MOS: Complementary Metal-Oxide Semiconductor) are used in advanced digital electronic applications. All this means in practice is that NMOS and PMOS devices have been produced in the same substrate. The devices are connected together within the substrate to generate basic digital 'building blocks' called **logic gates**.

Activity 15.3

Unlike the JFET (and BJT), MOSFETS require careful handling because they can be destroyed by static electricity. Produce a short report detailing the safeguards that should be taken when handling and using MOSFET devices.

Activity 15.4

Operate the JFET amplifier in Do This 15.2 from a 9 V power supply and connect to the BJT amplifier of Do This 15.1. (Just use one 10 μF coupling capacitor from the output of the JFET stage to the input of the BJT stage.) You now have a **two-stage amplifier**. Use an input signal of 10 mV peak-to-peak and measure the gain over a frequency range of 100 Hz to 12 kHz. (If the stage gains are A_{v1} and A_{v2} then the overall gain $A_v = A_{v1} \times A_{v2}$.) You should find that A_v should be at least 200. Experiment by adding negative feedback to the stage(s).

Assignment 15

This assignment provides evidence for:
Element 13.2: Investigate small signal amplifiers
and the following key skills:
Communication 3.2
Information Technology 3.1, 3.2, 3.3

The evidence indicators for this element require you to report on an investigation into a small-signal amplifier.

Report

Your report should include:

- a description of the d.c. operation of the amplifying device(s)
- the characteristics of the amplifying device, e.g. input and output characteristics
- the small-signal characteristics of the amplifier, i.e. gain, frequency response, input and output impedance
- small-signal equivalent circuits
- a record of the determination of the operational features.

If possible, include the use of CAD software which may be available at your Centre so that you can compare measured and predicted responses.

Chapter 16: Logic elements and digital circuits

All electronic systems can be broadly divided into two types – **analogue systems** and **digital systems**. In analogue systems (for example one of the simple amplifier circuits you have already investigated) the voltage and current signals can be smoothly varied over the range of values for which the system has been designed. In contrast, the voltage and current signals in digital systems can only have *two possible values – no intermediate levels can ever occur in a correctly operating digital circuit*. These two possible states are usually called **'logic 1'** and **'logic 0'** or '1' and '0' for short although sometimes the terms **'high'** and **'low'** are used instead. (The reason the word **'logic'** is adopted will become clearer as you progress through the element.)

In this element you will be solely concerned with understanding and using devices and circuits which are digital in nature.

When you complete this element you should be able to:

● describe combinational logic elements using truth tables
● simplify combinational logic circuits using Boolean algebra and Karnaugh maps
● describe the operation of sequential logic elements using truth tables
● describe the operation of sequential logic circuits
● monitor the operation of combinational and sequential logic circuits.

Analogue and digital signals

For comparison, two very simple circuits which can produce analogue and digital voltage waveforms are shown in Figure 16.1.

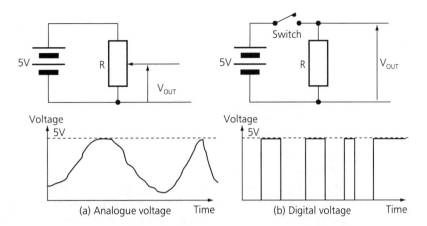

Figure 16.1 Analogue and digital waveforms

The analogue circuit can produce an infinite range of voltages V_{out} between 0 V and 5 V by varying the position of the wiper or knob of the potentiometer. On the other hand, the digital circuit will only generate two output voltages – either 0 V or 5 V depending on whether the switch is open or closed. The digital voltage has two definite 'states' called logic '1' (5 V) and Logic '0' (0 V). There are no intermediate states.

Combinational logic gates

The basic circuits that process these logic '1' and logic '0' states are called combinational logic gates or circuits and are the fundamental building blocks of all digital electronic systems.

Each logic gate will have only one output but may have several inputs. The term 'logic gate' is used because the state of the gate output is determined by certain mathematically logical conditions that exist on the inputs. Thus the output is at '1' or '0' depending on the logical combination of '1's and '0's applied to its inputs.

The AND gate
The operation of an AND gate can best be understood by examining the simple switching circuit in Figure 16.2(a).

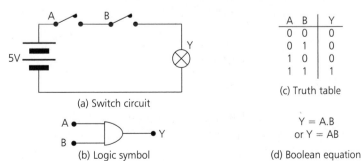

A	B	Y
0	0	0
0	1	0
1	0	0
1	1	1

(c) Truth table

(a) Switch circuit

(b) Logic symbol

$$Y = A.B$$
$$\text{or } Y = AB$$

(d) Boolean equation

Figure 16.2 The AND gate showing digital (ON/OFF) inputs from switches A and B to give a digital operation (ON/OFF) of lamp Y

The switches A and B and the lamp Y are two-state devices. The switches can only be open or closed and the lamp can only be on or off. It is easy to see that the lamp will only be *on when both A AND B are closed at the same time.*

If the convention '0' and '1' is used to represent the switch being OPEN and CLOSED, and '0' and '1' to represent the lamp being OFF and ON, then the logical operation of the circuit is described by the table in Figure 16.2(c). This table is called a **truth table**. It shows the state of the output Y for all combinations of the inputs A and B. For the AND gate, *Y is ON or '1' only when A and B are '1'.*

The logic symbol for an AND gate is drawn in Figure 16.2(b). Any electronic circuit which conforms to the truth table shown is by definition an AND gate and is represented by the logic gate symbol given.

A more concise way of describing the AND gate circuit operation is to use a form of algebraic equation called a **Boolean equation**. For the AND gate the Boolean equation is:

$$Y = A \cdot B$$

Here the dot '·' means AND but sometimes the dot is omitted and is written as $Y = AB$.

Progress check 16.1

Draw the truth table for an AND gate that has three inputs A, B and C. Write the Boolean expression.

The OR gate

Another basic logic gate is the OR gate illustrated in Figure 16.3.

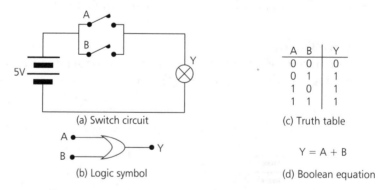

A	B	Y
0	0	0
0	1	1
1	0	1
1	1	1

(a) Switch circuit

(c) Truth table

(b) Logic symbol

$$Y = A + B$$

(d) Boolean equation

Figure 16.3 The OR gate

The lamp Y will light when switches A or B or both are closed. Thus *Y is ON or '1' when A or B or both are '1'.* The Boolean equation for an OR gate is:

$$Y = A + B$$

The '+' sign means OR in Boolean algebra not 'plus' as in normal arithmetic.

The NOT gate

This is the simplest of the basic gates and has a single input and output. The output is always *the inverse of the input* and for this reason the gate is often called an **inverter**. As shown in Figure 16.4, in Boolean algebra a NOT is indicated by a bar '⁻'. Thus the inverter equation is read as 'Y equals NOT A' or 'Y equals A bar':

$$Y = \overline{A}$$

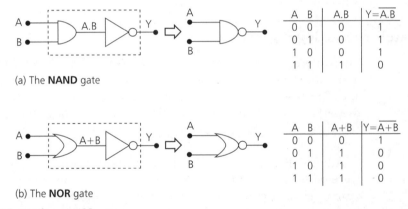

A	Y
0	0
1	1

$$Y = \overline{A}$$

(a) Logic symbol   (b) Truth table   (c) Boolean equation

Figure 16.4 The NOT gate or inverter

NAND and NOR gates

Two more very useful logic gates can be constructed by combining AND, OR and NOT gates. NAND and NOR gates are shown in Figure 16.5.

A	B	A.B	Y=$\overline{A.B}$
0	0	0	1
0	1	0	1
1	0	0	1
1	1	1	0

(a) The **NAND** gate

A	B	A+B	Y=$\overline{A+B}$
0	0	0	1
0	1	1	0
1	0	1	0
1	1	1	0

(b) The **NOR** gate

Figure 16.5 NAND and NOR gates

NAND and NOR gates are simply AND and OR gates followed by inverters. Notice how the small circle is used in the output of the logic symbol to indicate NAND and NOR gates, which is derived from the NOT symbol.

For a NAND gate, the output Y is the inverse of an AND gate. Hence:

$$Y = \overline{A \cdot B}$$

For a NOR gate, the output Y is the inverse of an OR gate. Hence:

$$Y = \overline{A + B}$$

You will see later in this element that it is possible to construct all logic gates by using either just NAND gates or just NOR gates. This will often simplify the design of more complicated logic networks.

**Progress
check 16.2**

Can you work out how you might construct a NOT gate from a two-input
NAND gate?

The exclusive OR gate
The truth table for an exclusive OR gate and its circuit symbol are drawn in
Fig 16.6.

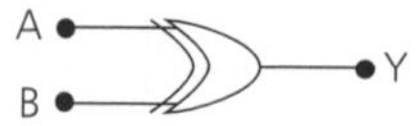

A	B	Y
0	0	0
0	1	1
1	0	1
1	1	0

Figure 16.6 Exclusive OR gate

It differs from a normal OR gate in that the last line of the truth table is different.
The Boolean equation is:

$$Y = A \oplus B$$

An exclusive OR gate will be used in a later case study.

Integrated circuit logic gates

The different logic gates (AND, OR, NOT, NAND, NOR) can be constructed in
many different ways. Initially circuits using individual resistors, thermionic valves
and, later, semiconductor diodes and transistors were used and wired together
to perform the logic gate function. This *discrete component approach*, as it was
called, was eventually superceded in the 1960's when integrated circuits (ICs)
began to be developed.

The first IC logic gates circuits were rather primitive by today's standards and
were basically the older discrete component circuits fabricated in the form of individual '**silicon chips**'. However, the technology rapidly improved and IC logic
devices have now evolved into several advanced **logic families**. Of these the most
popular are:

- Transistor–transistor logic (TTL) family
- Complementary metal-oxide semiconductor (CMOS) family

TTL logic gates use normal bipolar junction transistors whereas CMOS gates
are fabricated from MOSFET devices.

Historically, TTL was the first really successful IC logic family and was marketed
in the mid 1960s by Texas Instruments. The first TTL devices were used for mili-
tary applications (the 54 series) but became commercially available with a reduced
temperature operating range as the 74 series. The original standard TTL has been
continuously developed and now there are several sub-families available.

CMOS devices were available by the late 1960s and like TTL have been continuously improved and refined since then. The most attractive feature of CMOS
is its very low **power consumption** and high **packing density** (lots of gates on a
small area of silicon).

Both TTL and CMOS have their own advantages and disadvantages and the choice of which technology to use for a particular application will depend on many factors. As an example, the designers of a digital watch would use devices from the logic family which has the minimum power requirements. This is CMOS. On the other hand, where **operating speed** might be the decisive factor, one of the TTL sub-families would probably be used.

Specifying TTL and CMOS integrated circuits

The characteristics of logic gates can be classified under the following main headings:

- logic voltage levels
- fan-in/fan-out
- operating speed
- power dissipation.

We will examine each in turn.

Logic voltage levels

All logic gates are **two-state devices** and in an ideal world the logic '0' and logic '1' levels would correspond to two and only two distinct voltage levels in all digital circuit systems. Due to the nature of electronic components, however, this is not the case and bands of voltages are allocated to logic '0' and logic '1'. Thus, although nominal levels are specified by the manufacturer, a range of values is given within which there will be acceptable operation. Providing the two voltage levels in a digital circuit are within these ranges, they will be reliably recognised as logic '0' or logic '1'. This is illustrated in Figure 16.7.

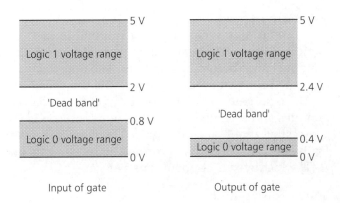

Figure 16.7 Logic levels for TTL

TTL circuits are designed to operate from a power supply voltage of 5V and the nominal logic '0' and logic '1' levels are:

- logic '1' ~ 3.6 V
- logic '0' ~ 0.2 V

In correctly operating TTL circuits, logic voltages within the ranges shown will exist – no levels within the 'dead bands' will occur.

The most common CMOS ICs are available in the 4000 series and can be operated from a range of supply voltages from 3 V to 18 V. As a consequence the logic voltages will depend considerably on the supply voltage used. Generally, any voltage greater than about two thirds the supply will be recognised as a logic '1' and any voltage less than about one third will be taken as a logic '0'.

Progress check 16.3

For a TTL digital circuit, decide whether the following voltages will be recognised as a logic '1' or logic '0': 3.4 V; 5 V; 0.3 V; 2.5 V.

Progress check 16.4

A CMOS IC is operated from a 5 V supply. What will be the approximate voltage ranges allowed for logic '1' and logic '0'?

FAN-IN and FAN-OUT

In real logic circuits the output of a gate will be connected to the inputs of other gates. We say that the gate output 'drives the following gate inputs'. There is a limit on the number of gate inputs that can be connected (loaded) to a gate output before the logic voltage levels (and currents) fall outside of their specifications. Manufacturers describe this characteristic by using the term fan-out. For standard TTL ICs the fan-out is limited to 10 which means that up to 10 gate inputs can be driven from one gate output. For CMOS gates the fan-out is about 50. The fan-in of a logic gate is the number of inputs that can be accepted by the gate.

Operating speed

The output of any gate will not respond immediately to changes at its input. A period of time will always elapse between an input change and the corresponding output change becoming apparent. The time delay is called the **propagation delay**. It is the propagation delay that limits the speed at which a gate can operate.

All logic gates are 'fast' by everyday standards (where time is measured in minutes and seconds) and delays of nanoseconds (10^{-9} s) are typical. CMOS is the slowest at about 30 ns per gate. There are several sub-families in the TTL range with standard TTL giving around 10 ns per gate and a new variation called Fast TTL which works at about 3 ns per gate.

Power dissipation

Even the simplest digital system will be assembled from several gates, and it is obviously important to know the overall power requirements so that the power supply for the circuit can be correctly chosen.

Power dissipation is closely related to speed and generally *the faster the gate the more power it needs*. CMOS requires far less power than TTL and this is a major advantage in many applications.

Working with logic ICs

A TTL IC from the 7400 series is shown in Figure 16.8. The IC outline is called a **dual-in-line package** (DIL) because there are two sets of connection pins and

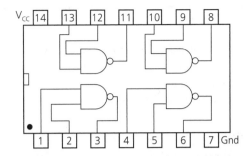

Figure 16.8 A TTL integrated circuit

each set are in-line with one another. Notice carefully how the pins are numbered anti-clockwise when looking from above. The 'dot' on the top indicates pin 1. This IC contains four NAND gates. The power supply is connected to pins 7 and 14, the remaining pins are used for input and output connections. Internal connections (not shown) are made from pins 7 and 14 to each of the four NAND gates to power them correctly.

Activity 16.1

Obtain data sheets for the 7400 series and the 4000 series of logic ICs.

Use your data sheets to identify the gates that the following chips contain:

(a) 7410   (e) 4001B
(b) 7420   (f) 4011B
(c) 7411   (g) 4023B
(d) 74LS06   (h) 4012B

Activity 16.2

Manufacturers specify their digital ICs in terms of the number of gates they contain. Find out what is meant by the terms SSI, MSI, LSI and VLSI.

Designing, building and testing combinational logic circuits

Combinational logic circuits or systems are constructed from collections of various basic gates interconnected or wired together. The output (or outputs) of the combined circuit at any instant *depends only on the combination of inputs that are applied at that time*. Past input combinations have no effect on the present output state.

The first step in any combinational circuit design is to *produce a truth table for the system requirements*. A **truth table**, you remember, gives the required state of the output for all combinations of the inputs.

WORKED EXAMPLE 16.1

Digital temperature sensors are located at three locations in a factory. If the temperature exceeds a certain level, a sensor will give a voltage level representing a '1', otherwise a '0' is produced. Design a logic circuit which will give a '1' output if any two (or all) of the sensors are at '1'.

Solution

1 *Allocate the variables.* Since there are three sensors, there are three variables and we could label them A, B and C.

2 *Draw the truth table.* For three variables there are eight input permutations or combinations ($2^3 = 8$). For each combination we decide whether the output Y is to be a '1' or '0'. The complete truth table in this case is:

A	B	C	Output Y
0	0	0	0
0	0	1	0
0	1	0	0
0	1	1	1
1	0	0	0
1	0	1	1
1	1	0	1
1	1	1	1

This demonstrates the requirement that an output '1' only occurs when two or more inputs change to a '1'.

3 *Write the Boolean equation.* There are four possibilities where Y = 1. The first is $\overline{A}BC$, the second is $A\overline{B}C$ etc. Therefore Y = 1 for: $\overline{A}BC$ OR $A\overline{B}C$ OR $A\overline{B}C$ OR ABC. So

$$Y = \overline{A}BC + A\overline{B}C + AB\overline{C} + ABC$$

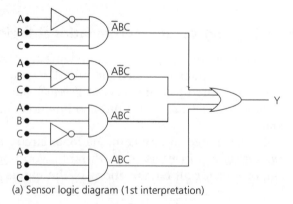

(a) Sensor logic diagram (1st interpretation)

Figure 16.9 Sensor circuit logic diagram.

4 *Draw logic diagrams and choose chips.* The circuit diagram is drawn in Figure 16.9(a). To build the circuit you would need three inverters, four 3-input AND gates and one 4-input OR gate. You could build this circuit now and providing no wiring errors are made it would work.

Practical aspects

In practice *we always try to simplify the Boolean equation* in some way. If simplification is possible it will:

- reduce the number of gates (and hence chips) required
- reduce circuit board space needed and circuit complexity
- reduce power consumption
- reduce weight (important in satellite applications for example)
- minimise cost

Working with Boolean equations

Two methods can be used to simplify Boolean equations. One uses the mathematical technique called **Boolean algebra**. The second employs a pictorial technique called **Karnaugh mapping** and is usually the best method to use in practice.

Progress check 16.5

What commercially available TTL AND, OR and inverter ICs could be used to construct the sensor circuit of Figure 16.9?

(Hint: You cannot get a 4-input OR gate. The 7432 has four 2-input OR gates. Can you make the 4-input OR gate from these?)

Simplification of Boolean equations using Boolean algebra

Mathematical simplification of Boolean equations makes use of a series of **logic rules** (or **identities**), the most important of which are:

1. $U + \overline{U} = 1$
2. $U + U = U$
3. $U \cdot U = U$
4. $U \cdot \overline{U} = 0$
5. $U + 0 = U$

6. $U + 1 = 1$
7. $U \cdot 1 = U$
8. $U \cdot 0 = 0$
9. $V \cdot (U + \overline{U}) = V$
10. $U + U \cdot V = U$

11. $U \cdot (\overline{U} + V) = UV$
12. $\overline{U + V} = \overline{U} \cdot \overline{V}$
13. $\overline{U \cdot V} = \overline{U} + \overline{V}$

Note 1: U and V can be single variables like A, B or C etc. or combinations of variables, e.g. AB, A$\overline{B}$C etc.
Note 2: Rules 12 and 13 are called De Morgan's theorem.

The validity of the rules can be checked by comparing the truth tables produced by both sides of the identity. If they are the same then the identity is true. Rules 12 and 13 (De Morgan's theorem) are probably the most widely used identities in Boolean algebra and logic generally. Please study the next worked examples.

WORKED EXAMPLE 16.2

Prove the identity $A + \overline{A} = 1$.

Solution
This is the proof of rule 1 where now U is the single variable A. We need to determine the output of an OR gate with inputs A and $\overline{A}$. The truth table is:

A	$\overline{A}$	$Y = A + \overline{A}$
0	1	1
1	0	1

The output is '1' when A is '0' or '1' and so the identity is proved.

WORKED EXAMPLE 16.3

Prove De Morgan's theorem $\overline{A+B} = \overline{A}\cdot\overline{B}$

Solution
The truth table is constructed for both sides of the equation and compared.

A	B	A+B	$\overline{A+B}$	$\overline{A}$	$\overline{B}$	$\overline{A}\cdot\overline{B}$
0	0	0	1	1	1	1
0	1	1	0	1	0	0
1	0	1	0	0	1	0
1	1	1	0	0	0	0

COMPARE

Since they are the same, then the theorem is proved.

The next worked examples show how some of the rules are used to simplify Boolean equations.

WORKED EXAMPLE 16.4

Simplify $Y = A + A \cdot B$

Solution

$Y = A \cdot 1 + A \cdot B$ (rule 7 A.1 = A)

$Y = A \cdot (1 + B)$ (take out common factor as in normal algebra)

$Y = A \cdot 1$ (rule 6 B + 1 = 1)

$Y = A$ (rule 7 A.1 = A)

(This is also a proof of rule 10 U + U.V = U)

WORKED EXAMPLE 16.5

Simplify the Boolean equation for the sensor circuit worked Example 16.1:

$Y = \overline{A}BC + A\overline{B}C + AB\overline{C} + ABC$

Solution

Remember the Boolean rules apply not only to single variables but also to combinations. We will use rule 1 ($U + \overline{U} = 1$) and rule 2 ($U + U = U$) to simplify the equation in stages:

$Y = \overline{A}BC + A\overline{B}C + AB\overline{C} + ABC + ABC$ (from rule 2: U + U = U)

$Y = BC(\overline{A} + A) + A\overline{B}C + AB\overline{C} + ABC$ (taking common factor BC from first and last term)

$Y = BC + A\overline{B}C + AB\overline{C} + ABC$ (rule 1: U + U = 1)

$Y = BC + A\overline{B}C + AB\overline{C} + ABC + ABC$ (rule 2 again)

$Y = BC + AC(B + \overline{B}) + AB(C + \overline{C})$ (take common factor)

$Y = BC + AC + AB$ (rule 1 again)

Phew! That was quite tricky. Boolean algebra takes lots of practise and it can still be quite difficult to 'see' your way through to a final solution.

Do this 16.1

Sensor circuit construction

Components needed:

- 1 × TTL (IC1) 7408
- 1 × TTL (IC2) 7432

Procedure:

1 Construct the logic diagram shown in Figure 16.10(a) by using the two components IC1 and IC2 in the circuit shown in Figure 16.10(b).

2 Take A, B and C to logic '0' and logic '1' in turn and complete the truth table. Notice carefully how a 3-input OR gate has been made from 2 × 2-input OR gates.

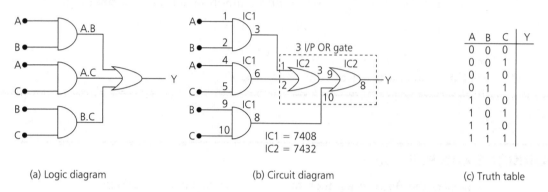

A	B	C	Y
0	0	0	
0	0	1	
0	1	0	
0	1	1	
1	0	0	
1	0	1	
1	1	0	
1	1	1	

(a) Logic diagram   (b) Circuit diagram   (c) Truth table

Figure 16.10 Sensor circuit construction

Progress check 16.6

Rather than connecting inputs A, B and C to '1', allow them to 'float', i.e. leave disconnected. What logic level does a floating TTL input behave as?

Karnaugh maps (K maps)

Karnaugh maps are used to simplify Boolean equations. Generally, it is easier to 'K-map' than to simplify an equation using Boolean algebra only. Two, three and four-variable K-maps are shown in Figure 16.11.

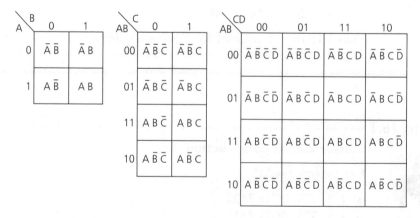

Figure 16.11 Two, three and four-variable K-maps

Each map consists of a number of **cells** with each cell representing a particular combination of the variables. Two variables can have four combinations, three variables have eight combinations and four variables have sixteen combinations.

In general, the number of cells is equal to 2^n where n is the number of variables. Notice carefully that *only one variable changes as you move horizontally or vertically between cells*. You can think of a K-map as a pictorial form of truth table. Each cell represents one line of the truth table.

K-mapping simplification is best understood by studying some worked examples.

WORKED EXAMPLE 16.6

Simplify $Y = \overline{A} B + AB$ with a two-variable K-map.

Solution

1 A '1' is placed in the correct cell for each term in the equation for Y. See Figure 16.12(a).

2 Adjacent 1's are looped to form groups of 2, 4, 8 etc. (In this example we have just one group of 2.) Now Y stays equal to '1' for this group even though A may be '0' or '1', i.e. A is **redundant** for this group.

3 For this group of 2, Y = 1 for B = 1, therefore Y = B. No further simplification is possible.

(It is, of course, easy to simplify this using Boolean algebra because $Y = B(\overline{A} + A) = B.1 = B$).

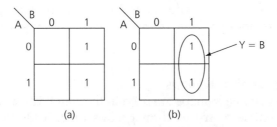

Figure 16.12 Worked example 16.6

WORKED EXAMPLE 16.7

Simplify $Y = ABC + AB\overline{C} + A\overline{B}C + \overline{A}\,\overline{B}\,\overline{C}$

Solution

1 The '1's are placed in the correct squares on the K-map. See Figure 16.13(a).
2 Two loops of 2 can be made but one term $(\overline{A}\,\overline{B}\,\overline{C})$ cannot be grouped with an adjacent '1'. Therefore:

$$Y = \overline{A}\,\overline{B}\overline{C} + AB + AC$$

is the most simplification possible.

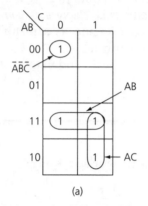 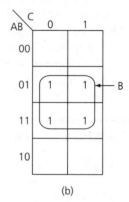

Figure 16.13 Worked examples 16.7 and 16.8

WORKED EXAMPLE 16.8

Simplify $\overline{A}\overline{B}\overline{C} + \overline{A}BC + AB\overline{C} + ABC$.

Solution
1   A group of 4 can be made on this map. See Figure 16.13(b).

For this group Y stays at '1' and B stays at '1', even though both A and C change from '0' to '1'. Thus A and C are redundant and Y = B.

4-variable K-maps
4-variable K-maps are used in the same way as 2 and 3-variable maps. The requirement is to try to form groups of 2, 4, 8, 16. Forming the largest groups will produce the greatest simplification.

WORKED EXAMPLE 16.9

Using the 4-variable map, simplify $Y = \overline{A}\,\overline{B}\overline{C}\overline{D} + \overline{A}\,\overline{B}\overline{C}D + ABCD + ABC\overline{D}$.

Solution
The map and loopings are shown in Figure 16.14(a).

Two groups of two can be formed, so:

$$Y = \overline{A}\,\overline{B}\overline{C} + ABC$$

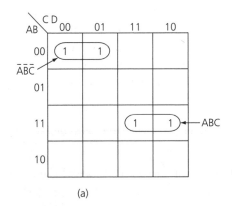
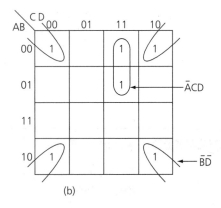

Figure 16.14 K-map for Worked example 16.9 and 16.10

WORKED EXAMPLE 16.10

Using the 4-variable map, simplify $Y = A\overline{B}C\overline{D} + \overline{A}\,\overline{B}CD + \overline{A}BCD + \overline{A}\,\overline{B}C\overline{D} + \overline{A}\,\overline{B}\,\overline{C}\,\overline{D} + A\overline{B}\,\overline{C}\,\overline{D}$.

Solution

The map and the loopings are shown in Figure 16.14.

Two groups can be formed so:

$$Y = \overline{A}CD + \overline{B}\,\overline{D}$$

Notice carefully the corner looping.

Universal gates

You have already seen that an inverter can be created simply by tying all the inputs of a NAND (or NOR) together. In fact, *all the logic gates can be made just by using NAND gates or NOR gates only*. For this reason, NAND and NOR gates are often called UNIVERSAL gates.

The use of NAND and NOR to construct the basic gates is shown in Figure 16.15. Notice how De Morgan's theorem is used to prove the equivalence for the AND and OR.

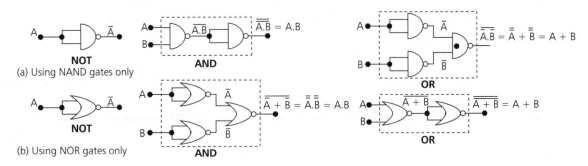

Figure 16.15 Universal gates

309

At first sight, it might appear that this approach is 'wasteful' in gates. In real circuit design, however, it is often found that the gates used as inverters cancel each other out. Additionally, from a cost point of view it means that you only need to stock either NAND gate chips or NOR gate chips to construct all of your combinational logic circuits – not a complete range of logic chips.

WORKED EXAMPLE 16.11

Draw the logic diagram for the sensor circuit (Worked Example 16.1) using NAND gates only.

Solution
The final logic diagram for the sensor circuit is redrawn in Figure 16.16(a). As shown in Figure 16.16(b), each gate is replaced by its NAND equivalent. Notice how the inverters 'cancel out'.

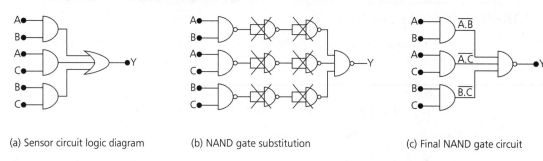

(a) Sensor circuit logic diagram   (b) NAND gate substitution   (c) Final NAND gate circuit

Figure 16.16 Worked example 16.11

For complicated logic designs this technique of gate replacement and 'inverter' removal can be quite tedious. *There is a simpler method which involves De Morgan's Theorem.* You do not even have to draw the logic diagram using AND, OR and NOT. The Boolean equation for Y is:

$$Y = AB + AC + BC$$

If we *invert Y twice*, nothing will change, i.e. we still have Y:

$$Y = \overline{\overline{Y}} = \overline{\overline{AB + AC + BC}}$$

Splitting the first inversion gives:

$$Y = \overline{\overline{AB} \cdot \overline{AC} \cdot \overline{BC}}$$

Now Y is entirely in NAND gate form. That's all there is to it and we can draw the final NAND gate diagram as in Figure 16.16(c).

WORKED EXAMPLE 16.12

Sensor circuit of Worked Example 16.1 with just NOR gates.

Solution
As for the NAND implementation, each AND and OR gate could be replaced by its NOR equivalent. However, the simplest method is to use De Morgan's theorem again.

For the sensor circuit $Y = AB + AC + BC$ and if we invert each term twice nothing will change, i.e. $\overline{\overline{AB}}$ is still AB. Therefore:

$$Y = \overline{\overline{AB}} + \overline{\overline{AC}} + \overline{\overline{BC}}$$

Applying De Morgan's theorem to 'split the lower line' gives:

$$Y = \overline{\overline{A} + \overline{B}} + \overline{\overline{A} + \overline{C}} + \overline{\overline{B} + \overline{C}}$$

Each term is now a NOR; but these are combined by an OR. If now Y is inverted, then $\overline{Y}$ is entirely in NOR form. Putting an inverter on the output of the circuit will invert $\overline{Y}$ to give Y as required. The NOR version of the sensor circuit can now be drawn as in Figure 16.17. It's a little more complex than the NAND.

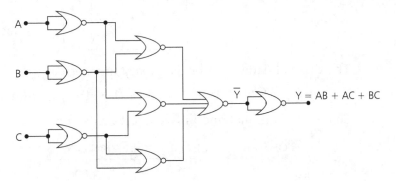

Figure 16.17 NOR version of the sensor circuit (Worked example 16.12)

Before you attempt the next case study, a brief revision of decimal and binary numbers is needed.

Decimal and binary numbers

Because humans have four fingers and one thumb on each hand, it is natural that a method of counting using ten symbols has evolved. We say that our **decimal system** has a **base of ten**. The symbols (or digits) chosen are:

9, 8, 7, 6, 5, 4, 3, 2, 1, 0

To express quantities greater than 9, collections of the symbols are used with each symbol having a greater 'weight' as we move from right to left. Thus, for example, the number 137 is

	HUNDREDS	TENS	UNITS

$$(1 \times 100) \quad + \quad (3 \times 10) \quad + \quad (7 \times 1)$$

$$\text{or} \quad (1 \times 10^2) \quad + \quad (3 \times 10^1) \quad + \quad (7 \times 10^0)$$

The number 7 has a weight of 1 (10^0) and is called the **least significant digit**. In this example, 1 is the **most significant digit** and has a weight of 100 (10^2).

The binary system has only two symbols but provides another way of being able to count. The symbols chosen are '0' and '1' and are called **bits**. The binary system has a **base of two**. Since logic gates are two-state devices, they are able to process binary numbers and can be assembled to carry out all of the normal mathematical operations. As an example, a computer performs all its operations in binary numbers and at the basic level is constructed from large quantities of logic gates described above. This is why digital circuits are so important.

Comparing with decimal, the position of the '1's and '0's in a binary number gives its weight (in powers of 2) as we move from right to left. Thus, for example, the binary number 1011 is equivalent to the decimal number 11 as follows:

most significant bit →	1	0	1	1	← least significant bit
	2^3	2^2	2^1	2^0	← binary weight
	8	4	2	1	← decimal equivalent weight

The decimal equivalent of the binary number is:

$$(1 \times 8) + (0 \times 4) + (1 \times 2) + (1 \times 1) = 11 \text{ decimal}$$

Some other decimal and binary equivalents are:

COUNT	DECIMAL	BINARY
ZERO	0	0000
ONE	1	0001
TWO	2	0010
THREE	3	0011
FOUR	4	0100
FIVE	5	0101
SIX	6	0110
SEVEN	7	0111
EIGHT	8	1000
NINE	9	1001
TEN	10	1010
ELEVEN	11	1011
TWELVE	12	1100
THIRTEEN	13	1101
FOURTEEN	14	1110
FIFTEEN	15	1111
SIXTEEN	16	10000
SEVENTEEN	17	10001
	etc.	

The next case study describes the design of the basic binary addition circuit called a HALF ADDER. It can add together two binary bits.

Case study

Design a circuit which will add two binary bits together to produce the **sum** and the **carry digits**.

Produce circuit diagrams using (i) AND, OR and NOT gates, (ii) an exclusive OR gate and (iii) just NAND gates.

The rules for binary additions are as follows:

> 0 plus 0 = 0
> 0 plus 1 = 1
> 1 plus 1 = 0 with a carry of 1

For example, '2 plus 2' in binary is:-

Binary		Decimal
10	=	2
10	=	2
100	=	4

The answer 100 is the binary sum with a carry digit '1'

1 *Produce the truth table.* We'll label the two binary inputs A and B and the two outputs that the circuit has to have as sum and carry (Figure 16.18(a)). The truth table is shown in Figure 16.18(b).

2 *Use K-maps to try and simplify.* The Boolean equations for the SUM and CARRY are:

$$SUM = \overline{A} \cdot B + A \cdot \overline{B} \quad CARRY = A \cdot B$$

The K-maps are drawn in Figure 16.18(c). No loops can be made – no further simplification is possible.

3 *Draw circuit diagram using AND, OR and NOT gates.* As shown in Figure 16.18(d), three ICs are needed.

4 *Using an exclusive OR.* Study the sum part of the truth table carefully. This is the exclusive OR function. A 7486 chip contains four 2-input EX-OR gates and simplifies our design. We now only need two chips and it could give us an edge over our competitors! The EX-OR implementation is shown in Figure 16.18(e).

5 *Obtaining a NAND circuit diagram.* De Morgan's theorem and a 'trick' with the K-map will help us here. So far we have only put '1's on a K-map for when Y = 1. Obviously, however, when Y is not '1' it must be '0'! Therefore, we could put '0's in the spaces and looping the '0's will give a $\overline{Y}$. Using the map in Figure 16.18(f) gives:

$$\overline{Sum} = \overline{A}\ \overline{B} + AB = \overline{\overline{A}\ \overline{B} + AB}$$

Inverting gives:

$$\text{Sum} = \overline{\overline{A} \cdot \overline{B} + A \cdot B}$$

Splitting the line with De Morgan's gives:

$$\text{Sum} = \overline{\overline{A}\ \overline{B}} \cdot \overline{AB}$$

De Morgan's on the first term gives us:

$$\text{Sum} = (\overline{\overline{A}} + \overline{\overline{B}}) \cdot \overline{AB} = (A + B) \cdot \overline{AB} = A \cdot \overline{AB} + B\overline{AB}$$

If we 'double invert', nothing changes:

$$\text{Sum} = \overline{\overline{A \cdot \overline{AB} + B \cdot \overline{AB}}}$$

'Splitting the inner line' gives the sum in NAND gate form:

$$\text{Sum} = \overline{(A \cdot \overline{AB}) \cdot (B \cdot \overline{AB})}$$

Double inverting the CARRY equation gives:

$$\text{CARRY} = \overline{\overline{A \cdot B}}$$

The final NAND circuit is drawn in Figure 16.18(f).

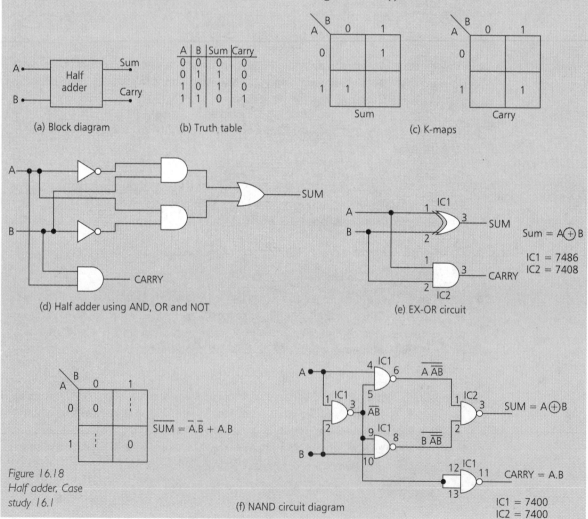

Figure 16.18
Half adder, Case
study 16.1

(a) Block diagram (b) Truth table (c) K-maps

(d) Half adder using AND, OR and NOT (e) EX-OR circuit

(f) NAND circuit diagram

Binary coded decimal (BCD)

If you look at the previous table, you can see how decimal numbers can be represented by binary bits.

Four bits are needed to represent the decimal digits 0 to 9. Even though a digital system (computer etc.) operates only in binary, whenever a result has to be indicated to us (or a number has to be input by us), decimal digits are used. It would be very confusing if the time on your digital watch was displayed in binary code, wouldn't it?

When four bits are used to represent a decimal number from zero to nine, the system is called **binary coded decimal** (BCD for short). Referring to the table shows, however, that 4 bits can have 16 combinations. BCD code is restricted to the range 0000 to 1001. The other six combinations (1010 up to 1111) cannot occur under normal operating conditions. It is an interesting exercise to design a logic circuit which would check whether a particular combination of 4 bits is a legal or illegal BCD code. You might be given the specification for the circuit something like the following Case study.

Case study

Design a **BCD code checker** circuit to show whether a particular BCD code is allowed or disallowed. If the code is allowed, the output of the circuit should be '1'. If the code is illegal, the output of the circuit should be '0'. Use TTL NAND gates to implement the final design.

1 *Draw the truth table* as follows.

BCD CODE D C B A	BOOLEAN EXPRESSION	OUTPUT OF CIRCUIT (Y)
0 0 0 0	$\bar{D}\,\bar{C}\,\bar{B}\,\bar{A}$	1
0 0 0 1	$\bar{D}\,\bar{C}\,\bar{B}\,A$	1
0 0 1 0	$\bar{D}\,\bar{C}\,B\,\bar{A}$	1
0 0 1 1	$\bar{D}\,\bar{C}\,B\,A$	1
0 1 0 0	$\bar{D}\,C\,\bar{B}\,\bar{A}$	1
0 1 0 1	$\bar{D}\,C\,\bar{B}\,A$	1
0 1 1 0	$\bar{D}\,C\,B\,\bar{A}$	1
0 1 1 1	$\bar{D}\,C\,B\,A$	1
1 0 0 0	$D\,\bar{C}\,\bar{B}\,\bar{A}$	1
1 0 0 1	$D\,\bar{C}\,\bar{B}\,A$	1
1 0 1 0	$D\,\bar{C}\,B\,\bar{A}$	0
1 0 1 1	$D\,\bar{C}\,B\,A$	0
1 1 0 0	$D\,C\,\bar{B}\,\bar{A}$	0
1 1 0 1	$D\,C\,\bar{B}\,A$	0
1 1 1 0	$D\,C\,B\,\bar{A}$	0
1 1 1 1	$D\,C\,B\,A$	0

2 *Draw the K-map and simplify.* The K-map can be drawn directly from the truth table and looped, as shown in Figure 16.19. Make sure you understand the 'edge' loopings.

3 *NAND implementation.* The simplified expression for Y is

$$Y = \overline{B}\,\overline{C} + \overline{D}$$

Inverting twice gives:

$$Y = \overline{\overline{Y}} = \overline{\overline{\overline{B}\,\overline{C} + \overline{D}}}$$

Splitting the inner line with De Morgan's theorem gives:

$$Y = \overline{\overline{\overline{B}\,\overline{C}} \cdot \overline{\overline{D}}} = \overline{\overline{\overline{B}\,\overline{C}} \cdot D}$$

Y is now in NAND gate form. The circuit diagram in Figure 16.19 shows that only a single chip is required – a 7400 quad 2-input NAND gate.

Construct the circuit and check its operation against the original truth table.

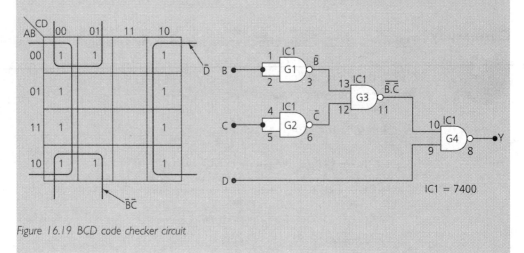

Figure 16.19 BCD code checker circuit

Sequential logic circuits

So far, only combinational logic circuits have been investigated. Their output(s) only depend on the combination of inputs applied at that instant. Sequential logic circuits have output(s) whose value is controlled not only by present input states but also by *past input conditions*. These circuits are still constructed from the basic logic gates, but because of in-built feedback connections, they have the ability to 'remember'. The basic sequential logic elements are called **latches** and **flip-flops**. Their operation is, as usual, described by a truth table.

Latches and flip-flops

Latches and flip-flops are **bistable circuits**. This means that the output (called Q) changes to and remains in one of two possible states. These states are called **set** and **reset**. The circuit symbols and truth tables for two bistables are shown in Figure 16.20.

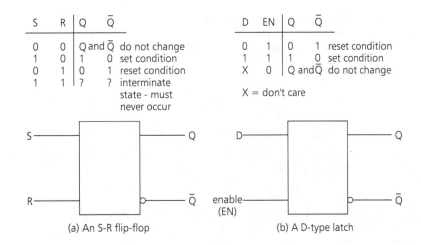

Figure 16.20 Two bistables

Notice that the output of the circuit is called Q. Manufacturers also make the inverse of Q ($\overline{Q}$) available on another pin of the IC as well. Note the '0' symbol. The truth tables are 'read' as follows. Consider the **S-R flip-flop**. When both S and R are at '0', the output Q will remain in its previous state (either '1' or '0'). When S is momentarily taken to '1', Q will SET to '1'. Q remains at '1' until R is taken to '1'. When this happens, Q will reset to '0'. The condition R = S = '1' is an invalid or disallowed condition and must be avoided in any R-S flip-flop application. In practice, S-R flip-flops are rarely used on their own; generally they are incorporated as internal sub-systems of more complicated digital IC's.

The **D-type latch** is common in digital systems, (we'll use it in a case study). The D stands for data and is so called because the value of D is stored at Q when the enable line (EN) is activated. Thus when EN = '1', whatever value is applied to D appears at Q. If the EN is inactive ('0' for the latch in Figure 16.20(b)) Q remains the same and is unaffected by changes on the D-input.

Activity 16.3

Obtain a TTL data book and locate the data for a 7475 and 74LS75 IC. Briefly describe the operation of the IC.

Progress check 16.7

The waveforms shown in Figure 16.21 are applied to the data and enable pins of one of the latches in a 7475 latch. If Q is initially reset, complete the Q waveform sketch.

Figure 16.21 Progress check 16.7

Edge-triggered flip-flops

Flip-flops differ from latches in that they have two data inputs. The output Q will only change state at a particular point on a **triggering input signal** called the **clock**.

The term **edge-triggered** is used to describe those flip-flops whose output can only change on the rising and falling edge of the clock signal. Both positive and negative edge-triggered flip-flops are commercially available.

The circuit symbols and truth tables for the most important edge-triggered devices are illustrated in Figure 16.22. In the diagrams, the label 'CP' is the **clock pulse input**.

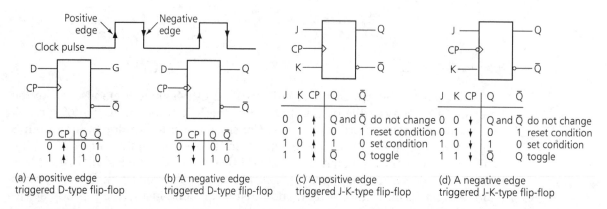

Figure 16.22 Some important edge-triggered flip-flops

Study the diagrams carefully and note the following:

- the wedge symbol on the clock input generally indicates edge triggering
- negative edge triggering is indicated by the sign 'O' in front of the wedge
- the arrow in the truth tables shows which edge is the active one.

Of all the flip-flops, the **J-K flip-flop** is the most useful and widely used. The first 3 lines of the truth table show that the device functions in the same way as an S-R flip-flop. The final line is the condition that is disallowed for an S-R flip-flop. When operated with J = K = '1' the flip-flop is said to **toggle**. Thus if Q = '0' then when the active clock edges arrives Q sets to '1'. The next active edge will reset Q to '0' again and so on. It is this toggling facility that makes the J-K so useful and versatile.

WORKED EXAMPLE 16.13

The waveforms shown in Figure 16.23 are applied to a negative edge-triggered J-K flip-flop whose Q output is initially RESET. Draw the waveform for Q.

Solution

1 Since the flip-flops are negative edge-triggered any changes will only occur on that edge.

2 At the end of the first clock pulse, J = '1' so Q will set to '1'.

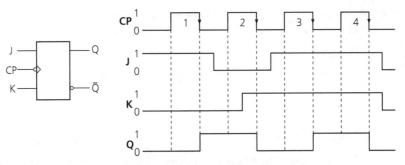

Figure 16.23 Waveforms for Worked example 16.13

3   At the end of the second pulse, J = '0', K = '1' so Q will reset to '0'.

4   At clock edge 3, J = K = '1' so the flip-flop will toggle and Q sets to '1'.

5   When the fourth active clock edge arrives, J and K are still equal to '1' so the flip-flop toggles again and Q will reset to '0'.

Progress check 16.8

The clock waveform shown in Figure 16.24 is applied to a negative-edge triggering J-K flip-flop which has J and K permanently wired to '1'. Assuming that Q is initially RESET, draw the output waveform. Can you suggest a use for the J-K flip-flop connected in this way?

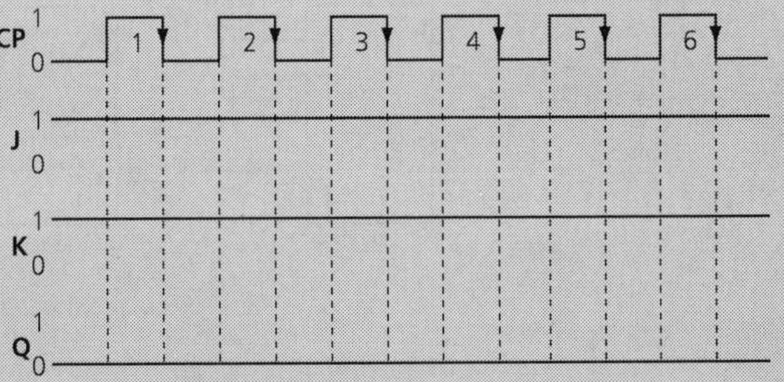

Figure 16.24 Waveforms for Progress check 16.8

Preset and clear inputs

Most commercial flip-flops have one or two other inputs which override the clock and the J-K inputs and allow the output to be set or reset at will. Such inputs are called **asynchronous** because they do not have to be synchronised with the clock to be effective.

In the maufacturer's data, these inputs are labelled S_D for **direct set** and R_D for **direct reset**. Some producers, however, use the terms PRE for **preset** and CLR for **clear**.

Activity 16.4

Use a TTL data book to obtain information for 7476 and 74LS76 J-K flip-flops. Produce a report highlighting the basic differences between the two devices.

Counter circuits

The necessity to have circuits with the ability to count is of supreme importance in many digital systems. Counters are collections of flip-flops arranged in a series so that the clock pulses cause the outputs of one or more of the flip-flops to set and reset others in a binary pattern related to the number of pulses applied.

Counters are classified according to the way in which they are clocked. In **asynchronous counters**, the flip-flops are wired in cascade so that the clock input of the first flip-flop is driven by the pulses to be counted but subsequent flip-flops have their clock inputs driven by the Q (or $\overline{Q}$ output) of the previous flip-flop. Consequently, the flip-flops do not change state at exactly the same time and for this reason the counters are called asynchronous.

With **synchronous counters** the pulses to be counted drive the *clock inputs of all the flip-flops simultaneously*. As a result of this, if the flip-flops change state they will do so at the same time under the control of the applied pulses.

At one time counters were constructed by wiring individual flip-flops together. This was a fairly laborious and error-prone exercise and nowadays a wide range

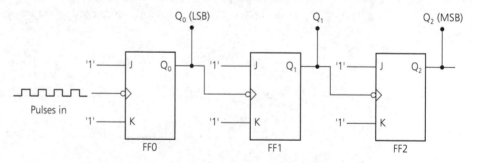

(a) 3-bit counter logic diagram

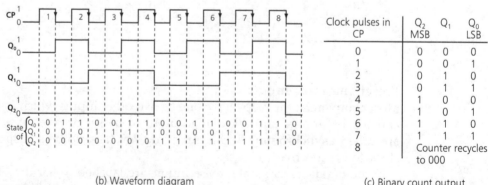

Clock pulses in CP	Q_2 MSB	Q_1	Q_0 LSB
0	0	0	0
1	0	0	1
2	0	1	0
3	0	1	1
4	1	0	0
5	1	0	1
6	1	1	0
7	1	1	1
8	Counter recycles to 000		

(b) Waveform diagram

(c) Binary count output

Figure 16.25 A 3-bit binary counter

of pre-wired medium-scale integration (MSI) counter circuits are available off-the-shelf. In your practical work you will use some of these devices. We will start off, however, by describing an asynchronous counter made with three flip-flops. It illustrates most of the important facts about asynchronous counters and their operation.

The 3-Bit asynchronous counter

When a J-K flip-flop is connected in the toggle mode, two pulses at the clock input generate one complete pulse at the Q (or $\bar{Q}$) output. (You would have spotted this if you did Progress Check 16.8 correctly.) It is this observation that forms the basis of asynchronous counter design.

A 3-bit counter can be constructed as in Figure 16.25(a). It is called a 3-bit because it uses 3 flip-flops and has 3 outputs. This one uses negative edge-triggered flip-flops. Notice how J and K are tied to '1'.

The counter works like this:

- Assume that all flip-flops are initially reset ($Q_0 = Q_1 = Q_2 = $ '0').
- On the negative edge of the first pulse, Q_0 will set to '1'. This '1' is connected to the clock input of FFB.
- At the end of the second pulse, FF0 will reset and FF1 will set.
- On the negative edge of the third pulse, FF0 sets again and FF1 remains set.
- On the negative edge of the eighth pulse, all the flip-flops reset and the whole cycle starts all over again.

Please note the following important points:

1 This counter uses three flip-flops and has eight distinct states (i.e. $Q_0 = $ '0', $Q_1 = $ '0', $Q_2 = $ '0' up to $Q_0 = $ '1', $Q_1 = $ '1', $Q_2 = $ '1'). *The number of states is often called the modulus of the counter.* Generally, if 'n' flip-flops are connected in this way, then:

$$\text{number of states} = 2^n$$

Thus, for example, if n = 5 the five bit counter would have a modulus of 32.

2 Each flip-flop *divides the frequency of the incoming pulses by a factor of two*:

One flip-flop divides by two

Two flip-flops divide by four

Three flip-flops divide by eight

One common use therefore of a counter is as a **frequency divider**.

As shown in Figure 16.25(c), if Q_0 is taken as the least significant bit, Q_0, Q_1 and Q_2 produce a *binary count of the input pulses*, i.e. the counter cycles from 000 to 111 and then recycles. The counter cycles through eight distinct states. It is called a **divide-by-eight counter**.

Activity 16.5

Use a TTL data book to obtain information about the 7490, 7492 and 7493 series of counter chips. Briefly compare their logic diagrams, their count sequences and the modes in which they can operate.

Case study

In a factory, finished articles move one after another along a conveyor belt. The articles pass a sensor which produces a pulse as each moves by. A pulse is generated every two seconds. Design a **decade counter circuit** which will count these pulses in batches of ten and display the count on a seven-segment LED display (i.e. the display will show 0 to 9 continuously). Add a facility to 'freeze' the display while the count continues.

1 *Choosing the display.* Seven-segment LED displays can be of the common anode or common cathode type. For the common anode, the anodes of the seven LEDs are tied together internally and this common connection needs to be connected to a positive voltage (V_{cc} = 5 V for our circuit). A segment is 'turned on' by taking its cathode LOW. A resistor needs to be placed in series with each LED to limit the current to a safe value – 270R resistors are sufficient. The pinout for the display is shown in Figure 16.26(a).

2 *Driving the display.* The display needs an interface circuit between itself and the logic system driving it. This interface is called a **decoder/driver** and performs two vital functions. It must:

● convert the binary input signals used in the logic system into the correct codes required by the segments
● be able to 'sink' current from each LED when it is turned on.

The counter needs to have ten states (0 to 9) and produce the BCD codes from 0000 up to 1001. We need a standard decoder/driver to convert from BCD to the codes required to drive the seven segments. An MSI combinational logic device with type number 7447 is available in the TTL range. The complete display section is drawn in Figure 16.26(b).

3 *Choosing the counter.* The 7490 is a dedicated **decade counter** (0 to 9). It has separate 'divide by 2' and 'divide by 5' sections. For a decade counter Q_0 (pin 12) must be externally connected to pin 1 so that the two sections are connected in series.

4 *Checking circuit operation.* When the circuit is complete, pulses need to be applied to the counter to check that it operates correctly. Many circuits can be used as **pulse generators**. A 555 timer chip and two resistors and a capacitor is an easy and popular way of producing pulses. The circuit is shown

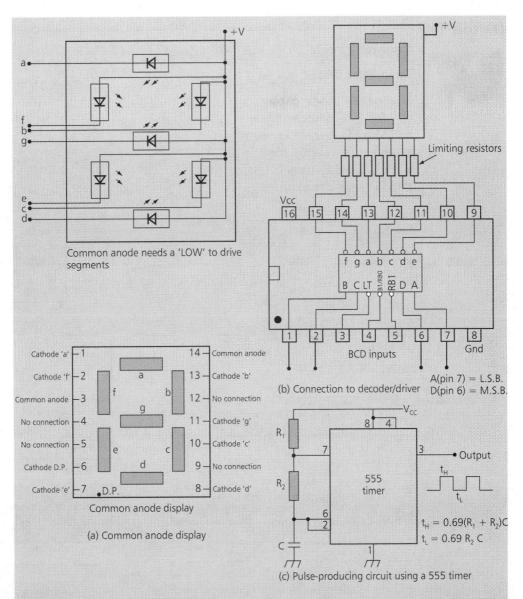

Common anode needs a 'LOW' to drive segments

Limiting resistors

Vcc

(b) Connection to decoder/driver

BCD inputs

A(pin 7) = L.S.B.
D(pin 6) = M.S.B.

(a) Common anode display

Cathode 'a' — 1   14 — Common anode
Cathode 'f' — 2   13 — Cathode 'b'
Common anode — 3   12 — No connection
No connection — 4   11 — Cathode 'g'
No connection — 5   10 — Cathode 'c'
Cathode D.P. — 6   9 — No connection
Cathode 'e' — 7   8 — Cathode 'd'

Common anode display

(c) Pulse-producing circuit using a 555 timer

$t_H = 0.69(R_1 + R_2)C$

$t_L = 0.69\,R_2\,C$

Figure 16.26 Display system

in Figure 16.26(c). R_1, R_2 and C determine the time for which the output from pin 3 is HIGH (t_H) and LOW (t_L) using the equations shown.

We need a negative edge every two seconds to increment the 7490 counter. Choosing $t_H = t_L = 1$ s will achieve this. To make t_H very nearly equal to t_L means R_2 must be much bigger than R_1. For $R_2 \gg R_1$ then $t_H \approx 0.69\,R_2C$. Choosing C = 10 μF gives:

$$t_H = 1 = 0.69 \times R_2 \times 10 \times 10^{-6}$$

$$R_2 = 144.9\text{K (preferred value 160K)}$$

R_1 can be 1K or 2K.

The final circuit with 'freeze' facility is constructed in the next practical activity.

323

Do this 16.3

Counter circuit

Components needed:

- common anode display
- 7447 TTL decoder/driver
- 7490 TTL decade counter
- 16-pin dil resistor module (270R)
- 7475 quad D-type TTL latch
- 555 timer circuit

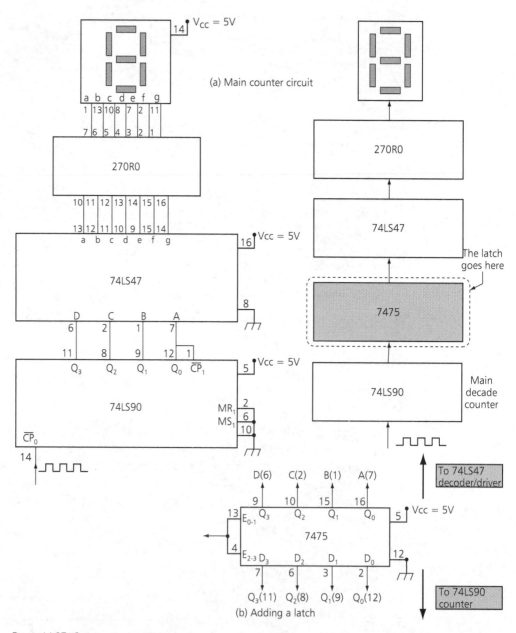

Figure 16.27 Counter circuit with latch

- 160K resistor
- 2K resistor
- 10 μF electrolytic capacitor

Procedure:

1 Assemble the circuit in Figure 16.27(a). (Leave space between 7447 and 7490 to add another chip later.)
2 Connect the 555 timer output to the clock input of the 7490. The count should continuously cycle.
3 *Freezing the display* is called **latching**. A 7475 quad D-type latch can be used. When the enable pins are at '1', data into the latch appears at the outputs and the display changes with the counter. If at '0', the outputs will not change – they are latched. Figure 16.27(b) shows the wiring of the latch. Put a flying lead off pins 13 and 4 and check operation.

Shift registers

Since the output of a flip-flop can be set to '1' or reset to '0', then it can act as a 1-bit memory device. If 'n' flip-flops are used then an n-bit binary word can be stored. Such a device is called a register. If the data in the register can be 'moved' then it is called a shift register.

A familiar example of a shift register application is the display system on your pocket calculator. As you enter digits in from the keyboard, the numbers previously entered are shifted to the left. When you have finished data entry, the register holds the data temporarily until further operations are performed.

Shift registers are available in a variety of MSI packages with various storage capabilities. Some basic register devices and operations are now described.

A serial-in-serial-out register (SISO)
The simplest shift register accepts data serially on a single input and, under the control of clock pulses, shifts the data in and subsequently out on a single output. A 4-bit SISO using D-type, positive edge-triggered flip-flops is illustrated in Figure 16.28. Notice that the Q output of the flip-flops feeds the data input of the following flip-flops.

Assume that the register is initially cleared and the data to be applied is 1000 with the MSB first. The sequence of events is illustrated in Figure 16.28(b).

1 On the positive edge of CP1, the '1' on the data line will be clocked into FF0.
2 When CP2 positive edge arrives, the data input is '0'. This is clocked into FF0. At the same time the '1' at the input of FF1 is clocked into it.
3 The sequence of events continues until, after the leading edge of CP4, the register contains 0001. Another four positive clock edges would totally move the original 1000 data word out of the register.

Other shift registers
Shift registers are available with various bit lengths, loading methods and shifting capabilities. The other important types are illustrated in Figure 16.29.

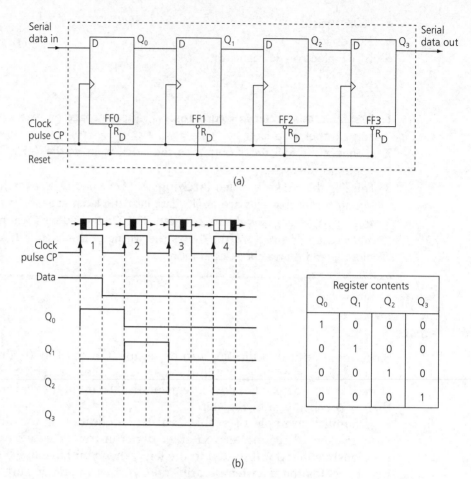

Figure 16.28 Serial-in-serial-out register (SISO)

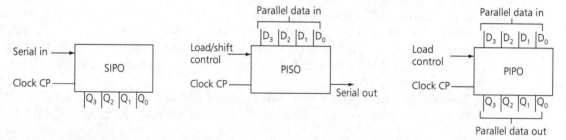

Figure 16.29 Other types of registers

In the **serial-in-parallel-out register** (SIPO), data is entered serially under control of the clock into the register and appears on the outputs of each internal flip-flop. For the 4-bit shown, the serial input information would appear on Q_0 to Q_3 after four clock pulses. The **parallel-in-serial-out register** (PISO) uses a **load/shift control** and the clock to enter data into it in parallel. The data is then shifted out serially under clock pulse control. In the **parallel-in-parallel-out register** (PIPO), data is entered in parallel from the data inputs. This data then appears in parallel on the outputs.

Activity 16.6

Use a TTL data book to obtain information about the 7495 4-bit shift register. Produce a brief report describing the facilities it provides.

Tri-state outputs

In microprocessor-based systems and computers generally, several devices such as memory and input/output chips share common communication pathways known as **buses**. A bus is simply a collection of copper tracks on a printed circuit board and data moves from chip-to-chip via the bus. Only one device should have access to the bus at any one time and to achieve this they are manufactured with what are known as three-state or **tri-state outputs**. Figure 16.30(a) shows the circuit symbol for a tri-state buffer.

Input ▷ Output   '1' ▷ '1'   '0' ▷ '0'   '1' or '0' ▷ High impedance

Enable   '1'   '1'   '0'

     (a)   (b)

Figure 16.30 A tri-state buffer

As well as an input and output, there is a control line known as the **enable**. The one shown has an *active high enable*. When the enable is active ('1'), the output will follow the input. The '0' and '1' states are the normal TTL levels. When the enable is not active ('0' in this case) the buffer operates in a third state known as the *high impedance state*, in which the buffer acts like an open switch and the output is completely disconnected from the input.

Many devices such as RAM and ROM chips etc., equipped with tri-state outputs, can be connected to a common bus. Enabling the output buffers of only one chip at a time will ensure that only that chip has access to the bus.

Assignment 16

This assignment provides evidence for:
Element 13.3: Investigate logic elements and digital circuits
and the following key skills:
Communication 3.2, 3.3
Application of Number 3.1, 3.2, 3.3

The evidence indicators for this element require you to provide a record of solutions to a given combinational logic circuit problem. You should include:

(a) A description of the logic elements used with truth tables.
(b) Simplification using Boolean algebra and K-mapping.
(c) Simplification for NAND gate or NOR gate realisation.
(d) Circuit diagrams of the final circuits.
(e) A record of tests carried out on the final circuits.

Report

Produce a report on a sequential logic circuit (a counter, shift register or register). Your report should include:

- explanations of the sequential logic elements used with truth tables and internal gate design.
- a description of the circuit operations with timing diagrams
- a circuit diagram of the circuit
- a record of tests carried out on the final circuit.

General notes should also be included for other sequential logic circuits not included in your report.

PART SIX: ENGINEERING INSTRUMENTATION AND CONTROL

Chapter 17: Selecting and testing measurement systems
Chapter 18: Engineering control systems
Chapter 19: Final control elements in engineering control
systems

- Modern industrial manufacturing processes require ever more complex systems to carry out accurate measurements and for control. In some cases extremely accurate positioning is required and this can only be achieved using a properly tuned control system. It is therefore important to gain an understanding of measurement and control techniques and to be able to specify and select suitable systems and devices.
- Part six introduces basic measurement techniques, simple control theory and investigates final control elements.

Chapter 17: Selecting and testing measurement systems

Measurement systems are used in a wide range of applications in laboratories, on the shop floor and in the field. They are often used for collecting and recording data on a wide variety of measurements such as temperature, pressure, flow rate, speed and displacement. It is important to understand the basic requirements for such a system and the different types of sensor used for the measurements. Most modern measurement systems make use of computers to record, analyse and display measured data. It is therefore useful to have an understanding of how to apply computer technology in this context.

When you complete this element you should be able to:

- describe and illustrate the functional elements of engineering measurement systems
- perform maintenance and test procedures on engineering measurement systems
- record maintenance and test data
- evaluate the performance parameters of engineering measurement systems using data.

System requirements

A typical measurement system is constructed from a number of sub-systems:

- sensing
- signal processing
- data acquisition/recording
- calibration/test.

Sensing

The first sub-system contains the device or devices used to actually make the measurement. The devices used are called **sensors** or **transducers** and convert one type of physical phenomenon, such as temperature, pressure or strain into another. Most of the commonly used sensors convert physical quantities into electrical quantities, such as voltage, current or resistance.

Signal processing

The second sub-system is associated with signal processing or signal conditioning. The output from a typical sensor may be in a form which cannot be directly processed such as resistance. The sensor output would therefore need to be converted into say a voltage. On the other hand the output from the sensor may be in the correct form but too small and would therefore require amplification. Conversion and amplification are examples of signal processing or conditioning and is one of the most important parts of any measurement system.

Recording/data acquisition

The third sub-system is associated with the recording, storing, displaying and analysing the measured data. In a simple system this sub-system may be just a display device indicating say temperature in degrees centigrade or in a more complex system it may be a computer which can display, store, record and analyse the measured data.

Calibration/test

The final sub-system is related to testing and calibration and will usually consist of what are called sensor simulators. This is usually a box containing very accurately calibrated voltage, current and resistance sources which are used in place of the actual sensors to calibrate and test the measurement system. In addition to this the sensors themselves will need to be calibrated. This is much more difficult because accurate sources of temperature and pressure etc. are required. This is often done by sending the sensors back to the original manufacturers on a regular basis for calibration.

Progress check 17.1

State the purpose of the sensing element in a typical measurement system.
Why is the signal processing/conditioning sub-system so important?
With reference to the recording/data acquisition sub-system what is the main advantage of using a computer?
Why does a measurement system need to be calibrated?

Measurements

Prior to investigating the various component parts of a typical measurement system we need to give some consideration to the actual physical quantities that need to be measured.

Temperature

This is a measure of the level of heat produced by a particular process and is typically measured using one of the following scales of temperature:

- Centigrade
- Fahrenheit
- Kelvin.

Centigrade
This scale is probably the most commonly used and has a range defined by 0° representing the freezing point of water and 100° representing the boiling point of water.

Fahrenheit
This scale was very popular at one time but has largely been superseded by the use of the Centigrade scale. It has a range defined by 32° representing the freezing point of water and 212° representing the boiling point of water.

Kelvin
This scale is popular in scientific circles with 0° representing absolute zero (the temperature at which all molecular movement ceases) and 273° representing our normal freezing point of water. It has a common relationship with the centigrade scale in that 1°C = 1°K.

Temperature scale conversions
It is possible to convert from one scale to another by applying a simple formula as follows:

- Given a temperature in centigrade convert to Fahrenheit using:

$$T°F = 9/5 \ T°C + 32°$$

- Given a temperature in Fahrenheit convert to centigrade using:

$$T°C = 5/9(T°F - 32°)$$

- Given a temperature in centigrade convert to Kelvin:

$$T°K = T°C + 273°$$

Progress check 17.2

Which two temperature scales are most commonly used?
Convert the following temperatures:

- 20°C to degrees Fahrenheit
- 100°K to degrees centigrade
- −10°C to degrees Kelvin

Pressure

The pressure acting on a surface is defined as the perpendicular force per unit area of surface. The unit of pressure is the pascal, (Pa), where 1 pascal is equal to 1 newton per square metre. Thus:

Pressure, $P = F/A$ (pascals)

where F is force in newtons acting at right angles to a surface of area A square metres.

WORKED EXAMPLE 17.1

As an example if a force of 20 N acts on an area of 4 m², then the pressure on the area is given by:

$P = 20/4 = 5$ Pa

The air above the Earth's surface also exerts pressure and this is known as **atmospheric pressure** and has an approximate value of 100 kilopascals (kPa). Because of this pressure two terms are commonly used when measuring pressure:

● Absolute pressure. This means the pressure above that of absolute vacuum (zero pressure).
● Gauge pressure. This is the pressure above that normally present due to the atmosphere. Thus:

absolute pressure = atmospheric pressure + gauge pressure.

For example, a gauge pressure of 50 kPa is equivalent to an absolute pressure of $(100 + 50)$ kPa = 150 kPa.

Another unit of pressure, used in particular for atmospheric pressure, is the **bar**, where 1 bar = 100 kPa

Activity 17.1

Find out what other unit is sometimes used in relation to atmospheric pressure. Having found out what this other unit is called, find the relationship between it and the bar and the pascal.

Progress check 17.3

Explain the difference between absolute and gauge pressure.
Calculate the pressure acting upon an area of 10 m² when a force of 50N is applied.

Flow rate

The measurement of flow rate is of great importance in many industrial processes and is also one of the most difficult measurements to carry out as the flow behaviour of a fluid depends on a great many variables concerning the physical properties of the fluid and the pipe within which it flows.

The basic unit of volumetric rate of flow V is cubic metres per second. However this is a very large unit and therefore in practice a smaller unit, litres per second is used. For very slow rates of flow litres per minute may be used.

In the case of gaseous fluids, such as air, mass flow rate is often used which has the basic unit of kg per second.

Activity 17.2

Find out what units are used by your local water board to measure volumetric flow and how much they charge per unit.

Using this information estimate how much water your household uses in a year and how much it costs.

Fluid level

The measurement of fluid level is very important in industries using and storing liquid chemicals as well as water-based products. The requirement may simply be to know when a tank is full or empty or to know exactly how much liquid is in a tank or similar container.

The first requirement is simply a measurement of the height of the liquid in a tank and units of millimetres would normally be used in most cases. The second requirement implies a more accurate measurement and involves the unit of volume which would be the litre.

Speed

Linear speed or velocity

This is the measurement of distance covered in a given amount of time in a straight line and would apply to any object whether it be a vehicle or part of a machine.

Speed is calculated using the equation:

$$S = D/T$$

where D = distance in metres or kilometres and T = time in seconds, minutes or hours.

For example if an object travels 30 km in one hour it is said be travelling at a speed of 30 km/h.

Average speed or velocity

There are situations where an object may travel at different speeds over a given period of time; in this case the idea of average speed is used.

WORKED EXAMPLE 17.2

An object travels at a speed of 10 m/s for 2 s and then travels at 20 m/s for 5 s. To calculate the average speed we need to work out the total distance travelled and the total time taken to travel that distance.

Total time $= 2 + 5 = 7$ seconds

Distance travelled during the first 2 s $= 2 \times 10 = 20$ m

Distance travelled during the next 5 s $= 20 \times 5 = 100$ m

Total distance travelled $= 20 + 100 = 120$ m

Average speed $=$ total distance travelled/total time $= 120/7 = 17.14$ m/s

---

Activity 17.3

The next time you go on a long car journey and you get bored, try this: make a note of the car speed at regular intervals and try to calculate the average speed of the car for the whole journey.

Rotational speed or velocity

This type of measurement applies to objects which rotate about a central axis such as a wheel on an axle or a pulley on the end of a shaft connected to an electric motor. The units used are either revolutions/second or radians/sec. The unit of revs/sec requires the counting of complete revolutions in a given time; for very high speeds of rotation it is more convenient to use the unit revolutions/minute or r.p.m.

Angular speed or velocity

Another way of looking at rotation is to consider a pulley which has a white line painted on it from its centre to its circumference.

As the pulley rotates the white line will move through an angle of 360° for each revolution of the pulley. If the pulley rotated once every second its angular velocity would be 360 degs/sec. At higher rotational speeds this unit of measurement is too small and therefore a larger unit is required. The unit adopted for this purpose is the radian, one radian is defined to be $360/(2\pi) = 57.3°$. There are therefore 2π radians in one revolution. The pulley above would therefore have an angular velocity of 2π rads/s.

Progress check

Calculate the angular velocity of a wheel which rotates 50 times/sec.

Displacement

Linear displacement

This is the measurement of how far an object is moved in a straight line and is particularly important when defining the limits of movement for parts of machinery used in manufacturing. The units used for this are either millimetres or metres.

Rotational displacement

This relates to an object which moves through an arc describing a circle. It is assumed that the object rotates through less than one revolution and can move clockwise or anti-clockwise. The movement of the object often has to be measured accurately and therefore units of degrees, minutes and seconds are used.

Examples of rotational displacement or movement would include a gun turret, an aerial dish or an operating arm in a piece of machinery, the movement of which would require accurate measurement or positioning.

Activity 17.4

Find three examples of linear displacement and three examples of rotational displacement and explain why the displacement requires to be measured in each example.

Functional elements

Sensing

Thermocouple

A thermocouple is created whenever two dissimilar metals touch and the contact point produces a small open circuit voltage or e.m.f. that varies as a function of temperature. The thermoelectric voltage is known as the **seebeck voltage** and

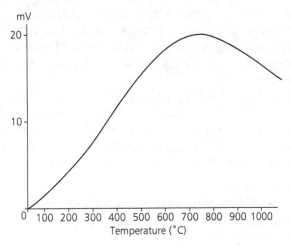

Figure 17.1 Graph showing variation in seebeck voltage with temperature

actually varies in a nonlinear way with temperature as shown on the graph in Figure 17.1.

As there is clearly a relationship between this voltage and temperature the device can be used as a temperature sensor. It is probably one of the most popular types of temperature sensor used in industrial circles and can measure temperatures over a wide range.

Several types of thermocouples are generally available and are designated by capital letters that indicate their composition according to American National Standards Institute (ANSI). For instance a J-type thermocouple consists of one iron wire and one constantan wire, where constantan is a mixture of copper-nickel alloy. This type of thermocouple can measure temperature from –200°C to about 850°C and is typically used in paper and pulp mills, reheat and annealing furnaces and chemical reactors.

Activity 17.5

Find out what other types of thermocouples are available and compare their characteristics in terms of output voltage and temperature range.

Pressure gauge

As its name suggests it is a device for measuring pressure and comes in a number of different forms. The most obvious form is that of a mechanical pressure gauge such as the Bourdon pressure gauge, the main components of which are a coiled-tube which moves under pressure.

When pressure is applied to the curved phosphor bronze tube, which is sealed at the pivot end, it tends to straighten out. The movement rotates the pointer across the scale. This type of pressure gauge can be used to measure large pressures both above and below atmospheric pressure. This type of gauge indicates 'gauge pressure' and is widely used in industry for indicating pressure.

Semiconductor pressure gauge

The other popular type of pressure gauge used is a semiconductor pressure sensor. This device is used where more accurate measurements of temperature are required and the value is to be electronically processed. This type of device will have a working range of 0 to 600 kpa and over this range will produce an output voltage from 0 to 10 V d.c. and can be used to measure either absolute or gauge pressure.

The device itself is actually what is referred to as a hybrid integrated circuit, this is because it contains various components which are mounted on a silicon chip. These components include the pressure sensor, temperature compensating components and a signal conditioning system. The physical construction of a typical encapsulation is shown in Figure 17.2.

Piezoresistor

The heart of the semiconductor pressure gauge device is the pressure sensor which is based on a bridge arrangement of piezoresistors. A piezoresistor is a resistor which changes its value of resistance when pressure is applied to it. Two of these are then connected into a bridge circuit, as shown in Figure 17.3.

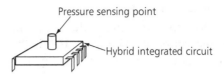

Figure 17.2 Physical structure of semiconductor pressure gauge

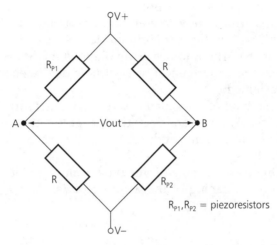

R_{p1}, R_{p2} = piezoresistors

Figure 17.3 Piezoresistor bridge circuit used in pressure gauges

The principle of operation is that at zero pressure the resistance of both arms of the bridge are the same and therefore no voltage appears across terminals A and B. As pressure is applied the value of the piezoresistors R_{p1} and R_{p2} increase, causing an imbalance in the bridge giving rise to a voltage across terminals A and B. The voltage produced is quite small and therefore the signal conditioning circuitry within the integrated circuit is used to increase its value in such a way that the final output voltage will vary over the range 0 to 10 V.

The semiconductor pressure gauge finds applications in motor vehicle diagnostics, machine tool control, hydraulics and pneumatics and oceanographics.

Activity 17.2

Modern day cars make use of engine management systems which control the operation of the car engine by measuring a large number of different parameters. Try to identify why a semiconductor pressure gauge would be used as part of this system and what the measurement would be used for.

Flow meter

The turbine flow meter is the simplest and most popular type and uses some form of multi-vane rotor which is driven by the fluid being investigated. There are two basic types: the rotary vane positive displacement meter and the turbine flow meter. The rotary vane flow meter measures the flow rate by indicating the quantity of fluid flowing through the meter in a given time. The device consists of a chamber into which is placed a rotor containing a number of vanes. Fluid entering the chamber turns the rotor and a known amount of fluid is trapped and carried round to the outlet. If X is the volume of fluid displaced by one blade then for each revolution of the rotor, the total volume displaced will be 6X. The rotor shaft may be coupled to a mechanical counter or electronic device which may be calibrated to give flow volume. This type of meter is widely used for measurement of water consumption, measurement of petrol in petrol pumps and measurements in the process industries.

The **turbine flow meter** contains a rotor to which blades are attached which spin at a velocity proportional to the velocity of the fluid which flows through the meter. The number of revolutions made by the turbine blades may be determined by a mechanical or electronic device enabling flow rate or total flow to be determined.

This type of flow meter is very compact, durable and has good pressure capabilities. Applications include the volumetric measurement of both crude and refined petroleum products and in the water, power, aerospace, process and food industries.

One disadvantage of this type of flow meter is that it requires periodic inspection and cleaning of the working parts.

Level gauges

Float-and-arm

The simplest form of level gauge is one based on the float-and-arm system. The angular movement of the arm can be measured by either mechanical or electronic means either of which can be calibrated to indicate the volume of fluid remaining in a container. Although cheap and simple it is less attractive if the fluid is corrosive as component parts would then have to be made from more expensive corrosion-resistive materials.

Air-compression level gauge

This type of gauge makes use of a capillary tube and pressure gauge, as shown in Figure 17.4.

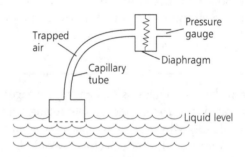

Figure 17.4 Air-compression level gauge

A tube is immersed in the liquid so that air is trapped inside the tube. A change in fluid level outside the tube will cause the air to be compressed or expanded which in turn causes a change in pressure which is sensed by the pressure gauge. Gauges similar to this are often used in washing machines to sense the water level in the drum.

Capacitance gauge

Where the fluid is corrosive a non-contact form of measurement is best. One such device which can be used in this situation is the capacitance gauge the construction of which is shown in Figure 17.5.

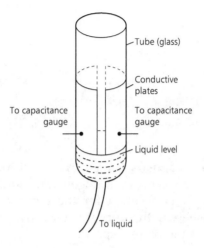

Figure 17.5 Capacitance gauge

The two conducting plates form a capacitor which will have an approximate value when no liquid is present given by the equation:

$$\text{Capacitance} = \varepsilon_o \varepsilon_r \frac{A}{D} \quad \text{(Farads)}$$

where ε_o = permittivity of free space, A = cross-sectional area and D = distance between the plates.

ε_r = the **relative permittivity** of the fluid and therefore as the fluid level rises between the plates so ε_r will increase in value and hence the value of capacitance will increase. The effective increase in capacitance can be measured electronically and when the system is suitably calibrated will give an accurate indication of fluid level or volume.

Progress check 17.5

What type of level gauge is used to measure water level in a domestic washing machine?

What type of level gauge would be best used with a corrosive fluid?

Tachometer

The most accurate way to measure angular velocity is to use a device called a tachometer or tachogenerator. The tachogenerator is a very small and precision-built a.c. generator which has rotating magnets with the output taken from stator coils to avoid the use of slip-rings. This type of construction means that the amount of power required to mechanically rotate the generator is minimal. The tachogenerator is normally mechanically coupled to a revolving shaft and the

341

electrical output is signal conditioned to produce a d.c. voltage proportional to angular velocity. A typical setup is shown in Figure 17.6, where the d.c. voltage is used to drive a meter calibrated in r.p.m.

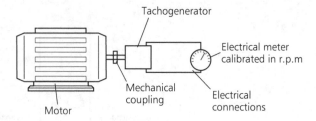

Figure 17.6 Tachogenerator setup

The main drawback for some applications is the need to make a mechanical coupling between the tachogenerator and the shaft.

The main application for the tachogenerator is that of electric motor speed control whereby the output is used by a control system to ensure that the motor remains at the correct speed.

Potentiometer

The potentiometer is a resistor which can be varied in value depending on the position of a shaft and a wiper connected to it as shown in Figure 17.7(a).

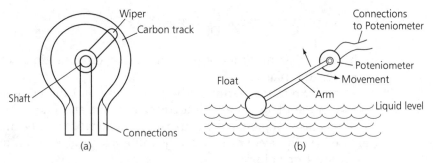

Figure 17.7 Potentiometer construction and application

The shaft is connected to the wiper which is in contact with a carbon track. The resistance measured between the wiper and one of the carbon track contacts will vary depending on the position of the wiper. Thus angular displacement can be measured in terms of a changing resistance value.

A typical application would be in relation to measuring fluid level by means of a float-and-arm system; this is clearly shown in Figure 17.7(b). The shaft of a potentiometer would be connected to the arm of the system; as the arm moves up and down so the wiper will rotate in the potentiometer causing the resistance value to change across the terminals. This resistance change can be processed electronically and with suitable calibration would give a true indication of fluid level or volume.

Activity 17.7

Think of another application for a potentiometer and explain how it could be used to make the measurement.

Strain gauge

Sensing strain involves the measurement of very small changes of length of an object. This is typically undertaken using a **resistive strain gauge**. The resistive strain gauge consists of a conducting material in the form of a thin wire or strip which is attached firmly to the material in which strain is to be measured. This material could be the wall of a building, part of a bridge or a turbine blade, anything in which excessive stress could indicate a serious problem. The physical structure of a simple strain gauge is shown in Figure 17.8.

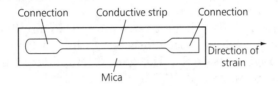

Figure 17.8 Simple resistive strain gauge

As the strain gauge is firmly attached to the material being measured any changes in material length will cause the strain gauge to stretch; this stretching causes the cross-sectional area of the resistive element to reduce which in turn causes the resistance of the gauge to increase. This is due to the fact that resistance is inversely proportional to cross-sectional area. The change in resistance can be processed electronically to produce a d.c. voltage which is proportional to strain.

Progress check 17.6

Could a resistive strain gauge be used to measure a structure which is in compression, rather than in tension as explained above? If so how would the strain gauge be mounted?

Signal processing

When sensors are used to measure parameters such as temperature, pressure, flowrate, level, speed and displacement, invariably the output produced by the sensor is in a form which cannot be used directly by a measurement system. The output from a sensor will therefore need to be conditioned or processed in some way so that it is compatible with the measurement system. Typically this will require amplification, compensation or conversion from one form of signal into another.

Converters

Pressure-to-current converter

In many sensing applications the actual sensor may be some distance away from the measurement system. This would normally require the sensor to be connected to the measurement system via very long leads. Typically the output from the sensor may be very small and therefore very susceptible to noise when sent over a long distance. Also, the sensor may need to be powered: this would require the addition of extra leads. To overcome these problems the output from the sensor is often converted into a current. If we take the example of measuring pressure using a semiconductor pressure gauge, the output from this type of device is typically in the range 0–10 V. This voltage range is converted into a current range of 4–20 mA using a voltage-to-current converter.

Voltage-to-current converter

Figure 17.9(a) shows a simplified circuit for a voltage-to-current converter. The output from the semiconductor pressure gauge would be connected to terminals 1 and 2. This voltage passes through to amplifier A the output of which is used to drive the base of transistor Q_B; this in turn causes an emitter current to flow through resistor Rx. The gain of amplifier A is adjusted so that over the input voltage range of 0–10 V the emitter current varies over the range 4–20 mA. Notice that the output current actually makes use of the power supply lines, which means that both the power for the sensor and the output can be connected to the measurement system using only two wires. We have thus overcome the problems associated with measuring small voltages remotely from the measuring system and reduced the connection requirement from four to two wires.

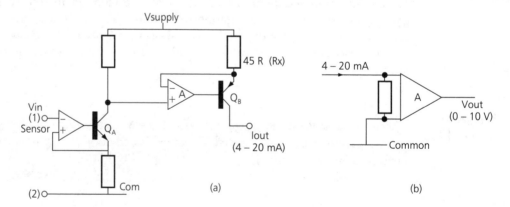

Figure 17.9 *Voltage-to-current and current-to-voltage converters used in remote sensing applications*

Current-to-voltage converter

Having converted the sensor output from a voltage to a current, most measuring systems deal with inputs in terms of voltage, so there is a requirement to convert the current back into a voltage at another stage in the system. This is achieved using a current-to-voltage converter as shown in Figure 17.9(b).

The current signal from the sensor is amplified by the differential amplifier A and converted into a voltage at its output terminals. The gain of the amplifier is adjusted to give a voltage output range of 0–10 V for an input current range of 4–20 mA.

Two-wire transmitter system

A system which uses the techniques discussed, to send sensor outputs to a measurement system, is often referred to as a two-wire transmitter system and is commonly used in industry to send back signals from sensors to remote measurement systems.

Many of the sensor manufactures now supply their sensors with voltage-to-current converters built into them with their outputs adjusted to the industrial standard range of 4–20 mA.

<table>
<tr><td>

Progress check 17.7

</td><td>

What are the main advantages associated with using a current signal rather than a voltage signal from a sensor output?
Why does a current signal normally have to be converted back into a voltage signal?

</td></tr>
</table>

Cold junction compensator

With certain types of sensor there is a need to electronically or otherwise compensate for the effects of temperature or small voltages generated within the sensor connection system. This is particularly true of the thermocouple: due to the fact that a thermocouple effect is produced when two dissimilar metals come into contact, an additional thermocouple is effectively created when the leads of the actual thermocouple come into contact with another metal, say a terminal block as illustrated in Figure 17.10.

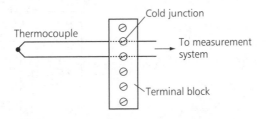

Figure 17.10 Generation of cold junction compensator

The voltage generated at this junction must be accounted for and is known as the cold junction. In a typical temperature measurement system the actual voltage measured by the system is not the actual thermocouple voltage and has to be calculated using the following equation:

$$Vtc = Vmeas + Vcj$$

where Vtc is the compensated thermocouple voltage, Vmeas is the actual measured voltage and Vcj is the cold junction voltage.

To calculate Vcj the temperature of the terminal block has to be known. It can be assumed to be at ambient temperature. However, for very accurate temperature measurements terminal blocks can be obtained with their own temperature sensors built in which can be used to accurately determine the temperature of the terminal block. Once the temperature is known using the table supplied by the thermocouple's manufacturer it is converted to a voltage (Vcj); this is then added to the measured value Vmeas to give the compensated thermocouple voltage Vtc. These days when using computer-based measurement systems all of this is taken care of automatically by the software.

Progress check 17.8	What is a cold junction and how is it created? What steps have to be taken to compensate for the effects of the cold junction?

Analogue-to-digital conversion

We have already identified the need to convert the output of a sensor into either current or voltage. However, it has become common practice to display the output from a sensor either on a digital display system or on a computer-based system neither of which will accept an input of varying (analogue) current or voltage. The problem with this type of system is that it can only deal with information in the form of 0's and 1's and therefore a system has to be devised to convert a voltage into a number of pulses consisting entirely of '0's and '1's. You should already know that the binary numbering system makes use of '0's and '1's to represent numbers, but we require a voltage to be represented by a binary number. This is exactly what happens in an analogue-to-digital converter: the input is an analogue voltage and the output is a binary voltage.

Figure 17.11 shows an 8-bit analogue-to-digital converter. The input voltage range is 0–2.55 V and the output is an 8-bit binary number.

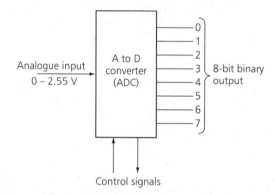

Figure 17.11 8-bit analogue-to-digital converter

With 8-bits we can represent numbers in the range 0–255. By dividing 2.55 by 255 we get the relationship between input and output in terms of the fact that the least significant bit represents 10 mV. This means that the smallest voltage which can be represented in this system is 10 mV. See also Chapter 22.

Progress check 17.9	What is the smallest voltage which can be measured using a 10-bit A to D converter?

Binary representation of voltage

If the input voltage was 10 mV then the binary output of the ADC would be 00000001; if the input was 50 mV then the binary output would be 00000101 and so on.

Most modern digital display or computer-based systems have an analogue-to-digital converter built in so that the voltage produced by a sensor can be measured directly. It is important to remember that analogue-to-digital conversion is not a continuous measuring system; the output jumps in steps representing 10 mV. Therefore if the input voltage was say 17 mV the output would jump to a number representing 20 mV and there would be an error of 3 mV in this case. This is illustrated by the graph shown in Figure 12, which shows the relationship between input and output of an 8-bit analogue-to-digital converter.

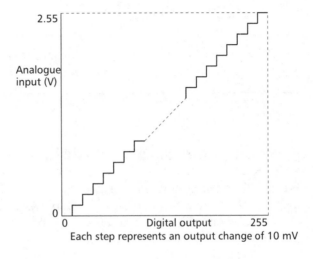

Figure 17.12 8-bit AtoD converter graph

The accuracy of a typical 8-bit analogue-to-digital converter can be no better than ± ½ LSB which represents a voltage of ± 5 mV.

Digital-to-analogue conversion

When a signal has been converted into digital form there is often a requirement to convert it back into analogue form. This is a requirement when a computer-based measurement system reads an analogue value from a sensor and then has

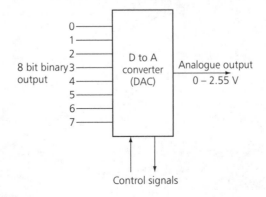

Figure 17.13 Basic digital-to-analogue converter

347

to output an analogue voltage in response to this measurement. A digital-to-analogue converter is used for this purpose. In this case the input is a binary number and the output is an analogue voltage as shown in Figure 17.13.

The process is the reverse of analogue-to-digital conversion, considering an 8-bit DtoA converter where the least significant bit represents 10 mV, a binary number of 00000101 would generate an analogue voltage at the output of 50 mV and so on. As with the ADC the accuracy is ± ½LSB. The resolution can be increased by increasing the number of bits used; for instance if 12 bits are used then a total 2 raised to the power of 12 voltage levels can be represented. In this example a total of 4096 voltage levels are available, if the maximum output voltage was say 4.096 V then the smallest voltage which could be output would be 4.096/4096 = 1 mV. See also Chapter 22.

Activity 17.8

Identify a piece of equipment which uses an analogue to digital converter and find out how many bits it uses and hence the smallest voltage it can measure.

Receiving, recording and data acquisition

Having considered how measurements can be made, consideration must be given to how the received information is displayed and recorded. Two basic systems are available: a dedicated display and recording system such as a chart recorder or a more versatile and flexible computer-based system.

Chart recorders

Chart recorders have been around for many years in various shapes and forms. Their main purpose is to give a visual display of the measured parameter and a paper record of variations over a set period of time. A typical chart recorder is shown in Figure 17.14.

In its simplest form a chart recorder consists of a moving drum which causes a sheet of graph paper to be moved at a steady rate under a pen which is attached

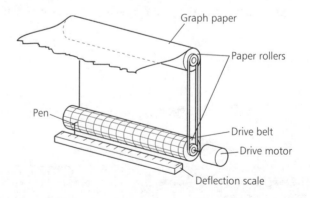

Figure 17.14 Typical chart recorder

to a pointer which is deflected by a voltage applied to it from a sensor. The linear movement of the pointer across a calibrated scale gives an indication of the value of the measured parameter and the pen on the graph paper produces a permanent record of the variations in the measured parameter over a given period of time. Modern chart recorders have the ability to record the output from many sensors at the same time and can be programmed to accept outputs from different types of sensor.

The chart recorder is a precision instrument and is quite expensive and tends to be used in dedicated situations, such as monitoring in power stations for example. Due to the limitations of the amount of paper which can be stored in the chart recorder, the time period over which it can record a particular parameter before a paper change, is limited. A 24 hr recording cycle would be fairly typical.

Advances in modern technology has meant that computer systems now form a large part of a typical measurement, recording and data acquisition/collection systems and to a certain extent have displaced devices such as chart recorders.

Computer-based data collection/display systems

Modern day computers can now be configured with special plug-in boards to interface directly with different types of sensor, carry out analogue-to-digital and digital-to-analogue conversions. Software can now be purchased which allows the computer to be setup as a complete measurement system. Information received from various sensors, can be displayed on the computer screen in a variety of different ways, recorded at regular intervals and stored away on the computer over periods of time from a few seconds to a few years. This type of computer-based system is often referred to as a **data acquisition system**; a typical setup is shown in Figure 17.15.

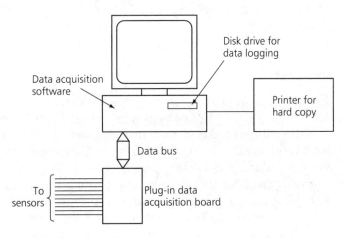

Figure 17.15 Computer-based data acquisition system

The versatility of the software allows the measured parameter to be displayed in any number of different forms, such as a line graph, bar graph, an analogue meter or digital meter. An example display screen from a piece of software called Labtech Notebook is shown in Figure 17.16.

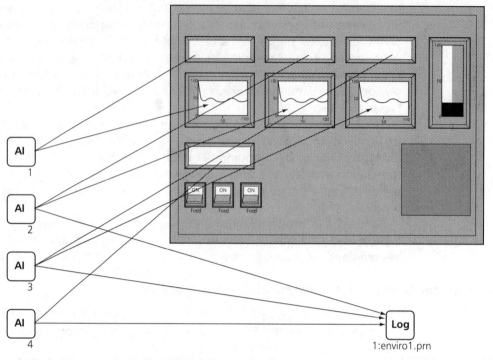

Figure 17.16 Display screen from Labtech Notebook

<table>
<tr><td>**Progress
check 17.10**</td><td>A measurement system is required to monitor 8 different temperatures and record variations over a period of 28 days. What would be the best type of recording system to use for this application?</td></tr>
</table>

Calibration and test

For any measurement system to operate correctly and to display the correct value for the measured parameter it must be correctly calibrated and regularly tested. For the computer-based measurement system this means applying known voltages representing the outputs of the various sensors to the inputs of the computer system's data acquisition plug-in board. A purpose-made calibration box is normally used for this purpose and is often referred to as a sensor simulator box. A typical calibration setup is shown in Figure 17.17.

The system is calibrated by adjusting the plug-in board, usually by means of software, until the displayed value is equal to the value of output from the simulator box. The measurement system is then said to be calibrated. This process may be carried out at regular intervals with the simulator box being sent away to the manufacturers every 12 months for recalibration. The various sensors associated with the measurement system also have to be tested at regular intervals. This is often carried out with dedicated sensor test boxes obtained from the sensor manufacturers. These boxes test the integrity of the sensor and ensure that it is working correctly.

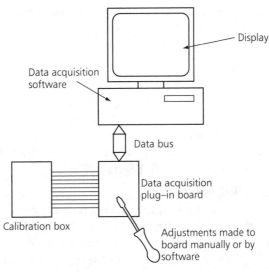

Figure 17.17 Calibrating a computer-based measurement system

Selection criteria

When selecting functional elements such as sensors or a complete measurement system it is important to give consideration to:

- fitness for purpose
- reliability
- accuracy
- sensitivity
- repeatability.

Fitness for purpose
When considering say a functional element such as a sensor for a given application we have to consider whether or not it is fit for the purpose intended. For instance if we have to select a sensor to detect changes in fluid level, an obvious selection would be a simple float-and-arm system, but upon further investigation we may find that the fluid involved is highly corrosive. Thus although our selection would indeed measure fluid level it is not suitable for that purpose in this case as clearly a non-contact method of measurement would be preferred.

Reliability
When selecting a sensor or measurement system it is important that it performs reliably in the given application. For instance choosing a sensor for a particular measurement is one thing but if its construction is such that it is very delicate and in a given application it is subjected to a great deal of mechanical stress it will very quickly break and therefore become an unreliable element within the system.

Accuracy
This is one of the most important parameters of a measurement system and has to be considered very carefully. The accuracy for the measurement would normally be specified for a given application; for example temperature may need

to be measured to an accuracy of ± 0.1°C. It would clearly be no good at all to set up a measurement system and then find that the temperature can only be measured to an accuracy of ± 0.5°C.

Sensitivity

This is very important especially when dealing with sensors which produce very small output voltages. This is especially true when using a computer-based system which make use of analogue-to-digital converters. For instance if the smallest voltage to be measured is 1 mV it would be a waste of time selecting a plug-in board for the computer which used an 8-bit AtoD converter which could only measure to the nearest 10 mV.

Repeatability

When the same measurement is taken repeatedly over a period of time it is important that the measurement system returns the same value each time; this is called repeatability. A measurement is non-repeatable if the measurement system introduces variations due to temperature effects, supply voltage variations and so on. When selecting elements of a measurement system be sure to check for such variations and where possible compensate for them.

Activity 17.5

A measurement system is required to measure pressure and temperature and to display both measurements and to record the variations of pressure and temperature over a period of 10 days. Both sensors will be mounted remotely from the actual measurement system. Using a block diagram identify each component part of a system to meet the above specification.

Assignment 17

This assignment provides evidence for:
Element 14.1: Investigate engineering measurement systems
and the following key skills:
Communication 3.2, 3.3
Information Technology 3.1, 3.2, 3.3, 3.4

You are required to assemble and test for functionality a computer-based measurement sytem to measure ambient temperature.

(a) The temperature variations are to be displayed as a line graph on the screen of the computer and the value of the temperature is to be recorded on disk every 10 s for a total of 2 minutes.
(b) A detailed block diagram of the whole system as well as a printout of the recorded file on disk and a screen dump of the line graph is required. Also, to be included are details of calibration and test methods used to prove correct operation of the system including details of any test data collected.

Report

The evidence indicators for this element suggest that a report is required, covering two engineering measurement systems, and should include:

- description of each systems functional elements, supported by appropriate diagrams or drawings
- a log of maintenance and test procedures performed on each system
- record of test and maintenance data
- evaluation of the performance parameters of each system, using maintenance and test data.

Chapter 18:   Engineering control systems

In today's competitive world, a company must be efficient, cost-effective and flexible if it wishes to survive. In manufacturing and process industries this has resulted in a greatly increased demand for industrial control systems in order to streamline operations in terms of speed, reliability, versatility and material throughput. It is important therefore to gain an understanding of control systems and their component parts and how they can be correctly applied.

When you complete this element you should be able to:

● identify and illustrate the functional elements of the basic types of engineering control systems
● describe the different control modes for closed loop systems
● describe controllers used in engineering control systems
● describe how controllers are configured
● record engineering control system performance data
● evaluate the parameters of control systems.

Functional requirements

A control system consists of a number of different elements, the general arrangement of these elements is shown in Figure 18.1(a).

Plant refers to the machine or process which is being controlled. The controller can be a person in which case the control system is said to be a **manual control system**. Alternatively, if the controller is a device such as an electronic circuit, computer, or mechanical linkage the system is referred to as an **automatic control system**. The final control element provides the interface between the controller and the plant and is usually some form of **actuator**, which may be pneumatic, hydraulic or electromechanical. The **measurement system** provides information regarding the status of the plant. All of these elements together form a complete control system; the way in which they interact depends on the type of plant being controlled and the accuracy of control required.

Measurement system

In modern automatic control systems the measurement system contains devices such as **sensors** and a **signal processing/conditioning** unit all of which have been described in some detail in Chapter 17. The main purpose of the measurement system is to feed back to the controller a signal which represents the status of the process being controlled which will in turn cause the controller to react to changes in the process.

Controller

This is the main component part of the control system. In most modern day control systems it would be a form of computer system. It is the controller which decides how to adjust the process being controlled to meet a particular requirement. For example if the controller receives a signal from the measurement system indicating that the temperature of the process is too high the action of the controller would be to turn off the source of heating.

Final control elements

The final control element is the actual device which directly effects the process. For example it could be an electrically operated valve which controls the flow of water which is required for a particular process. In this example the controller would output a signal which would turn on the valve causing say a tank to fill with water; when the tank was full the measurement system would send a signal to the controller to tell it to turn off the valve and so on.

Feedback

For a complete control system to operate effectively there must be what is called a feedback loop from the process to the controller. Feedback is said to occur in an automatic control system when the control action depends upon the measured state of the process being controlled. Feedback gives an automatic control system the ability to deal with unexpected disturbances and changes in the behaviour of the process being controlled. Using the idea of feedback, as shown in Figure 18.1(b), the output of the motor controller or control effort, as it is sometimes called, is determined by the difference between the actual motor speed measured

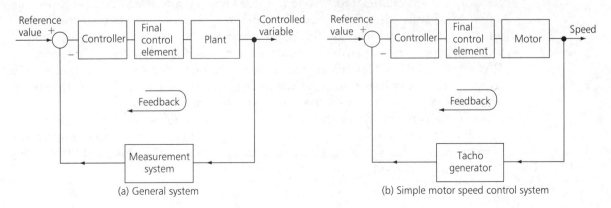

(a) General system   (b) Simple motor speed control system

Figure 18.1 Control systems

and the required speed. In the case of an electric motor, the speed would be measured using a tachogenerator the signal from which would be compared with a signal representing the required speed; the difference signal produced is then used by the controller through the final control element to control the actual speed of the motor.

Progress check 18.1

What do you understand by the term 'plant'?
Why is feedback so important in a control system?

Feedback control modes

Automatic control systems are normally used for either regulation or trajectory following.

Regulation
A control system for maintaining the plant output at the required value in the presence of any external disturbances is called a regulator. The regulator must apply control effect to attempt to maintain the plant output at the required value with minimum error.

Trajectory following
A control system is often required to make the plant output follow a certain profile or trajectory. Examples include the control of a machine tool or the control of a gun turret.

Regulation or trajectory following form the basic requirements for most types of control situations. To enable the control system to operate in this way the controller can be made to operate in a number of different modes.

On/off control

This is the simplest mode in which a controller can operate, it requires only a simple ON/OFF signal from the measurement system, which is used to generate corrective action by the controller. It is the mode quite commonly used in the control of domestic heating systems. It does have some disadvantages especially when the plant being controlled does not respond immediately to the control effort applied. Any delay in the plant response means that the process output will continue to rise even when control effort has been removed. This can actually cause wild oscillations in the process being controlled. The effect for a temperature control system is shown in Figure 18.2(a), where the actual temperature will rise and fall about the required temperature of 20°.

For a domestic heating system this variation is quite acceptable but for other systems a much more accurate method is required especially if the temperature is to be maintained within very small limits, e.g. ± 0.1°. For this kind of accuracy the proportional mode of operation is chosen.

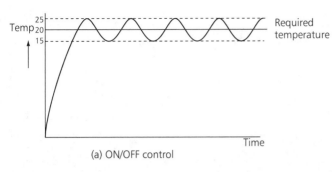
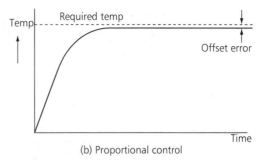

Figure 18.2 Temperature control systems

Proportional control

In this mode the amount of control effort applied to the plant is proportional to the difference between the required value and the actual measured value. This implies that the measurement system is required to supply a much more accurate and meaningful signal. The response of a typical proportional temperature control system is shown in Figure 18.2(b).

Notice the difference in the response compared to the ON/OFF controller: the wild variations have disappeared. However, there is still a small problem known as **offset error**; this is the difference between the required value of temperature and the actual temperature achieved. This error exists because for the controller to operate in proportional mode there must always be a small error signal present at the input of the controller to enable it to generate control effort.

To see how a proportional control system is used in practice let us consider the following case study.

Case study

A control system is required to control the **speed of an electric car** in such a way that when the speed of the car is set to a particular value, it remains at that speed irrespective of road gradient. We will now see if the proportional controller is suitable for this application.

Figure 18.3(a) shows a block diagram of a proportional control system for this case study, indicating its various component parts.

The power (p) supplied to the motor no longer depends on the driver's speed setting. It depends on the difference between the speed setting and the actual speed. To see how well the system works we need to develop some simple mathematical models for the controller and the electric car. If we say that speed is represented by (v), speed setting is represented by (s) and the feedback voltage by (a) and the output of the driver's controller block by (b) volts. We now have to sort out the relationships between these variables. Firstly, in this example (a) represents 1 V for every 10 mph of speed; (b) must also have the same relationship with (s).

We can therefore say that: $a = v/10$ and $b = s/10$.

By testing the car it is found that it requires 100 W of power for every 10 mph of speed, therefore $v = p/10$.

When the car reaches a gradient it is found that the speed reduces by 1 mph per degree of slope, therefore the speed equation becomes:

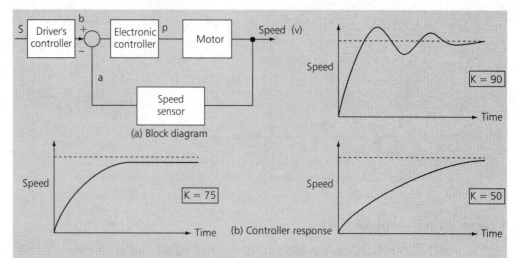

Figure 18.3 *Electric car proportional control system*

$$v = \left(\frac{p}{10} - \text{deg of slope} \right)$$

The power output of the controller is given by the equation:

$$p = K(b - a)$$

To use this equation we have to calculate a value for K. this is done by considering the speed of the car on the flat, using $v = p/10$. The expression for p is now substituted into the equation for v, hence:

$$v = \frac{K(b - a)}{10}$$

Now we substitute in the values for a and b as follows:

$$v = \frac{K\left(\dfrac{s}{10} - \dfrac{v}{10} \right)}{10}$$

which becomes:

$$v = \frac{K(s - v)}{100}$$

Rearranging gives:

$$\left\{ v + \left(K \cdot \frac{v}{100} \right) \right\} = \frac{Ks}{100}$$

which finally gives us:

$$v = \frac{\dfrac{k}{100}}{\left(1 + \dfrac{K}{100} \right) d} \quad \text{or} \quad v = \left(1 - \frac{1}{1 + \left(\dfrac{K}{100} \right)} \right) s$$

Offset or steady-state error
The equation for v shows that whatever value is chosen for K the actual speed (v) is never going to equal the required value of speed (s). This difference is known as the offset or steady-state error, the actual value being given by the term 1/(1 + K/100). This term has to be made as small as possible, but making K too large can cause the system to oscillate about the required value of speed. Therefore, a compromise value has to be chosen. Figure 18.3(b) shows the case study controller responses for different values of K; a suitable value is about 75 which gives a rapid but stable response.

The controller therefore cannot maintain the speed at the desired value due to the steady-state error although the effect on speed of a gradient will be significantly reduced compared with having no controller at all. This case study has shown that a proportional controller alone is unsatisfactory for this particular application. To remove the steady-state error another term has to be added to the control system, known as the **integral term**.

Progress check 18.2

What is the main problem with ON/OFF control?
Explain the term offset or steady-state error.

Proportional plus integral (PI) controller

The integral term is connected across the proportional controller as shown in Figure 18.4(a).

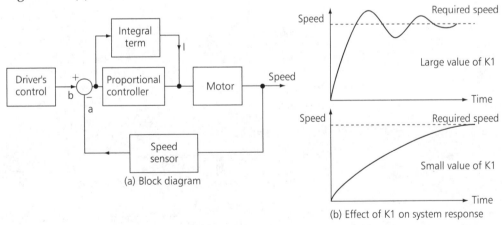

Figure 18.4 Proportional plus integral control

To see how the integral term effects the response of the controller a mathematical expression is required. The output of the integral controller is (I) and is given by the expression:

$$I = \{I + (K1(b - a))\}$$

The output from the integral controller will increase if a < b and will decrease if a > b and will remain constant when a = b. This means that the steady-state error will be completely removed. The proportional controller will therefore rapidly bring the actual speed close to the required speed and then the integral controller will slowly remove the steady-state error. The speed at which this steady-state error is removed depends on the chosen value for K1. Figure 18.4(b) shows how the overall response of the control system is affected by the value of K1.

Effect of K1 (integral gain) on system response

From Figure 18.4(b) it is clear that too large a value of K1 causes oscillations and too small a value produces a slow response. Therefore, as with the proportional controller a compromise value is required.

From this response we can now see that the control system will maintain the speed of the electric car at the required speed irrespective of road gradient.

When another term is added in this way to a proportional controller the overall control system is know as a **two-term controller**.

Proportional plus derivative (PD) controller

In the previous case study it was shown that a proportional plus integral (PI) controller was required to solve the problem. There are however, many different control situations and each one will need to be considered in terms of the best type of controller to use. It is easier to start with a proportional controller and then to simply add more terms until the required response is achieved.

Another term which is often used with a proportional controller is known as the **derivative term**. This can be added across the proportional controller in a similar way as in the PI controller and is shown in Figure 18.5(a).

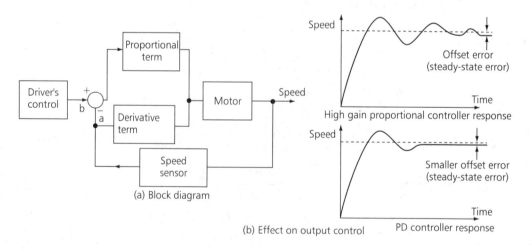

Figure 18.5 Proportional plus derivative control

To show the effect of derivative control we will use the same electric car control system as before. Notice that the input of the derivative controller, shown in Figure18.5(a) is connected to the (a) output which represents the actual speed of the electric car. The output generated by the derivative controller is actually proportional to the rate of change of the actual speed of the car. This is represented mathematically by the expression:

$$D = \frac{d(a)}{dt}$$

When discussing the characteristics of the proportional controller in the case study it was stated that if the proportional gain was increased to a large value the speed of response was increased and the steady-state error was reduced. The only problem was that the response oscillated about the required speed which was unacceptable. Therefore if a term could be added to the controller which would maintain the speed of response and small steady-state error but reduce the oscillations, we would end up with a controller which would react much more rapidly to a change in speed. This is in fact what happens when derivative control is added to the proportional controller. The response of a PD controller is shown in Figure 18.5(b) along with the response of a high gain proportional controller. It is clear to see that the rapid response is maintained but the oscillations have been effectively reduced or damped.

The response of a PD controller is much faster than that of a PI controller but as can be seen from Figure 18.5(b) the steady-state error remains. Ideally it would be nice to combine these two effects and get the best of both worlds. This is precisely what is done in a 3-term proportional plus integral plus derivative (PID) controller.

Proportional plus integral plus derivative (PID) controller

Figure 18.6(a) shows a typical PID speed control system where the benefits of both integral and derivative control are utilised.

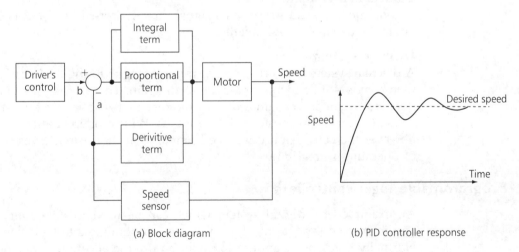

(a) Block diagram   (b) PID controller response

Figure 18.6 PID controller

The response of a PID controller is effectively the combined response of a PI and a PD controller and is shown in Figure 18.6(b).

Activity 18.1

For each of the following control systems select the most appropriate controller mode and explain the reasoning behind your choice.

Control system 1.
A control system is required to accurately control the temperature of an oven; speed of operation is not important. However, when a particular temperature is set the oven must achieve this temperature with an error of less than 0.1°C.

Control system 2.
A speed control system is required which will respond to rapid changes in speed due to changing load conditions. The accuracy of the actual speed achieved is not as important as the ability to maintain a constant speed under varying load conditions.

Controllers

Having investigated the various controller modes of operation it is now important to consider how a controller is implemented in practice.

A controller can be implemented using analogue circuit components, transistors, operational amplifiers etc. as covered in Chapter 15 or it can be based on digital circuit components also covered in Chapter 16.

Continuous system

A continuous system processes a signal directly in real time and is usually based on analogue circuit components.

Discrete system

A discrete system converts a signal from analogue to digital form for processing, then converts it back into analogue form again. The process of signal conversion does not occur in real time due to the fact that it takes time for the system to convert signals for one form into another. With modern digital control systems however the conversion times are so small that the system gives the appearance of operating in real time.

Programmable logic controllers

An example of a digital system which can be used for ON/OFF, proportional and in some cases 3-term PID control is a programmable logic controller (PLC). Normally this type of application will require the addition of analogue-to-digital and digital-to-analogue conversion modules. The basic operation of the PLC is covered in Chapter 23 with Figure 18.7 showing a PLC as a temperature controller.

The PLC in this example is performing the function of a proportional controller, with the required temperature being set by a thumb wheel switch.

Direct digital control systems

Although a PLC can be used as a controller in this way it under-utilises the capabilities of the PLC. It is much more likely that a dedicated digital system is used; these are available to perform ON/OFF, proportional, PI, PD and PID control

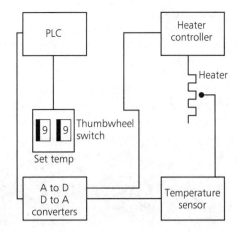

Figure 18.7 *PLC temperature controller*

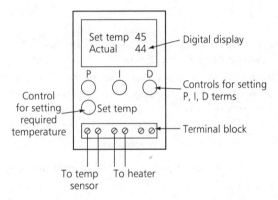

Figure 18.8 *Digital controller*

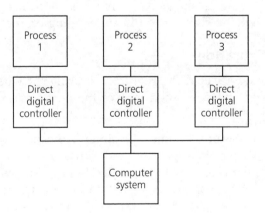

Figure 18.9 *Distributed control system*

functions and are referred to as direct digital control systems. Figure 18.8 shows a typical dedicated digital controller.

Distributed control systems

Where many different processes are required to be controlled a distributed control system is often used. In this application individual direct digital controllers are used to control each process but the overall process is monitored and controlled by a computer system. Figure 18.9 shows a typical distributed control system.

**Progress
check 18.3**

Explain the difference between a discrete system and a continuous system. How can a PLC be used as a controller?

Selection criteria

Fitness for purpose

To choose the correct controller type for a given application requires a detailed knowledge of the specification for the system. This specification must identify the required speed of response, the signal range required for inputs and outputs and give information regarding the process or plant to be controlled. Plant or process information is usually obtained by carrying out a series of tests to establish a relationship between plant/process input and plant/process output.

The controller type is selected by using the following guide:

1 Start by using a proportional controller and perform a simple mathematical analysis, as per the electric car case study.
2 Add additional terms of integral and/or derivative control until the correct output is obtained.
3 Adjust the system gain factors (K) until the correct speed of response is obtained.
4 Having established the type of controller required, select either a dedicated digital controller or PLC depending on the overall requirements of the system.

Controller configuration

Having selected the required controller it will need to be installed and correctly set up and fine tuned to give the desired response.

Setting
In the case of a dedicated digital controller this will require the initial setting of proportion gain, integral and derivative gain factors.

In the case of a PLC a suitable program will have to be written to enable the system to respond accordingly. Within this program, constants will be used to define the proportional, integral and derivative gain factors.

Tuning
Having selected the initial values for the gain factors the whole system is then usually tested and its response to changes graphed. The main difficulty is deciding on the initial gain values and also identifying the correct response.

Correct response

Consider the general system response curve shown in Figure 18.10, this represents the response to a sudden or step change in say the required speed.

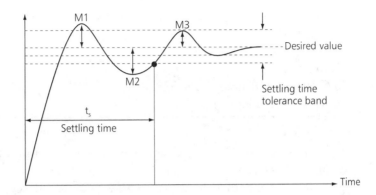

Figure 18.10 Step input response curve

The important terms to consider are:

- *The Decay ratio*. The Decay ratio which is defined as the ratio of two successive overshoots and is calculated using the expression:

$$\text{Decay ratio} = \frac{m3}{m1}$$

- *The settling time*. The settling time which defines the time it takes for the output to settle to within a specified tolerance band, typically ± 2%. This parameter will be specified by the application, the ideal decay ratio is considered to be ¼. In other words the second overshoot should be ¼ of the size of the first overshoot.

Tuning a multi-term controller

The best way to tune a multi-term controller is to start by turning off all of the terms except for the proportional term. The proportional gain is then set to a minimum value and the required output setting or set point value is increased by about 10%, the system response is then observed. The set point value is returned to its original setting; the proportional gain is then increased. This process is repeated a number of times until the system output oscillates continuously.

The value of proportional gain which caused this effect is noted. The proportional gain is then set to about half this value and then the gain factors for the other terms are adjusted to give a decay ratio of ¼. Small adjustments can then be made to meet the settling time requirement.

Assignment 18

This assignment provides evidence for:
Element 14.2: Investigate engineering control systems
and the following key skills:
Communication 3.2
Information Technology 3.1, 3.2, 3.3
Application of Number 3.1, 3.3

You are required to analyse the characteristics of the following position control system to establish the correct type of controller to be used, suitable values for gain factors and then to describe how you would set up and tune the system indicating the terms which you would use to judge the overall system performance. The system block diagram and component characteristics are shown in Figure 18.11.

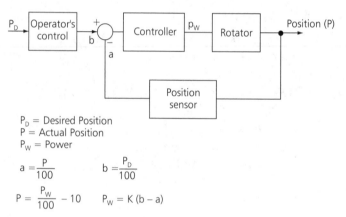

P_D = Desired Position
P = Actual Position
P_W = Power

$$a = \frac{P}{100} \qquad b = \frac{P_D}{100}$$

$$P = \frac{P_W}{100} - 10 \qquad P_W = K(b - a)$$

Figure 18.11 Position control system

If you have access to suitable practical resources then an attempt should be made to assemble an actual control system, as described above, and to then tune the system to obtain the desired response.

Report

The evidence indicators for this element suggests that a report is produced describing two engineering control systems, with different control modes, including:

● identification of the functional elements supported by appropriate illustrations
● description of the control modes of closed-loop systems
● description of controllers used in the control system and how they are configured
● a record of the performance data for the control system
● notes evaluating the parameters of the control systems.

Chapter 19:   Final control elements in engineering systems

This chapter covers:
Element 14.3:   Investigate final control elements in engineering control systems.

... and is divided into the following sections:
- Actuators and servos
- Valves
- Valve selection criteria.

Defining a control system for a particular application cannot be completed until the element which actually controls the final process is selected. The final control element as it is called, can be electrical, pneumatic or hydraulic, and is the subject of this chapter.

When you complete this element you should be able to:

- describe the final control elements in engineering control systems
- describe the operating principles of final control elements
- identify the final control elements for two given engineering applications.

Actuators and servos

The final control element is the device which a control system operates, to control or complete the actual process stage, and is normally an actuator or a valve.

An actuator is a final control element which applies a force. This force can be applied by the movement of say a piston or a mechanical arm or by the rotary movement of a shaft. In general actuators can be either **linear actuators** or **rotary actuators**.

Rotary actuators

Electric servo motor
A typical servo motor is basically a small d.c. motor. The operation of d.c. motors was covered in Chapter 6. The major difference between a servo motor and an ordinary d.c. motor is that it is physically smaller and made of lightweight materials and uses low friction bearings and the field is generated using permanent magnets. This gives the servo motor the following characteristics:

- rapid acceleration
- linear relationship between armature current and speed/torque

The speed of the servo can therefore be accurately controlled by electronically varying the armature current. Servo motors find applications in position control and speed control systems such as would be found in printers, plotters, CNC machines and robots.

Applications of servo motors

In position control system the servo motor is often used in conjunction with a gearbox as shown in Figure 19.(a).

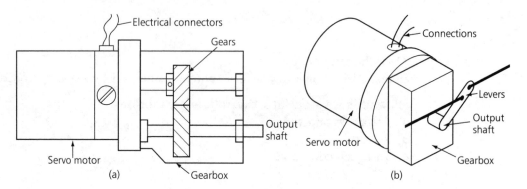

Figure 19.1 Servo motor with gearbox

Using a gearbox to reduce the speed of the output shaft means that rotational movement can be controlled very accurately. When used as part of a complete control system the rotary position of the output shaft would be measured using a **shaft encoder** which would give an output proportional to angular movement. A typical linear position control system is shown in Figure 19.1(b).

The system shown in Figure 19.1(b) operates due to the fact that when the output shaft turns clockwise the right-hand lever moves to the right as does the left-hand lever. When the output shaft turns in an anti-clockwise direction the levers move to the left. Hence linear movement is achieved and by the connection of additional levers various directions of movement can be obtained. For example the steering of the front wheels of a small vehicle can be controlled in this way as shown in Figure 19.2.

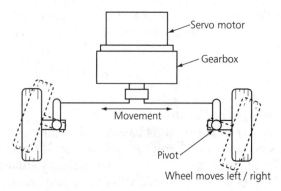

Figure 19.2 Servo motor with steering control system

The levers connected to the gearbox output shaft are connected to two levers which are in turn connected to the front wheel hubs. Left and right movement of the output shaft levers will cause the front wheels to turn to the right and to the left.

**Progress
check 19.1**

What are the main advantages of a servo motor?
Describe two applications for a servo motor.
How can the speed of a servo motor be controlled?

Stepper motor

A stepper motor to is a special type of d.c. motor with a permanent magnet rotor and a number of stator windings wound around the rotor to form magnetic pole pairs. The spacing between windings can be as little as 10°. Each winding is pulsed, in turn, with a digital voltage which causes the rotor to aline itself with that particular winding. The motor can therefore be either run continuously or the rotor can be stepped through a precise angle of rotation. The amount of rotation per pulse is referred to as the step angle and depends on the number and spacing of the stator windings.

Stepper motors are produced in sizes ranging from a few milliwatts to many kilowatts. Applications for this type of motor include drive motors for printers and plotters, numerically controlled machines (CNC) and very small versions are used in quartz analogue clocks and watches.

Linear movement actuators

Where greater linear movement is required with more force a direct linear movement actuator device is preferred.

Electric solenoid

The electric solenoid is an ideal linear movement actuator and is available in many different physical sizes to suit most applications. The amount of linear movement obtained is proportion to solenoid current. Figure 19.3 shows a typical solenoid used as a linear movement actuator.

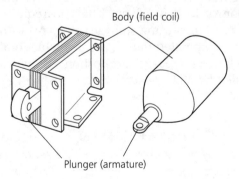

Figure 19.3 Solenoid as linear movement actuator

The solenoid does have some limitations however; the force available from the solenoid can be limited for applications requiring a large distance of linear movement. In this case it is best to use a pneumatic cylinder or hydraulic cylinder.

Pneumatic cylinder

A typical pneumatic cylinder is very simple in that compressed air supplied to one side of the cylinder will cause it to extend and when supplied to the other side will cause it to retract. The pneumatic cylinder is probably one of the most popular and widely used linear movement actuators and is used in most types of manufacturing processes. It has the advantage of being relatively cheap, easy to maintain and easy to control and comes in a wide range of physical sizes to suit most applications. The movement of the cylinder piston is often detected by either using lever-operated switches or proximity detectors. A typical setup is shown in Figure 19.4.

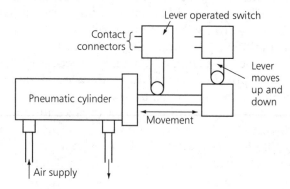

Figure 19.4 Pneumatic cylinder with position detectors

Hydraulic cylinder

For heavy-duty applications requiring high force levels the ideal actuator is the hydraulic cylinder the construction of which is similar to the pneumatic cylinder.

The operation of this device is the same as the pneumatic cylinder except that hydraulic fluid instead of compressed air is used to extend and retract the piston. Typical applications include presses and the control of steel rollers in a steel making mill. These devices are expensive and require a high pressure hydraulic fluid source to drive them. They do require regular maintenance as the rubber seals often fail causing fluid leakage. As with the pneumatic cylinder, associated electrically operated valves are available to allow them to be controlled electronically. In fact a large proportion of hydraulic and pneumatic cylinders are controlled by PLC systems as covered in Part 8 of this book.

Progress check 19.2

What are the main limitations of the solenoid as a linear actuator?
Explain why you would use a pneumatic cylinder rather than a solenoid for a given application.
Why would an hydraulic cylinder be used instead of a pneumatic cylinder for a given application?

The control of both pneumatic and hydraulic cylinders requires the control of air/fluid flow. This is achieved by using a variety of different types of valve.

Valves

Pneumatic

Valves for pneumatic control are available with four basic functions:

- directional
- non-return
- pressure control
- flow control.

Directional valves

These are used to control the way the air passes in terms of starting, stopping and changing the direction of air flow. Depending on the number of paths the air is allowed to take, directional valves are either two-way, three-way, four-way or multi-way. Typically a path is defined as an inlet connection, outlet connection or exhaust connection to the atmosphere.

The **two-ways valve** tends to be found in pneumatic control systems requiring a straight through function, that is starting or stopping the air flow. A ball-seat valve is often used for this purpose as shown in Figure 19.5.

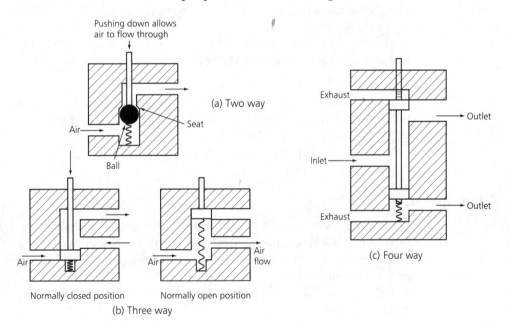

Figure 19.5 Directional valves

The **three-way valve** is used where during a control cycle air would need to be exhausted into the atmosphere such as when using a single-acting cylinder. It is interesting to note that a valve can be of a normally open or normally closed type as shown in Figure 19.5(b).

A **four-way valve** is normally the maximum number of ways used in a pneumatic system: **multi-way valves** with greater than four ways tend to be used in hydraulic systems. In a pneumatic control system a four-way valve would normally be used to control a double acting piston where the direction of air flow has to be changed when the piston is retracted. The basic operation of this type of valve is shown in Figure 19.5(c).

Designation of directional valves

There is an accepted designation for directional valves using two digits, the first to indicate the number of connections and the second to indicate the number of positions. Thus a 2/2 valve would indicate two connections and two positions, a 3/2 valve would indicate three connections and two positions and so on.

Valve design

Directional valves are designed mainly using either a lifting motion where they are known as poppet valves or a sliding motion known as piston or rotary valves.

Poppet valves

The flow of air through a poppet valve is controlled by the lifting of a plate, disc, ball or plug with flexible gaskets to seal the valve seat. A poppet valve features a rapid response coupled with few moving parts and therefore longer working life.

Ball-seat valve

This is the most inexpensive type of poppet valve and is normally made as either a 2/2 or 3/2 type valve.

This type of valve tends to be used in non-critical situations where some air leakage can be tolerated as the sealing of this type of valve is not perfect.

Disc-seat valve

This is a more expensive form of poppet valve which has much better sealing properties and is available as 2/2, 3/2 and 4/2 type valves. A typical disc-seat valve is shown in Figure 19.5(b).

Notice how all poppet type valves are sprung loaded and can be designed to be normally open or closed.

Valve operation

For a directional valve to operate some form of external effort must be applied to it and this can be mechanical, electrical, air pressure or fluid pressure. In the case of poppet valves the external effort is applied through air pressure. As more and more pneumatic systems are controlled using PLC's the most common form of valve activation is via a solenoid and this is most appropriate when used with a sliding type valve.

Sliding valves

The most common type of sliding valve is the piston or **spool valve** and these days is most commonly operated via electrical solenoids. The construction of this type of valve is quite complicated due the problems of sealing. The most common type of construction uses what are called 'O' ring seals which can be mounted either in the bore or on the spools as shown in Figure 19.6.

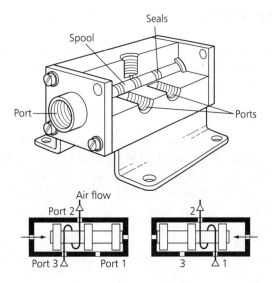

Figure 19.6 Spool valve construction

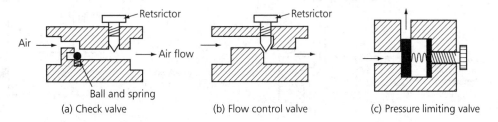

Figure 19.7 Valve types

Non-return valves

A non-return valve only allows air to flow in one direction. The simplest type of non-return valve is the **check valve**, shown in Figure 19.7(a).

This valve completely blocks the flow of air until the inlet pressure develops a force greater than that of the spring.

Flow control valves

These valves are used to control the flow of compressed air and therefore the speed of operation of pistons by using a restricting or throttling effect as in Figure 19.7(b). They are also available as non-return valves.

Notice that the throttling function is available in one direction only with the type of valve shown.

Pressure control valves

This type of valve prevents the pressure from rising above a maximum level and although they are used in compressed air generation systems they are rarely used in pneumatic control systems. Figure 19.7(c) shows a typical pressure limiting valve.

Valve selection criteria

Selecting the right type of valve for a given application is important and basically comes down to controlling single or double-acting pistons and their speed of operation.

Single-acting piston control

With this type of piston, once extended the air has to be exhausted form the piston assembly for it to retract. It is therefore a requirement in this application to choose a valve which will allow air to be exhausted when in one position. In this case the best valve to use would be a 3/2 type valve as shown in Figure 19.8(a).

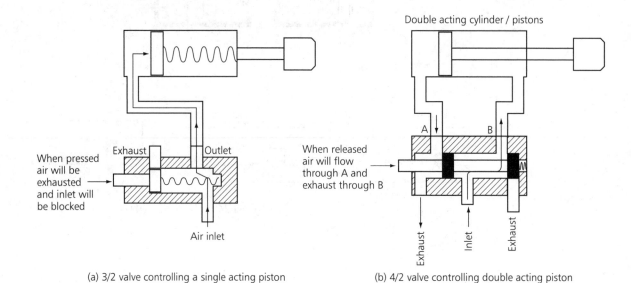

(a) 3/2 valve controlling a single acting piston   (b) 4/2 valve controlling double acting piston

Figure 19.8 Valve control systems

Activity 19.1

A pneumatic system is required which uses a single-acting piston and a 3/2 valve to control it. However, it is required that the piston operates at a lower speed on the return stroke. Produce a diagram of a suitable pneumatic system to meet this requirement.

Double-acting piston control

With a double-acting piston air pressure must be applied first in one direction to extend the piston and then in the other to retract it. Therefore a more complex directional valve arrangement is required. Typically a single 4/2 valve could be used as shown in Figure 19.8(b).

Activity 19.2

Think of another way of controlling a double-acting piston including speed control in both directions. Draw a suitable diagram to illustrate your answer.

Assignment 19

This assignment provides evidence for:
Element 14.3: Investigate the final control elements in engineering control systems
and the following key skills:
Communication 3.2, 3.3
Information Technology 3.1, 3.2, 3.3

A company called Manufacturing Processes Ltd requires a simple pneumatic control system to operate two pistons on a production line. The pistons are to operate in such a way that one piston extends and retracts and then the other extends and retracts and then the cycle repeats. The speed of the pistons may also need to be adjusted in either direction. The actual cycling is to be controlled by a PLC system which is already available. Your job is to specify a suitable pneumatic system including piston and valve types.

Report

The performance indicators for this element suggest that a report is produced on an investigation of the final control elements for a process control system and a servo control system and should include:

● a detailed description of the final control elements identified for each control system
● a description of their operating principles
● indentification of the final control elements for given engineering applications.

PART SEVEN: MICROELECTRONIC SYSTEMS

Chapter 20:   **Microprocessor-based systems**
Chapter 21:   **Software programming**
Chapter 22:   **Interfacing microprocessor-based systems**

- There is probably no one who is reading this textbook who has not had at least some experience of using a personal computer (PC). You have probably word processed your assignment material, for instance. In fact, it is hard to now imagine any facet of our lives which is not affected (or controlled!) by a computer of some sort. Closely linked to the growth of the use of the computer is the development of an electronic circuit built on a single piece of silicon called the **microprocessor**. Microprocessors first appeared in the early 1970's and were the result of developments in **silicon integrated circuit technology**. By that time it had become possible to integrate up to 10,000 transistors on tiny 'chips' of silicon crystal. The **silicon chip**, as it came to be known, could now be used for very large numbers of the basic logic gates (investigated in Element 13.3, Chapter 16) manufactured as one single micro-sized integrated circuit for processing large amounts of information in the form of digital signals. When such a microprocessor is combined with some additional integrated circuits it forms a general-purpose **microelectronic computing system**.
- In this part, the structure, programming and interfacing of microprocessors is described. When necessary the practical activities use a microprocessor-based system which uses a Z-80 microprocessor. Practical kits can be obtained from the supplier whose address is given at the beginning of this book.

Chapter 20: Microprocessor-based systems

This chapter covers:
Element 19.1:   Investigate microprocessor-based systems.

... and is divided into the following sections:
- Basic concepts
- Block diagram of a microprocessor-based system
- Hexadecimal numbers
- Internal structure of a microprocessor
- Microcomputer memories
- Microprocessor timing.

A microprocessor is an integrated circuit with thousands of logic elements that can be programmed to carry out arithmetic and logic operations following a predetermined sequence. They are truly versatile electronic devices that can be used in a huge variety of applications ranging from electronic games and toys to complex industrial control systems. Their versatility is due to the fact that the **standard logic arrays** can be programmed to perform different tasks. The program, or list of instructions, that each is required to follow will vary from application to application but since the microprocessor chip itself does not need to be customised they can be mass-produced so that manufacturing costs are kept low. Because of its adaptability, the microprocessor is often called the 'universal chip'.

Microprocessors form the heart of all microcomputer systems where it is used as the **central processing unit** (CPU) and operates in conjunction with other integrated circuits or 'chips' such as memory circuits and input/output devices (I/O) to achieve the desired functions. This chapter considers the main features of microprocessors and how microprocessor-based systems operate.

When you complete this element you should be able to:

- describe the features of microprocessing
- describe microcomputers in terms of their constituent elements using block diagrams
- describe the operation of microcomputer elements using block diagrams
- explain factors influencing microcomputer design.

This chapter begins by examining a very simple system. Although basic, it will illustrate many important concepts which apply to all microprocessors.

Basic concepts

The system in Figure 20.1(a) shows some of the essential components that would be needed in a 4-bit microprocessor. The system is imaginary but will illustrate many important points. Its overall function is to take in binary data from input switches, perform some operation on that data and then output to some lamps. Let us consider each block in turn.

The **accumulator** (A) is a 4-bit register. Registers were investigated in Element 13.3, Chapter 16 when the basic logic circuits were described. They are digital devices which can store bits. This register is 4 bits wide and can store the '1' or '0' signals from the input data. The **system controller** is a set of logic gates which manipulates the binary data in the accumulator depending on the pattern of bits held in the **instruction register** (IR). It also controls the rest of the system. The data in the IR in our case is set by some 'instruction switches'. The binary data into and out of A passes from the **input port** and on to the **output port**. For the moment we can think of the ports as being similar to registers. The operation of the system overall is controlled by the **timing control** block. At each step of the timing control circuit, the system controller carries out an instruction determined by what is held in the instruction register.

System operations

Since there are four instruction switches, there are $2^4 = 16$ patterns of '1's' or '0's' that can be set as **instruction conditions** on the switches. For any microprocessor the total number of instructions is known as the **instruction set**. Here the sixteen instruction **words** range from 0000 up to 1111. Suppose at the design stage of the system controller we have decided that some of the 4-bit words that can be set by the switches mean:

- 1111 = 'send (or input) data from input port to A'
- 0011 = 'send (or output) data from A to the output port'
- 1010 = 'shift contents of A one place to the right'.

We will suppose that the system has to perform the following operations in sequence:

- 'input data to A'
- 'output data to lamps'
- 'shift contents of A one place to the right'.

The sequence of operations that the system has to carry out (or **execute**) is called the **program**. The way it would be executed is shown in Figure 20.1 and 20.2

Figure 20.1 shows the initial condition. The registers will contain data 'bits' left over from previous operations and are shown here as '?'. The instruction switches in our example have been set to '1111' – the first instruction – and the input data switches are set at '1100'.

At the first step of the timing control circuit which we will call the **clock**, the instruction '1111' is placed in the instruction register and the data '1100' at the input is placed or clocked into the input port, Figure 20.2(a). At this stage the controller examines the instruction register and **decodes** the instruction. It now knows 'what it has to do'. While the controller is decoding the instruction, the

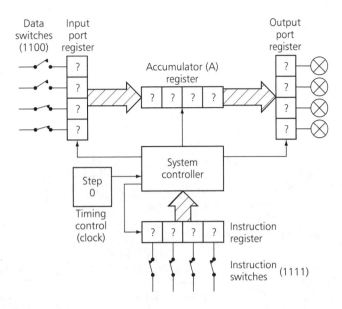

Figure 20.1 Simple (imaginary) 4-bit microprocessor

next instruction can be set on the instruction switches (0011 in this example) whilst the data input is maintained as before. The state of the system is shown in Figure 20.2(a).

At the next step of the clock, the controller transfers the input data to the accumulator and transfers the next instruction to the IR. Whilst the controller is decoding the instruction, the next instruction (1010) is set on the instruction switches. See Figure 20.2(b).

At the next step of the clock, the data in the accumulator is sent (output) to the lamps and the next instruction is read into the IR. For this program, no further instruction needs to be set on the switches. See Figure 20.2(c). Notice that the contents of A are not affected by the operation of transferring data to the output port. The controller decodes the instruction (1010) and is ready for the next operation.

At the completion of the next step of the clock, the controller moves the contents of A one place to the right. This operation has divided the original input data by two, i.e. 1100 in A has now become 0110. The system has performed a purely internal operation which has been arithmetic in nature unrelated to the input and output conditions. See Figure 20.2(d).

Before we highlight how improvements can be made to our system, you should note the following points:

- By changing the list of instructions (the **program**), a whole range of different tasks can be performed. This realisation is the essence of the importance of programmable microprocessors.
- Both the data and instructions are in binary code. Thus, for example, the word 0101 could be used to represent an instruction and data.
- The system can perform arithmetic operations (like the division in the example) and logical operations. For example, one of the instruction codes (say 0000) could instruct the controller to invert each bit in A, i.e. change '0' to '1' and vice-versa. (This is called 'complementing A').

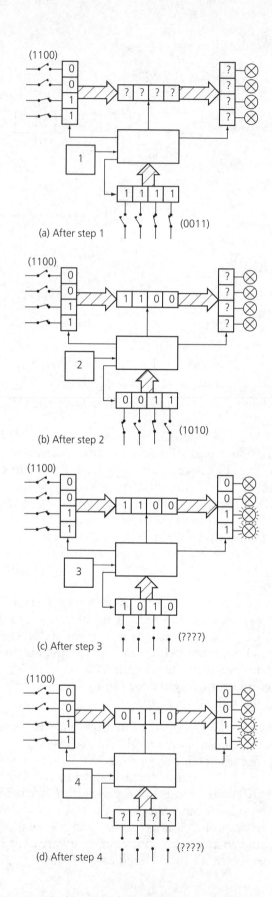

Figure 20.2 Basic
microprocessor
operation

- The overall operation is controlled by a timing circuit called the clock. You can think of the clock as providing the 'heartbeat' of the system ensuring that each part operates correctly in synchronisation with the other parts of the system one pulse at a time. In real microprocessor-based systems, the clock is derived from an oscillator circuit which is controlled by a quartz crystal.
- Our system always follows the same sequence of operations. It fetches an instruction, it decodes that instruction and finally it executes that instruction. This is called the **fetch/execute cycle** and will be examined in more detail later in the element.

Improving the system

The basic system can be made more versatile and powerful with some obvious improvements. Firstly, consider how instructions have been applied to the system. Rather than manually setting switches, the program instructions (and data) can be stored in some form of **memory**. The memory can be thought of as a series of separate locations with each location holding a 4-bit word. The 4-bit word may be an instruction that has to be executed or data that the microprocessor has to manipulate. Each memory location will have a unique **address** which will also be a binary number. Thus the microprocessor must be able to communicate with the memory and either read binary data from it or write binary data to it.

In microprocessor-based systems two types of memory devices are used. One is called **read only memory** (ROM). The microprocessor can only read the binary data that is held in ROM and cannot overwrite or alter it in any way. Another type of memory device called the **random access memory** (RAM) can be read from or written to under the control of the microprocessor. In microprocessor-based systems, ROM and RAM are separate integrated circuits with various storage capacities. We'll consider them in more detail later.

The second obvious improvement is to include more **work registers** within the basic microprocessor in addition to the accumulator A. In real microprocessors, these additional registers are labelled 'B', 'C', 'D' etc. and are used for the storage of data and other functions. In addition to these work registers other special-purpose registers will be included. We are now in a position to draw the block diagram of the basic elements that comprise a more realistic microprocessor-based microcomputer system and begin considering its operation in detail.

Progress check 20.1

> What would the contents of the accumulator be in our basic 4-bit microprocessor if the second instruction was 'complement A'?

Block diagram of a microprocessor-based system

A block diagram of a microprocessor-based system is shown in Figure 20.3(a). The individual devices (integrated circuits) that make up the system is usually called the hardware. Also shown in Figure 20.3(b) is the **pinout** (connection pins) of a Z-80 microprocessor. We will consider the microprocessor extensively in this part of the text book.

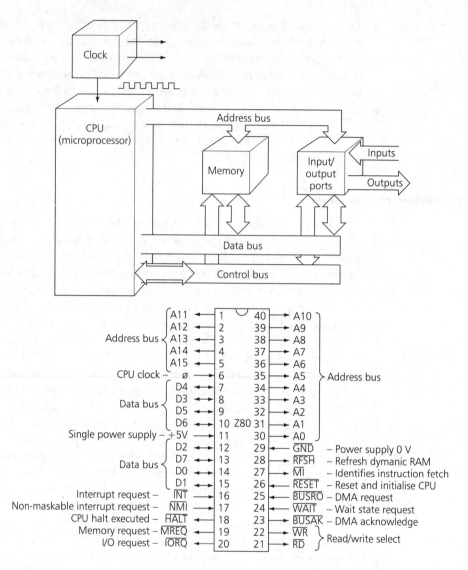

Figure 20.3 A microprocessor-based system and the Z-80 CPU

Central processing unit (CPU)

The CPU is the 'brain' of the system and carries out all of the arithmetic and logical operations as well as controlling how the other parts of the system operate together. As shown, it is fed with a stream of pulses produced by the clock circuit. The clock is used by the CPU to synchronise its own operations and all other operations throughout the system.

The control of, and communication with, the other components by the CPU is via the three buses called the **address, data and control buses**. In a real system the buses will be copper tracks on a printed circuit board to which the bus connection pins on the integrated circuits will be soldered.

The address bus

The address bus leaves the CPU and is connected to other parts of the system. Since it is 'one way' it is called **unidirectional**. During CPU operations, binary data will be placed on the address bus, i.e. groups of '1's' and '0's'. These bit patterns are called **addresses** and the address 'tells' the system which part is to be used at any instant. The Z-80 CPU in Figure 20.3(b) has 16 address lines labelled A_0 to A_{15}. The number of binary bit patterns possible is therefore $2^{16} = 65,536$, so the Z-80 CPU can produce 65,536 separate addresses. The Z-80 would be described as having a 16-bit address bus.

The data bus

The data bus is used to carry binary data from the CPU to the rest of the system or to carry data back to the CPU from that part whose address appears on the address bus. Thus the data bus is **bidirectional**. The Z-80 has an 8-bit data bus labelled D_0 to D_7 in Figure 20.3(b) and so data would be moved around a Z-80 system 8 bits at a time. Microprocessors with 8, 16, 32 and 64-bit data widths are available.

The control bus

The control bus carries signals which control and synchronise the overall operation of the system. Most control signals are produced by the CPU but a few will be passed into the CPU from other parts of the system. The number of control signals produced will depend on the particular microprocessor. The Z-80 CPU has 13 control bus lines, for example. Notice that most are generated by the CPU while others are accepted by the CPU as inputs.

Progress check 20.2

On what pins must the clock and the power supply be connected on a Z-80 CPU?

The memory

The memory is that part of the system that holds the instructions that the CPU must execute and any data that the CPU needs to process. Each location in the memory has a unique address which can be accessed by the CPU when the CPU puts that address on the address bus. Sometimes the CPU may need to 'read data' from the memory while at other times it may need to 'write data' to the memory. Data to and from the memory travels along the data bus. Considering the Z-80 for example: since it has an 8-bit data bus, each memory location will hold 8 bits of binary data. In computer jargon, 8 bits is called a **byte**. Since the Z-80 can produce 65,536 separate addresses on its address bus, then it can use a memory system which is 65,536 bytes long.

In general, the memory will store the CPU instructions and data sequentially in one address after another. The CPU will then read the instructions from the memory, one after the other and carry out the operations required by the program in sequence. As indicated previously, both ROM and RAM memory devices will be used as part of the overall memory of the system. We will examine these devices later in this chapter.

Input/output ports

The system provides communication with external devices such as printers, disk drives, VDU's, keyboards etc. These external devices are usually called **peripherals**. Under the control of the CPU, data to and from the CPU (or memory) from or to the peripherals passes through input and output **ports**. The ports are part of an integrated circuit called an input/output circuit or **I/O chip**. Most I/O devices have a number of separate ports. Generally, a port can be set up to take data in from a peripheral (say a keyboard) or transfer data out to another, for example a printer. Thus the I/O device providing the input and output ports allows for the correct **interfacing** of the microprocessor-based system with the outside world. The interfacing of microprocessor-based systems will be examined in Chapter 22.

Hexadecimal numbers

The relationship between decimal and binary numbers was examined in Chapter 16 in our work on logic circuits. The binary number system was the obvious choice to use for logic circuits since the inputs and outputs of the logic gates were either about 0 V ('0') or about 5 V ('1'). Since our microprocessor-based system is also a digital system involving processing and controlling of these digital voltages, then binary numbers are used again. The memory in the system, for example, may contain instructions and data in the following sequence:

Memory location 1 0 0 0 1 0 1 0 1

Memory location 2 0 1 1 1 1 1 1 1

Memory location 3 0 1 0 1 1 0 1 1

etc.   etc.

Each 8-bit word (a byte) is part of the program that the CPU has to execute.

Denary	Binary	Hexadecimal
0	0000	0
1	0001	1
2	0010	2
3	0011	3
4	0100	4
5	0101	5
6	0110	6
7	0111	7
8	1000	8
9	1001	9
10	1010	A
11	1011	B
12	1100	C
13	1101	D
14	1110	E
15	1111	F

Furthermore, each memory location has a unique address which is also a binary number. The Z-80, for example, has a 16-bit address bus. A particular address may be '010111100001011'. Remembering and understanding these long binary numbers is very difficult so a shorthand notation is used instead. It is called the **hexadecimal numbering system** or hex for short. Using hex is easy but you should never forget that the digital microprocessor system is always using binary patterns of approximately 0 V and 5 V signals in all of its operations. Thus, no matter how a program for a system is written or developed, it will always end up as a series of binary numbers that the microprocessor understands. These binary instructions are called the **machine code** for the microprocessor.

In the hexadecimal system, the binary numbers from 0 to 15 are represented by a single character. These characters are 0, 1, 2, 3, 4, 5, 6, 7, 8, 9, A, B, C, D, E, F. When long binary numbers need to be represented in hex, the number is divided into groups of 4 bits starting from the right and each group of four bits is replaced by the correct hexadecimal character. The decimal, binary and hexadecimal equivalents are shown in the table.

WORKED EXAMPLE 20.1

Convert the binary numbers (a) 1010111100010010 and (b) 1100000011101111 to hexadecimal. (c) Convert the hexadecimal number BCD7 to binary.

Solution

(a) Divide the binary number into groups of 4 bits starting from the right and convert in hex form.

1010	1111	0001	0010
A	F	1	2

Therefore, 1010111100010010 = AF12 in hex.

(b) Similarly:

1100	0000	1110	1111
C	0	E	F

(c) Each hex character is replaced by its binary equivalent:

B	C	D	7
1011	1100	1101	0111

WORKED EXAMPLE 20.2

The Z-80 CPU has a 16-bit address bus. What are the hexadecimal values of the lowest and highest addresses that the CPU can produce?

Solution

The lowest address is when all the address lines A_0 to A_{15} have a '0' (about 0 V) placed on them by the CPU. Thus the lowest address is 0000 hex. That is:

$$A_{15}\ A_{14}\ A_{13}\ A_{12} \quad A_{11}\ A_{10}\ A_9\ A_8 \quad A_7\ A_6\ A_5\ A_4 \quad A_3\ A_2\ A_1\ A_0$$

$$0\ \ \ 0\ \ \ 0\ \ \ 0 \quad\ \ 0\ \ \ 0\ \ \ 0\ \ \ 0 \quad\ \ 0\ \ \ 0\ \ \ 0\ \ \ 0 \quad\ \ 0\ \ \ 0\ \ \ 0\ \ \ 0$$

$$0 \qquad\qquad 0 \qquad\qquad 0 \qquad\qquad 0 \qquad \text{(hex)}$$

The highest address is when the CPU places a '1' (about 5 V) on each address line. The highest address is FFFF Hex. That is, when a '1' appears on all address lines.

The Z-80 therefore has an address range of 0000, 0001 etc. all the way up to FFFF. It can access 65,536 individual addresses.

Progress check 20.3

What is the range of data codes in hexadecimal that can exist on the one-byte wide Z-80 data bus? How many possible combinations is this?

Internal structure of a microprocessor

The internal structure of a microprocessor is often called its **architecture**. Each type of microprocessor from different manufacturers will have its own unique

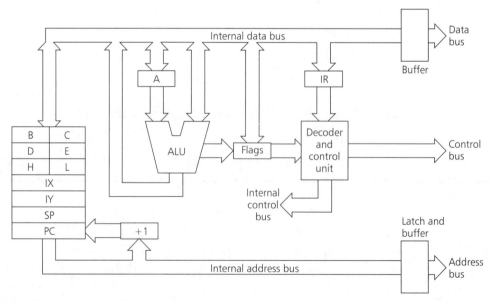

Figure 20.4 Simplified Z-80 internal structure

architecture. This may range from being relatively simple like the 8-bit Z-80, up to complicated structures like the Intel Pentium which has a data bus width of 64 bits. An understanding of the internal structure of any microprocessor is essential if it is to be constructively used in any practical application. We will concentrate on the Z-80 CPU but the lessons learnt here can be applied to other microprocessors. The simplified internal structure of the Z-80 is shown in Figure 20.4. Remember, in reality it is a 40-pin integrated circuit. We will consider each part in turn.

Registers

General-purpose registers
There are six 8-bit registers labelled B, C, D, E, H and L. These are connected to the internal data bus of the chip via a circuit, which is not shown in the diagram, called a **multiplexer**. The multiplexer ensures that only one register can be connected to the data bus at any instant. You have met the term multiplexer before in Part 3 of this book, in connection with time division multiplexing techniques. The multiplexer ensured that only one channel could have access to the common transmission medium at any one time.

Although registers B, C etc. are 8-bit, there are some instructions that the CPU performs that allow them to work together in pairs. Thus 16-bit operations are possible. The register pairs are BC, DE and HL.

16-bit registers
In the register block are four other registers that are 16 bits wide. These are the **index registers** (IX and IY), the **stack pointer** (SP) and the **program counter** (PC). The function of each is as follows.

Program counter (PC). We have discovered that the microprocessor carries out its operations by fetching the instructions from memory and then executing these instructions one after the other. When an instruction is needed from memory, the address of the memory location is placed by the PC onto the address bus. The data will then return from memory via the data bus. During the time this is happening, the address in the PC is **incremented by one**. (That's what the '+1' box indicates in the diagram.) The PC thus points to the next address in memory and the CPU can receive further instructions or data from it. The CPU therefore uses the program counter to step through the program held in memory one step at a time.

Stack Pointer (SP). When the CPU is executing a program it may store data temporarily in a part of the memory of the system. This area of memory is called the **stack** and operates on a **last in/first out** basis. (This means that the last byte of data in will be the first one to be taken out.) Data to be stored on the stack is placed into the memory location pointed to by the 16-bit SP. The SP then automatically decrements by 1 before the next byte of data is stored. If a byte of data is read off the stack the SP is incremented by 1. Thus the last byte in is the first one out.

Index registers (IX, IY). IX and IY are 16-bit registers which can be used to 'point' to any part of the memory. They are particularly useful when tables of data bytes are stored in memory and might need to be used by the program. Usually, an index register is loaded with the address of the base of the data table that is held.

To read, say, the third data value from the table, 03 Hex is added to the index register value and then the third data value can be used.

Instruction register/decoder control unit

When the microprocessor is running and executing a program, each instruction from the memory is first passed into the **instruction register** (IR). The pattern of 8-bits is examined by the IR and decoded. At this point the CPU 'knows' what is required by the instruction and the control unit sets up the necessary control signals to execute that instruction.

Accumulator and arithmetic logic unit

If an instruction requires an arithmetic operation (e.g. adding or subtracting two numbers) or a logical operation (e.g. ANDing or ORing two numbers), then the **arithmetic logic unit** (ALU) is used. Generally, the accumulator (A) will contain one of the numbers and will be used to hold the result of the operation. Like registers B, C etc., A is an 8-bit register but is more versatile. Data from input and to output ports, for example, use the accumulator.

Flagging system

The **flag** (or **status**) register is a collection of flip-flops which are affected when the ALU is used. Flip-flops were examined in Element 13.3 when we investigated logic circuits. Most microprocessors have five or six flags which can be either set or reset by operations that have just taken place in the ALU. For example, one important flag is the **zero flag**. If the result of an operation produces a result of zero, e.g. subtracting two data values that are identical, then the zero flag is set to '1'. It could be that when this flag is set, a particular set of program instructions needs to be executed but if it is reset then some other operations need to be performed. The CPU therefore uses the flag register to make decisions. Flags will be considered in more detail in Chapter 21 when the programming of microprocessors is investigated.

Alternative register set

The Z-80 CPU has an alternate register set which is not shown in Figure 20.4. These registers are labelled A', F', B', C', D', E', H' and L'. Either the main set or the alternative set may be used, *but not at the same time*. There are, however, some special instructions that allow the programmer to switch between each register set for some operations. The use of the alternate register set is specialised and will not be considered further in this book. Because of the large number of registers provided in the 8-bit Z-80, it is often called a register-orientated microprocessor.

Buffers

A buffer is an electronic circuit which allows its input and output to be isolated from one another. The data bus buffer isolates the CPU data bus from the external data bus when required.

The **latch** on the address bus allows the address on the external address bus to be held fixed, even though the address on the internal address bus might change.

Activity 20.1

Obtain a data sheet for a 6502 microprocessor. Produce a short report comparing it with the Z-80 in terms of the number of registers and bus widths.

Case study

Intel is a world leader in the manufacture and development of **Intel microprocessors**. Since their introduction in 1971 Intel, and other manufacturers, have been constantly improving microprocessor architecture with the resulting increase in speed and computing power. The first micrprocessor, the Intel 4004, was available in 1971 and was produced for a Japanese company manufacturing electronic calculators and is 'archaic' compared to present day processors. The table below shows some of the capabilities of some common processors.

	8086/88 (the 'XT')	80286 (the 'AT')	80386	80486	80586 (the Pentium)
Date	1978	1982	1985	1989	1993
IC package	40 pin	68 pin	132 pin	168 pin	273 pin
Number of transistors in the chip	29,000	120,000	275,000	1.2 million	3.1 million
Data bus width	8088 8-bit 8086 16-bit	16-bit	32-bit	32-bit	64-bit
Address bus width	20-bit	24-bit	SX 24-bit DX 32-bit	32-bit	32-bit
Size of memory that can be addressed	1 Mbyte	16 Mbyte	SX 16 Mbyte DX 4 Gbyte	4 Gbyte	4 Gbyte
Clock speed (MHz)	4.77/8	10/12/16/20	SX 16/20/25 DX 33/40	SX 20/25 DX 33/50 DX2 50/66 DX4 99	60/75/90/100 133/150/166/ 200

Notice the ever increasing clock speeds, data and address bus widths and the complexity of the integrated circuit.

The choice of which microprocessor (and hence microcomputer system) to use will in general depend on many factors. In some industrial control applications, for example, an 8-bit, 'slow running' microprocessor like the Z-80 may be perfectly adequate for the task in hand. If a microcomputer was only to be used for word processing, for example, then a 'low speed' machine would be suitable. In this application it is the typist's typing speed that is important and there is little advantage in having very fast processing and other powerful facilities when the overall speed depends on manual operation. On the other hand, the processing of complicated spreadsheets, databases, graphic software etc. will require powerful and fast microprocessors.

Activity 20.2

Investigate the computer network at your Centre and find out what microprocessors are used within the system.

Microcomputer memories

Two types of memory device must be provided within a microprocessor-based system. There must be fixed memory called **read only memory** (ROM) which cannot be changed or erased and alterable memory called **random access memory** (RAM) which can be altered at will by the user. Memory devices which retain their data when power is removed are called **non-volatile**. A device which loses its data when power is disconnected is called **volatile**.

Both ROM and RAM ICs are available in various types, as illustrated in Figure 20.5.

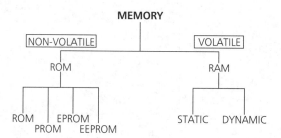

Figure 20.5 Types of memory

Types of memory

ROM

ROM is **read only memory** that is programmed once by the manufacturer. It is **non-volatile** and hence retains its program even when power is removed. It is usually only economical to produce ROMs when large quantities of identically programmed devices are required.

PROM

PROM is **programmable ROM** and is supled by the manufacturer in a blank state. A PROM device is produced with a large number of 'fusible links' which can be preset (blown) by the user using a special device called a PROM programmer. Once the program has been set into a PROM it cannot be altered and so PROMs are normally only used when a program is finalised and fully working.

EPROM

EPROM is **erasable PROM**. It is programmed in a similar way to a normal PROM but can be erased by shining ultra-violet light through a quartz glass window in the top of the IC for about 20 minutes. EPROMs are very useful for program development because of their ability to be reprogrammed and erased at will.

EEPROM

EEPROM is an **electrically erasable PROM**. This is similar to EPROM but is erased electrically rather than by ultra-violet light.

RAM

RAM is **random access memory** and is a type of memory that can be read from or written to. It is volatile in that whatever data/program it contains is lost as soon as power is removed.

Static RAM consists of individual 'cells' based on two transistors which can store a '1' or a '0' until power is removed.

Dynamic RAM consists of individual 'cells' based on a single transistor and a capacitor. The capacitor stores charge for a '1' and is discharged for a '0'. Because the stored charge leaks away relatively quickly, dynamic RAM chips need additional circuitry called **refresh** circuitry to 'top up' the stored charge every few milliseconds.

RAM devices can be written to, and read from, as often as required and if it is necessary to retain data when power is lost, special battery-backed RAMs are available. Normally CMOS RAM is chosen because of its low power consumption.

Memory capacity

The capacity of a memory device is the number of individual bits it can hold. As an example, Figure 20.6(a) shows a commercially available EPROM from RS Components. This one has a total capacity of 32,768 bits arranged as 4096 bits by 8 bits wide. In computer 'jargon' 8 bits is called a byte and 1024 bits is called '1K' – so this EPROM is a 4K by 1 byte device.

In general, the size of a memory in bytes is determined by the number of **address lines** connected and can be calculated using the formula:

As shown more clearly in Figure 20.6(b), this device has twelve address lines (A_0 to A_{11}) giving a size of:

$2^{12} = 4096 = 4K$

$$\text{Memory size in bytes} = 2^N$$

$$\text{where } N = \text{number of address lines}$$

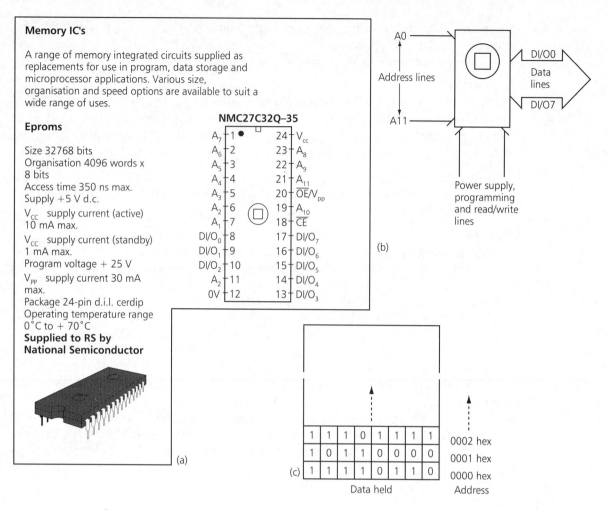

Figure 20.6 An EPROM (courtesy RS Components)

There are eight data lines labelled DI/O_0 to DI/O_7 and additional lines for programming, power supply and reading and writing.

Figure 20.6(c) is a 'model' you can use to understand how the memory is arranged.

Each group of 8 bits held has a **unique address**, usually given in hexadecimal. The lowest address is when A_0 to A_{11} are all zero volts, i.e. '0'(0000 hex) and this address holds the data 11110110. The next address is 0001 hex and holds data 10110000 and so on. If the EPROM is being used and address 0001 Hex is being read from, then the data 10110000 would be available on the Pins DI/O_0 to DI/O_7 after a time (for this device) of 350 ns. This is called the **access time**.

Constructing larger memories

Rarely can a memory requirement for a microprocessor-based system be provided by a single memory chip. Usually several devices are required. Since each chip of the same type number is identical, they need to be arranged to have different address ranges within the total address space that the microprocessor can have. The usual way of achieving this is to incorporate logic circuits called decoders

within the system. How decoders are used is illustrated with the following
Worked Example.

WORKED EXAMPLE 20.3

A Z-80 microprocessor-based system needs 8 Kbytes of EPROM memory. The EPROMs
need to occupy the bottom 8K of the 64K (65,536 addresses) memory space that the
16-bit address bus can address. Design a system to achieve this using a 2-line to 4-line
decoder.

Solution

(a) *Choosing the EPROMs.* The EPROM shown in Figure 20.6 is a 4K device so
two of these chips are needed for this design. The 12 address lines A_0 to A_{11}
need to be connected to each chip to access the 4K (2^{12} = 4096) memory locations.

(b) *Using a decoder.* A 2-line to 4-line decoder is shown in Figure 20.7(a).
Depending on the logic level at the inputs (00, 01 etc.), one of four outputs
will go to logic '0'. These outputs are connected to the **chip enable** pins of
the memory devices. In this application only two outputs are needed to select
either of the two EPROMs. The higher order address lines (in this case A_{12}

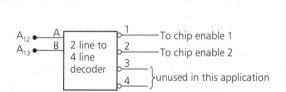

Truth table

Input		Outputs			
B	A	1	2	3	4
0	0	0	1	1	1
0	1	1	0	1	1
1	0	1	1	0	1
1	1	1	1	1	0

(a) 2 line to 4 line decoder

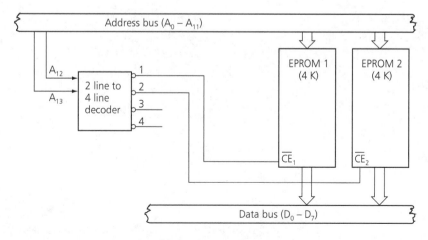

(b) Circuit connection

Figure 20.7 Worked example 20.3

and A_{13}) are used as the decoder inputs. The final circuit connections are shown in Figure 20.7(b).

Its operation can be considered by examining the following table:

A_{15} A_{14} A_{13} A_{12}	A_{11} A_{10} A_9 A_8	A_7 A_6 A_5 A_4	A_3 A_2 A_1 A_0	ADDRESS RANGE (HEX)	CHIP
X X 0 0	0 0 0 0	0 0 0 0	0 0 0 0	0000	
		to		to	1
X X 0 0	1 1 1 1	1 1 1 1	1 1 1 1	0FFF	
X X 0 1	0 0 0 0	0 0 0 0	0 0 0 0	1000	
		to		to	2
X X 0 1	1 1 1 1	1 1 1 1	1 1 1 1	1FFF	

Since A_{14} and A_{15} are not used in this application, they are 'don't care' (X) so we can assume they are '0'. When $A_{12} = A_{13} = 0$, the first output of the decoder is '0' so EPROM 1 is selected. EPROM 1 occupies the address range: 0000 hex to 0FFF hex (first 4K of memory). When $A_{12} = 1$ and $A_{13} = 0$, EPROM 2 is selected. EPROM 2 occupies the address range: 1000 hex to 1FFF hex (second 4K of memory).

Notice that if required, the other outputs of the decoder could be used to 'chip enable' other memory devices and so expand the memory capability of the system.

The **memory map** which shows how the available memory space is allocated to the memory chip is drawn in Figure 20.8.

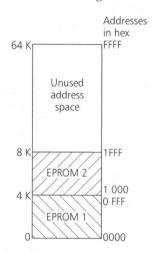

Figure 20.8 Memory map for Worked Example 20.3

In a typical system some RAM chip(s) would probably be added to store programs written by the user.

**Progress
check 20.4**

A RAM chip has 10 address lines (A_0 to A_9). What is its capacity?

The fetch-execute cycle

The implementation of every instruction by a microprocessor can be divided into two timing stages:

1 **Fetch** the instruction from memory and place in the instruction register.
2 **Execute** the instruction.

The fetching part of the cycle is illustrated in Figure 20.9.

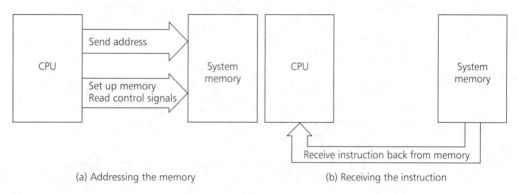

(a) Addressing the memory   (b) Receiving the instruction

Figure 20.9 Fetching an instruction

The CPU first sets up the necessary control signal to read the memory and its **program counter** sends the address of that memory location on the address bus. This is called '*addressing the memory*'. A short time later, depending on the **access time** of the device, the instruction which is usually called the **operation code** (or op code for short) is sent by the memory back along the data bus to the CPU. The CPU 'reads' the data bus and places the op code into its instruction register. Once it has decoded what the op code means, it sets up the necessary control signals to execute the instruction. Some instructions may require additional data to be fetched from memory while others may require only arithmetic and logic operation inside the CPU itself.

Types of instruction

The number of memory addresses required by an instruction will depend on the complexity of that instruction. A program stored in memory might look something like that shown in Figure 20.10.

The instructions are generally described as being 1 byte, 2 byte, 3 byte etc. and indicate the number of memory locations occupied. An important point to understand is the answer to the following question: 'How does the CPU know where the first instruction is in the memory?' The answer is to understand what happens when the CPU is reset. If you look back at Figure 20.3 you will see that the Z-80 has a RESET pin (pin 26). When reset, all of the internal registers are cleared, i.e. 'filled with '0's'. The program counter contains 0000 Hex at reset. Thus, the

397

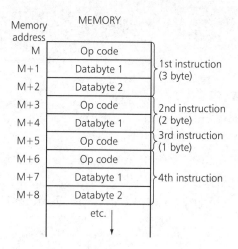

Figure 20.10 A possible program stored in memory

Z-80 *must find a valid op-code at this address*. If it doesn't, the CPU 'crashes'. Notice, however, that this instruction may be one that causes the Z-80 to 'jump to' another address that holds the main program.

Instruction set

At this stage it is necessary to have a simple appreciation of the types of instructions that the CPU will execute. From now on only the Z-80 microprocessor will be considered.

The range of instructions that a microprocessor recognises and executes is known as its instruction set. When stored in memory each instruction is, of course, just a pattern of binary bits. It would be very tedious and error prone to have to write these binary codes on paper when programs are being designed. A more convenient method is to use an abbreviated name, which is easily understood, which represents the op code. The abbreviated name is known as a **mnemonic**. The mnemonic is followed by variables which are called **operands**.

For example, the instruction

LD A, B
↑

mnemonic

means 'load register A from register B'. The mmnemonic is load (LD) and is easily understood. The actual binary code used to tell the Z-80 to perform the operation is 01111111 = 7 F in hex. The use of mnemonics greatly simplifies program writing and is an important skill to develop.

In general, the instructions for a microprocessor will fall into three categories called:

● data transfer instructions
● arithmetic and logic instructions
● test and branch instructions.

We will now describe a few instructions from each group so that you have a 'flavour' of typical microprocessor instructions.

Data transfer instructions

Example: LD A, FDH. This means 'load register A with FD hex'. It is called **immediate addressing** because the data to be loaded into A is in the memory location immediately after the op code.

Example: LD A, C. This means 'load register A from register C'. It is called **register addressing** because the data to be placed in A is found in CPU register C. Note that C will still hold the same data after the operation is completed.

Example: LD A, (HL). This means 'load A with the contents of the memory location whose address is held in the HL pair'. This is called **indirect addressing**.

Example: LD A, (1900H). This means 'load A with the contents of memory location 1900H'. This is called **extended addressing**. Notice that since the address is a 16-bit number, the instruction would occupy 3 bytes of memory in the order, op code, low byte of address (00H), high byte of address (19H).

Arithmetic and logic instructions

Example: ADD A, B. This means 'add the contents of B to the current contents of A'. When executed, the result of the addition is placed in A.

Example: DEC B. This means 'decrement by 1 the contents of B'. This type of instruction is often used for 'counting purposes'.

Example: AND B. This is a **logic instruction**. This means that the contents of B are ANDed with the contents of A. The result is placed in A. For example, suppose the contents of A and B are D8 hex and B3 hex. The result of executing the instruction AND B would be:

> (A) 1101 1000
>
> (B) 1011 0011

AND B = 1001 0000

This is the new contents of A.

Test and branch instructions

Example: CP OAH. This is the **compare instruction** which is often used when 'decisions' have to be made by the CPU. The CP instruction compares two values but does not alter either. It does, however, effect flags – for example the **zero flag**. In this example, the contents of A is compared to the immediate data value OA hex. If they are equal the zero flag is set, otherwise it is reset. The result of this test allows decisions to be made. For example, if the zero flag were set, the program might be designed to JUMP to another address and continue executing the program from that address.

Example: JP 1900H This is an unconditional jump instruction. When the microprocessor reads this instructions it 'jumps' to the address 1900 hex and continues executing the program from there.

Example: JP NZ, 1807H. This is a **conditional jump** instruction. If the result of the previous operation did not set the zero flag, (i.e. the result was not equal to zero), then jump to address 1807H and continue executing the program from there.

Microprocessor timing

All microprocessor operations are controlled by the clock circuit. The clock produces a steady 'stream of pulses' that are fed into the microprocessor (pin 6 on the Z-80). The microprocessor uses the clock to synchronise all of its operations. One cycle of the clock is known as a T-state. For example, if the clock frequency is 4 MHz, then the time per T-state is

$$\text{T-state time} = \frac{1}{4 \text{ MHz}} = 250 \text{ ns}$$

The number of T-states required to execute an instruction is not fixed but depends on the particular instruction.

The first operation the microprocessor must perform with every new instruction is to fetch the op code from memory and decode it to determine what it has to do. The sequence of events is called the *op code fetch cycle or machine one cycle* (M1). The designers of the Z-80 allocated 4, and only 4, T-states for this operation to be completed – nothing more and nothing less. Thus the M1 cycle is always 4 T-states long. If the op code 'tells' the CPU that it must read the next piece of data from the next memory location, then the designers have allocated a certain number of T-states for this operation. This one would be a normal memory read and takes 3 T-states. Each operation is precisely defined – there is no ambiguity whatsoever.

The precise timing diagrams will be examined later. For the moment we can get a good idea of how the Z-80 operates by using the simplified architecture diagram of Figure 20.4 to examine what happens when a typical instruction is executed.

WORKED EXAMPLE 20.4

The two-byte instruction LD A, 0AH is held in memory locations 1800H and 1801H in a Z-80 microprocessor-based system. Use the simplified architecture diagram to illustrate how the instruction is executed.

Solution

As you will discover in the next element, the hex code for the instruction LD A, 0AH is 3E 0A. These bytes of data would be stored in memory as follows:

Memory address	Data stored
1800	3E
1801	0A

The Z-80 must therefore read and decode the op code 3EH held in memory address 1800H and then load the data value 0AH into its accumulator. The first part is machine cycle 1 which takes 4 T-states. The final part is a normal memory read that takes 3 T-states. The sequence of events is as follows. The 7 T-states are illustrated in Figure 20.11.

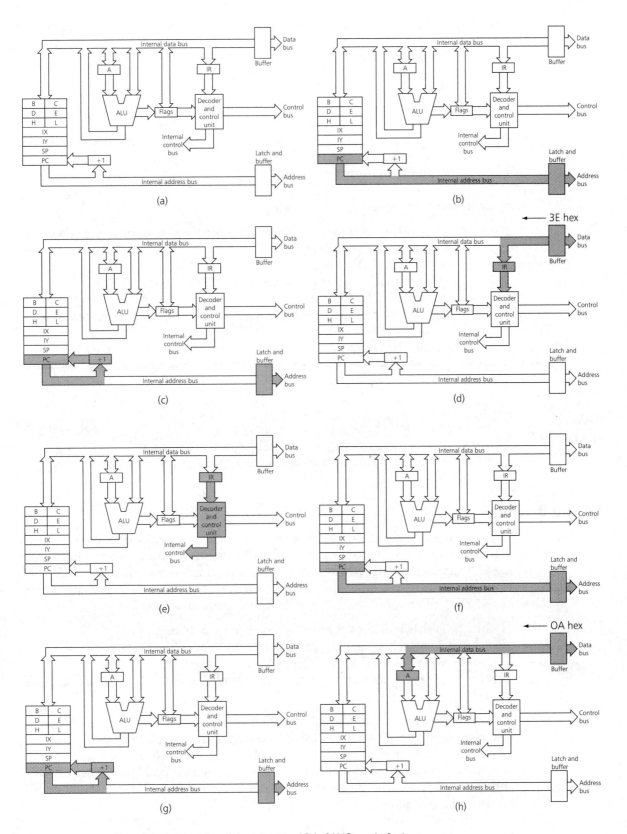

Figure 20.11 Execution of the instruction LDA, 0AHOo code fetch

T-state 1 Address 1800H transferred from PC to address bus. Control signals set up to read data from memory.
(Figure 20.11(b))

T-state 2 PC incremented to contain address 1801H.
(Figure 20.11(c))

T-state 3 Instruction 3E returns over data bus and placed in instruction register.
(Figure 20.11(d))

T-state 4 CPU decodes instruction and sets up necessary control signals for execution.
(Figure 20.11(e))

Normal memory read

T-state 1 Contents of PC (1801H) placed on address bus. Control signals set up to read data from memory.
(Figure 20.11(f))

T-state 2 PC incremented to contain address 1802H.
(Figure 20.11(g))

T-state 3 Data 0A hex returns from memory on data bus and is placed in accumulator.
(Figure 20.11(h))

Progress check 20.5

If the clock circuit for the Z-80 system in Worked Example 20.4 runs at 4 MHz, how long does it take to complete the LDA, 0AH instruction?

Timing diagrams

Manufacturers give full details of how their microprocessors operate. Of particular importance are the timing diagrams. These show which control lines from the CPU are active and when data and address bus values are stable and valid. Figure 20.12 shows the op code fetch cycle. If you refer back to the pinout diagram of the Z-80, the pins of the control lines that are active are:

$\overline{M1}$ pin 27 which indicates that an op code fetch is taking place.
$\overline{MREQ}$ pin 20 which is usually used to select memory decoders.
$\overline{RD}$ pin 27 which is used to read from memory chips.
$\overline{RFSH}$ pin 28 which is used to refresh dynamic RAMs if they are used in the system.

Note that on the rising edge of T_3 the op code on the data bus is read into the instruction register. Note that $\overline{MREQ}$ becomes active again during T_3 and T_4 to access the refresh address that the Z-80 places on the address bus.

A normal memory read diagram is shown in Figure 20.13. Notice only $\overline{MREQ}$ and $\overline{RD}$ are active. The write line $\overline{WR}$ (pin 22) stays high. Data enters the CPU on the falling edge of T_3.

If Figures 20.12 and 20.13 were placed one after the other, then we would have the precise timing sequence for Worked Example 20.4.

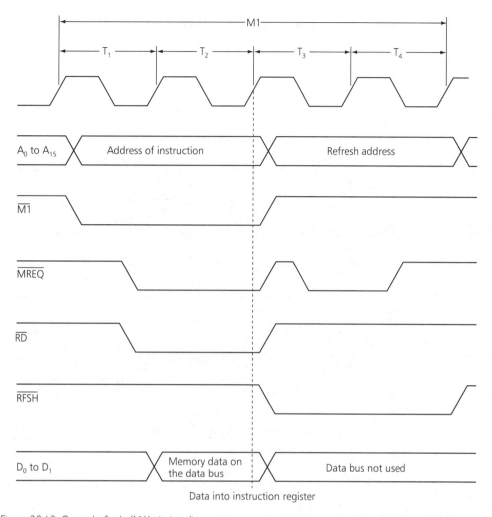

Figure 20.12 Op code fetch (MI) timing diagram

The memory write timing diagram is shown in Figure 20.14. This takes 3 T-states as well. Note that when writing to memory from the CPU, the $\overline{RD}$ pin is inactive and stays high. $\overline{MREQ}$ and $\overline{WR}$ are the active controls.

Timing diagrams similar to those in Figures 20.13 and 20.14 are given by the manufacturer for reading and writing to input and output ports. In this case $\overline{IORQ}$ would be the main control line.

Case study

Throughout this element the CPU we have focused on has been the Z-80 manufactured by Zilog. This CPU has formed the basis of several training systems to develop both hardware and software skills in microcomputer technology. In this case study we will examine some of the facilities provided by a Z-80 **microcomputer training system** called the Micro-professor. A block diagram of the system is drawn in Figure 20.15. Only the more important connections are shown.

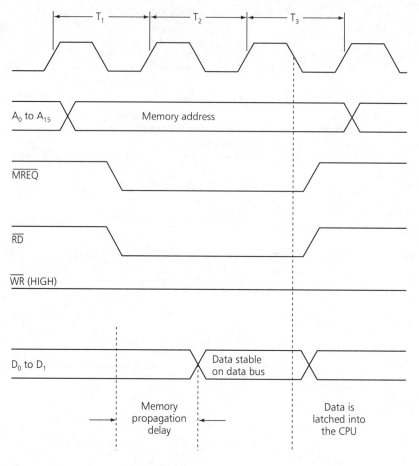

Figure 20.13 Normal memory read timing diagram

Many of the features provided by the Z-80 have already been described but the more important ones can be summarised as follows:

The Z-80 CPU
16-bit address bus allowing 64K of memory to be addressed.

- 8-bit data bus.
- Control line for memory addressing, $\overline{MREQ}$.
- Input/output addresses are resricted to the lower 8-bits (A_0–A_7) of the address bus so 256 I/O ports could be used if required.
- Separate input/output control line, $\overline{IORQ}$.
- Separate read and write control lines, $\overline{RD}$ and $\overline{WR}$.
- Several internal 8-bit registers.
- 16-bit stack-pointer register so that, in theory, the stack could occupy the whole of the 64K of addressable memory.
- Single clock input.

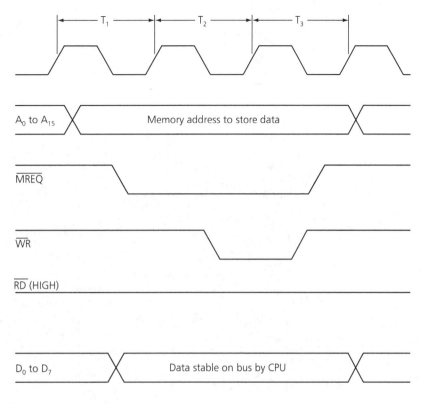

Figure 20.14 Memory write timing diagram

ROM
ROM is provided in the form of two 4K EPROMs with address ranges 0000H to 0FFFH and 2000H to 2FFFH. The EPROMs contain the **monitor program** which ensures that the system operates correctly when it is switched on. Notice that the system has a keypad and some seven-segment displays. One function of the monitor program is to continually 'scan' the keyboard to see which keys are pressed. It is also responsible for outputting data to the displays. The monitor program will be examined a little later after the rest of the system has been described.

RAM
The system has a single 2K RAM occupying the address range 1800H to 1FFFH. The programs that you, the user, want to execute are entered into the RAM memory. The writing of programs and their execution is the subject of the next element.

The PPI
The **programmable peripheral interface** is a single integrated circuit that is used as an input/output device. It has three ports (A, B and C) and correctly interfaces the keypad and the displays to the system.

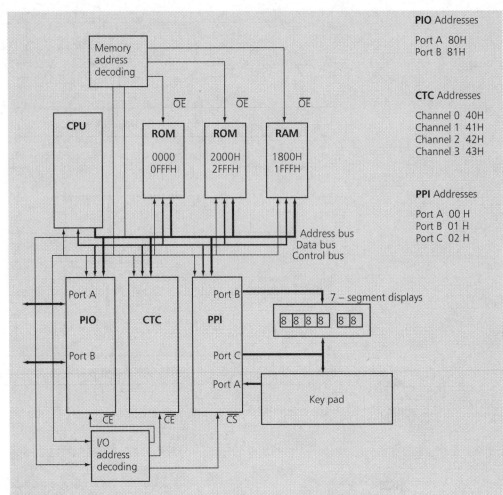

PIO Addresses

Port A 80H
Port B 81H

CTC Addresses

Channel 0 40H
Channel 1 41H
Channel 2 42H
Channel 3 43H

PPI Addresses

Port A 00 H
Port B 01 H
Port C 02 H

Figure 20.15 Block diagram of the Micro-professor

The CTC

This is the **counter timer chip** and is designed to perform counting and timing operations. The CPU itself can be programmed to carry out timing and counting operations but the use of a separate CTC chip allows the CPU to be used for other tasks while the CTC takes over routine counting and timing.

The PIO

This is the **programmable input/output** chip and is a 40-pin single integrated circuit. It allows external electronic circuits to be connected to and be controlled by the system. The PIO has two ports (A and B) that can be set up under the control of the program. Each port is 8 bits wide and bytes of data are sent or received from the ports in parallel. This means that 8 bits of data can be sent to, for example, Port A at the same time and appear on the 8 data pins of Port A simultaneously. (This is in contrast to some serial interface chips that allow data to leave or enter a port one bit after another.)

Decoding

Decoder chips are used for memory and I/O decoding. The memory address decoding circuitry is selected by the $\overline{\text{MREQ}}$ line from the CPU whenever a memory operation is taking place. The decoding system ensures that the correct chip is selected from the three memory chips.

The I/O address decoding circuitry is selected by the $\overline{\text{IORQ}}$ line from the CPU whenever input/output (or CTC) operations are required.

The monitor program

The monitor program stored in the EPROMs is fairly complex and is often called the **operating system**. It provides the following main facilities:

- Scanning and reading the keyboard.
- Displaying information on the seven-segment displays.
- The user can enter instructions and data into the system which the monitor translates into binary and stores them in the correct memory locations.
- Data in memory and in the CPU registers can be examined and changed.
- A useful facility called a 'single step' facility which allows you to execute one instruction at a time.

Although your Centre may use a different learning system, possibly based on a different microprocessor, the facilities offered will be fairly similar to those provided by the Micro-professor.

Case study

Many industrial and domestic control operations that used to be performed by specifically designed electronic circuits have now been replaced by **microprocessor-based control systems**. A good example is the incorporation of a 'microchip' into a domestic washing machine. A simplified block diagram of a typical system is shown in Figure 20.16.

As shown, the system needs to accept signals from various inputs and produce outputs to drive other parts of the washing machine. Most of these signals will be digital, such as switching the motor on and off, starting and stopping the pump and opening and closing valves, etc. The temperature sensor, however, which measures the water temperature is an analogue device which produces an output voltage proportional to the water temperature. Thus, an analogue-to-digital converter (ADC) will be needed to convert the analogue voltage into a binary number that the CPU can process. (The ADC may be incorporated in the microchip itself – see below.)

At switch-on, the program that is held in the ROM examines the binary code that is produced by the wash programme selector switch. The code 'tells' the system exactly what it has to do – for example 'heat to 80°C', 'spin for 3 minutes' etc. The program in ROM will direct the microprocessor to send out signals to the heater and motor etc. and receive data back from the various sensors. This data will be held in RAM.

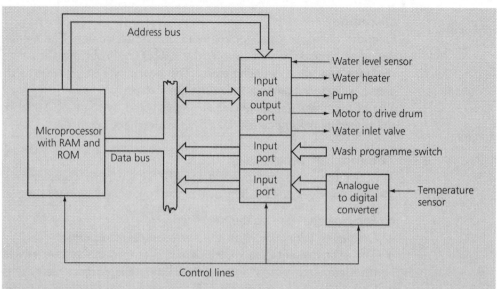

Figure 20.16 Washing machine control with a microprocessor

It is important to realise that if the manufacturer needs to upgrade or modify the machine to improve its performance, then it may only be necessary to alter the program that is stored in ROM.

In an application such as this, a powerful microprocessor (like a Pentium) will certainly NOT be needed! Furthermore, only a relatively small amount of ROM and an even smaller amount of RAM will be required. It is unlikely that the CPU, ROM and RAM and I/O devices will be in separate chips as in the previous case study. Rather, they will be combined in a single integrated circuit called a **microcontroller**. A range of microcontrollers are produced by all microprocessor manufacturers for applications such as this. The table below shows some 8-bit controllers that are available from Intel and their basic features.

Chip number	ROM	RAM	I/O lines	Clock	ADC	CTC
8048	1 Kbyte	64 bytes	27	Yes	No	No
8748H	1 Kbyte of EPROM	64 bytes	27	Yes	No	No
8049H	2 Kbytes	128 bytes	27	Yes	No	No
8022	2 Kbytes	64 bytes	28	Yes	Yes 2×8-bit	Yes
8050AH	4 Kbytes	256 bytes	27	Yes	No	No

Assignment 20

This assignment provides evidence for:
Element 19.1: Investigate microprocessor-based systems
and the following key skills:
Communication 3.1, 3.2, 3.4
Information Technology 3.1, 3.2, 3.3

The evidence indicators for this element require you to report on the design of two contrasting microprocessor-based systems.

You may decide to compare the microprocessor training systems used at your Centre, for example, with a system (or microcontroller) used for a simple control application.

Report

Your report should include:

- the features of the microprocessors used
- a description of the constituent elements that make up the microcomputer systems
- a description of the operation of the microcomputer elements
- an explanation of the factors which influenced the design.

Chapter 21: Software programming

This chapter covers:
Element 19.2: Write software to solve problems.

. . . and is divided into the following sections:
- Introduction to software development
- Assembly language programming
- Subroutines and using the stack
- Using an assembler.

The main advantage that a microprocessor-based system has over conventional digital circuits built from logic gates etc. is that it can be programmed to perform a wide range of tasks. The system is, of course, useless without a program – just as a CD or cassette player is useless without a CD or tape. The list of instructions that defines the task is called the **program software** or just software. The microprocessor is only capable of interpreting these instructions when they are presented to it as a series of binary coded words which are retrieved from the memory one after the other. This binary form is known as the **machine code**. If programmers had to directly write out these 'strings' of binary codes then it would be an extremely tedious and error-prone exercise to produce even the most basic of programs. Fortunately, techniques have been developed to avoid having to write at machine code level.

In this chapter some of the techniques of software development and design will be examined and described. The programs will be restricted to a Z-80 microprocessor system – the Microprofessor – whose hardware was described in the previous chapter. The techniques, however, are applicable to all microprocessor-based systems.

When you complete this element you should be able to:

- specify facilities to be provided by software to solve given problems
- write software using structured design techniques
- write instructions using an appropriate language
- check that software operates to specification.

Introduction to software development

In order to create a program that a microprocessor will run, a logical and well thought out procedure must be followed right from the outset. The method that is frequently chosen follows a series of logical steps which will now be described.

Defining the problem

Right at the beginning, the exact nature of the problem to which a program solution is required must be clearly stated. This cannot be over-emphasised. If the problem is incorrectly defined, no matter how good the final software is, solving the wrong problem is of no use at all. Some factors that need to be identified and stated are:

- a precise statement of what the program is expected to do
- full details of the number and types of inputs and outputs needed, etc.
- memory requirements and restrictions, etc.
- speed of operation
- cost
- timescale
- special testing requirements

Program design

Unless the programs are very simple, it is usually not possible to go straight from the program definition to the final machine code in one step. Usually it is necessary to break down the main program task into smaller sub-tasks that can be solved independently. These sub-tasks can be further broken down if necessary into a set of basic modules, as shown in Figure 21.1. The program solutions for the basic modules can then be tested to see if they work correctly. Firstly, the program solutions for all the tasks can be linked together to provide the final solution. This structured technique is generally known as **top-down design**.

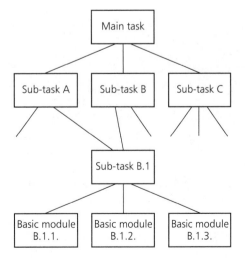

Figure 21.1 Top-down design technique

Another useful technique is to use an **algorithm**. This is a sequence of steps written in English which define the method used to solve the problem. It is possible for a problem to have more than one algorithm which means that there is more than one way of solving that particular problem. Some of the programs we will develop later in this element will use algorithms as a method of solution.

One way of representing an algorithm is to use a graphical method called a **flowchart**. A flowchart uses a set of simple standard symbols and each step in the algorithm is represented by one or more of the symbols. Each symbol is labelled with a description of the action to be performed and they are linked together to show the 'flow' of the program. The more important flowchart symbols are shown in Figure 21.2. Flowcharts will be used in some of the sample programs that follow.

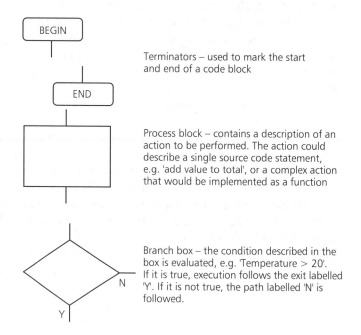

Terminators – used to mark the start and end of a code block

Process block – contains a description of an action to be performed. The action could describe a single source code statement, e.g. 'add value to total', or a complex action that would be implemented as a function

Branch box – the condition described in the box is evaluated, e.g. 'Temperature > 20'. If it is true, execution follows the exit labelled 'Y'. If it is not true, the path labelled 'N' is followed.

Figure 21.2 Some flowchart symbols

Program coding

Program coding (or coding for short) is the process of translating the algorithm or flowchart into the binary words (the **machine code**) that the microprocessor understands. As pointed out above, this is tedious and error prone so the machine code programs are written in hexadecimal form or a notation known as **assembly language**. Assembly language uses instruction **mnemonics** to make the process easier. For example, the instruction *'load the accumulator with a number'* for the Z-80 CPU is:

0011 1110 ← Machine code in binary

3E ← Machine code in hex

LD A ← Assembly language mnemonic

Of the three, the mnemonic form (remember the word mnemonic means 'an aid to memory') is the simplest to remember and is derived from the English language representation of the instruction. Thus LD A is derived from the instruction LoaD the Accumulator. The instruction mnemonics are all carefully chosen to be two or three letter commands that are easy for the programmer to understand and to remember.

Programming in assembly language using the mnemonics also means that each mnemonic must be translated into the machine code for the microprocessor to understand and remember. There are two ways of doing this. One way is to convert each assembly instruction into its machine code by hand. An **instruction set** for the microprocessor is needed in order to do this which lists all of the mnemonics and the corresponding machine code is hexadecimal. Some **hand coding** is important when you first start programming microprocessors because it gives you a 'feel' for what is happening.

The second method uses a separate computer program called an **assembler** which automatically produces the machine code from the assembly language mnemonic. Besides relieving the user of the task of looking up each code, the use of an assembler also has many other advantages over hand assembly. Both techniques will be used in this element.

Assembly language is called a **low-level language**. Languages like COBOL, PASCAL, BASIC etc. are called **high-level languages**. COBOL, for example, looks almost like a list of instructions written in English. When using COBOL, details which you would need for assembly language programming such as address, where to store data in RAM, etc. are not required. A program called a **compiler** is used when the program is complete to convert the instructions into the machine code of the microprocessor. In general, high-level languages are not used for small microprocessor-based systems, mainly because more memory is required and the 'execution times' of programs written in a high-level language are usually longer than those written in assembly language. For small microprocessor-based systems, assembly language is normally used.

Debugging and testing

Once written and converted into the machine code of the microprocessor, the program is tested to ensure that it operates correctly. Except in the case of simple programs, it is unlikely that it will run correctly first time. Several errors may occur and these '**bugs**' have to be identified and corrected. The process of tracking these errors down and correcting them is called 'debugging'.

Documentation

It is very important that any program should be properly described and documented. Customers, for example, may need to know how the program works and how to use it. Good documentation also simplifies the task of upgrading and modifying at a future date – particularly if this is undertaken by a programmer who is not the original one.

Assembly language programming

The majority of instructions consist of two parts: the **op-code** and the **operand**, i.e.:

Part 1: Op-code	Part 2: Operand
The op-code specifies the operations to be carried out	This specifies either: (a) the actual data value or (b) the address of the data

For example, to load the accumulator A from register B, the instruction is: LD A, B, where LD = operation mnemonic, A is the destination and B is the source. As another example, LD (1900H), A, means 'load memory address 1900H from A'. LD = operation mnemonic, A is the source and address 1900H is the destination.

An example of a simple assembly language program is:

Example program

Label	Mnemonic	Comment
	ORG 1800H	; start address
START:	LD A, (1900H)	; load A with contents of 1900H
	LD B, A	; load B from A
	LD A, (1901H)	
MAKESUM:	ADD A, B	
	LD (1902H), A	; load address 1902H from A
	HALT	; halt the cpu

The program is designed to add the two bytes of data 02H and 03H stored at memory locations 1900H and 1901H, store the result at address 1902 and then halt the CPU. Study the program and note the following:

- The ORG statement or ORIGIN address is used to refer to where you want the program to start from. Here the address 1800H has been chosen. This is the address at which the program will be located when it is converted into machine code.
- The labels START and MAKESUM refer to addresses in the final machine code program. At this stage, however, these addresses are not known so symbolic addresses or 'labels' are used. Notice the colon that follows the label.
- The program consists of a series of mnemonics written in the order that they are to be executed.
- Some lines have comments added. Notice that a semicolon precedes each comment. Comments have nothing to do with the program as such, but they are an aid to how the program works. Commenting is optional but good comments greatly aid the understanding of 'program flow'.

Hand assembly

Now that the program has been written, we need to convert the mnemonics into the machine code for the Z-80 CPU. This means looking up the hex code for each mnemonic from the INSTRUCTION SET for the Z-80 CPU. We will examine Z-80 instructions in detail a little later. For the moment, just accept that the hex codes for each mnemonic are the correct ones. Our machine code program looks like this:

Address	Hex code		Mnemonic	Comments
			ORG 1800H	; start address
1800	3A 00 19	START:	LD A, (1900H)	; load A with contents of 1900H
1803	47		LD B, A	; load B from A
1804	3A 01 19		LD A, (1901H)	
1807	80	MAKESUM:	ADD A, B	
1808	32 02 19		LD (1902H), A	; load address 1902H from A
180B	76		HALT	; halt the CPU

Notice that:

- as well as adding the correct hex code for each instruction, the address occupied by each byte has to be worked out
- when an address is part of the instruction, e.g. LD A, (1900H) the address part has to be entered 'low byte' then 'high byte', i.e. 00 then 19 NOT 19 then 00
- the address of the label START is 1800H and MAKESUM is address 1807H.

This program is now ready to be tested. The memory addresses used in this program have been chosen so that it would run on the *Z-80 microprocessor system described in the previous element*. This is called the **target system**. The monitor program resident in the EPROMs allows the hex codes to be entered into the RAM chip which starts at address 1800H. The monitor is then used to run and execute the program. For this example the data bytes 02H and 03H are entered into addresses 1900H and 1901H before the program is run. When the CPU halts, the monitor is used again to examine the contents of 1902H. In this case it would contain the result 05H. If the address of RAM is different for the target systems used at your Centre, then the addresses would have to be changed accordingly.

Progress check 21.1

What affect will the following instructions have:

(a) LD A, (1850H)
(b) LD H, L
(c) LD (190AH), A

The instruction set

Different microprocessors of course have different instruction sets and the manufacturers of a particular microprocessor will supply an instruction set applicable to the device concerned. In this section we will describe only some of the instructions available for use with a Z-80 microprocessor and, in particular, those mainly used in the program examples described in the remainder of this element. In general, the instructions that a microprocessor can execute can be divided into the following groups:

- Data transfer instructions
- Arithmetic and logic instructions
- Program flow instructions
- Input/output instructions
- Miscellaneous instructions.

We will examine each group of instructions as the element progresses. Case studies will be used to illustrate their use.

In addition to the type of instruction that the CPU executes, there are several methods by which the CPU locates the data to be operated on when executing that instruction. These are called **addressing modes** and the main ones are summarised in the table. Their meaning will become more obvious when we examine the instruction set in some detail.

Addressing modes

Z-80 addressing mode	Example	Meaning
Immediate	LD B, 12H	The byte of data (12H) is immediately after the op code in memory. Thus mode is used to place data into a register.
Extended	LD BC, 1902H LD A, (1900H)	Extension of immediate addressing. The two bytes following the op code in memory i.e. the operand.
Register	LD A, B	This mode allows data to be copied from one register to another.

Register indirect	LD A, (HL)
	
	LD (HL), A
		This mode is used to copy data from/to a memory location and a register.
		The HL pair contains the memory address in this example.

Implied	AND B	The op code implies one or more of the Z-80 registers as containing the operands. In the example the accumulator A holds one of the operands and the result will be stored in A.

Data transfer instructions

These instructions are used to transfer data between registers and memory locations or vice versa and between registers. They are probably the most commonly used instructions in programming. The 8-bit and 16-bit load groups are shown in Figure 21.3.

The 8-bit load group
Consider the 8-bit group first of all. How the chart is used is shown in the next worked example.

8-bit load group

Source

		Implied		Register							Reg indirect			Indexed		Ext. addr	IMME.
		I	R	A	B	C	D	E	H	L	(HL)	(HL)	(DE)	(IX+d)	(IY+d)	(nn)	n
Register	A	ED 57	ED 5F	7F	78	79	7A	7B	7C	7D	7E	0A	1A	DD 7E d	FD 7E d	3A n n	3E n
	B			47	40	41	42	43	44	45	46			DD 46 d	FD 46 d		06 n
	C			4F	48	49	4A	4B	4C	4D	4E			DD 4E d	FD 4E d		0E n
	D			57	50	51	52	53	54	55	56			DD 56 d	FD 56 d		16 n
	E			5F	58	59	5A	5B	5C	5D	5E			DD 5E d	FD 5E d		1E n
	H			67	60	61	62	63	64	65	66			DD 66 d	FD 66 d		26 n
	L			6F	68	69	6A	6B	6C	6D	6E			DD 6E d	FD 6E d		2E n
Reg indirect	(HL)			77	70	71	72	73	74	75							36 n
	(HL)			02													
	(DE)			12													
Indexed	(IX+d)			DD 77 d	DD 70 d	DD 71 d	DD 72 d	DD 73 d	DD 74 d	DD 75 d							DD 36 d
	(IY+d)			FD 77 d	FD 70 d	FD 71 d	FD 72 d	FD 73 d	FD 74 d	FD 75 d							FD 36 d
Ext. addr	(nn)			32 n n													
Implied	I			ED 47													
	R			ED 4F													

Destination (label at left of table)

16-bit load group with 'LD' 'PUSH' and 'POP'

Source

		Register							Imm. ext.	Ext. addr.	Reg. indir.
		AF	BC	DE	HL	SP	IX	IY	nn	(nn)	(SP)
Register	AF										F1
	BC								01 n n	ED 4B n n	C1
	DE								11 n n	ED 5B n n	D1
	HL								21 n n	2A n n	E1
	SP				F9		DD F9	DD F9	31 n n	ED 7B n n	
	IX								DD 21 n n	DD 2A n n	DD E1
	IY								FD 21 n n	FD 2A n n	FD E1
Ext. addr.	(nn)		ED 43 n n	ED 53 n n	22 n n	ED 73 n n	DD 22 n n	FD 22 n n			
Reg. ind.	(SP)	F5	C5	D5	E5		DD E5	FD E5			

Destination (label at left of table)

Push instructions → (pointing to (SP) row)

POP instructions ↑ (pointing to (SP) column)

Figure 21.3
8-bit and 16-bit load groups

Note: The Push and Pop instructions adjust the SP after every execution

WORKED EXAMPLE 21.1

What are the hex codes for the following instructions?

(a) LD A, 03H   (b) LD B, C

(c) LD A, (1900H)   (d) LD A, (HL)

Solution

(a) LD A, 03H. This means 'load register A with the data byte 03H'. This is immediate addressing. Look at the 8-bit load group in Figure 21.3 with the column headed 'IMME' as the source. Locate the intersection of that column with the row labelled A as the destination. The instruction is '3E n' when in this case n = 03H. Therefore, LD A, 03H is coded 3E 03H.

(b) LD B, C. This means 'load B from C' and is register addressing. Find the intersection of the column headed C as Source with the row B as Destination. The instruction is 41H. Therefore, LD B, C is 41H.

(c) LD A, (1900H). This means 'load A with the contents of address 1900H'. This is extended addressing. Find the intersection of the column headed 'Ext Addr' as the source with the row labelled A as the destination. The instruction is 3A n n. The 'n', 'n' means two bytes, giving the address with the low byte written first. So, LD A, (1900H) is 3A 00 19.

(d) LD A, (HL). This means 'load A from the address that is held in the HL pair'. This is register indirect addressing. Obviously, the HL pair must have been *previously loaded with an address* so that this instruction would load the contents of that address into A. Find the intersection of the column labelled (HL) as source, with the row labelled A as destination. So, LD A, (HL) is 7EH.

The 16-bit load group

In this group the most important are the ones that are used to load a register pair with a 16-bit number. In Figure 21.3 these instructions are highlighted with thick lines. For example,

LD BC, 1850H   is   01 50 18

LD DE, 190AH   is   11 0A 19

                  etc.

Arithmetic and logic instructions

The Z-80's mathematical ability is limited to addition, subtraction, increment and decrement. Consequently, to multiply and divide, for example, short programs have to be written using these basic functions. The instructions for the 8-bit arithmetic and logic group are shown in Figure 21.4.

With 8-bit arithmetic operations, one of the operands is always in the accumulator (A) and the result is placed back into A. For example, the two addition

Source

	Register addressing							Reg. Indir.	Indexed		Immed.
	A	B	C	D	E	H	L	(HL)	(IX+d)	(IY+d)	n
'ADD'	87	80	81	82	83	84	85	86	DD 86 d	FD 86 d	C6 n
ADD w carry 'ADC'	8F	88	89	8A	8B	8C	8D	8E	DD 8E d	FD 8E d	CE n
SUBTRACT 'SUB'	97	90	91	92	93	94	95	96	DD 96 d	FD 96 d	D6 n
SUB w carry 'SBC'	9F	98	99	9A	9B	9C	9D	9E	DD 9E d	FD 9E d	DE n
'AND'	A7	A0	A1	A2	A3	A4	A5	A6	DD A6 d	FD A6 d	E6 n
'XOR'	AF	A8	A9	AA	AB	AC	AD	AE	DD AE d	FD AE d	EE n
'OR'	B7	B0	B1	B2	B3	B4	B5	B6	DD B6 d	FD B6 d	F6 n
Compare 'CP'	BF	B8	B9	BA	BB	BC	BD	BE	DD BE d	FD BE d	FE n
Increment 'INC'	3C	04	0C	14	10	24	2C	34	DD 34 d	FD 34 d	
Decrement 'DEC'	3D	06	0D	15	1D	25	2D	35	DD 35 d	FD 35 d	

Figure 21.4 8-bit arithmetic and logic group

instructions are 'add' and 'add with carry' – ADC. The first is a straight binary addition and the second includes the contents of the carry flag as well. (Flags will be examined later.) Thus ADD and ADC mean: ADD A, S, and ADC A, S, where 'S' is the register to be added to A or the correct mnemonic for the other addressing modes. For example, the instruction for ADD A, C is 81H and ADD A, 05H is C6 05 etc.

To add or subtract 1 to a register of memory location the increment and decrement instructions are used, e.g. INC D is 14H, DEC C is 0DH etc. The 8-bit logic instructions shown in the chart will be examined later.

16-bit arithmetic instructions

These are shown in Figure 21.5. The instructions are 'read' in the usual way. For example, ADD HL, DE is 19H; INC HL is 23H, etc.

We now have experience with a sufficient number of instructions to write some useful programs.

		Source					
		BC	DE	HL	SP	IX	IY
'ADD'	HL	09	19	20	30–		
	IX	DD 09	DD 19		DD 39	DD 29	
	IY	FD 09	FD 19		FD 39		FD 29
ADD with carry and set flags 'ADC'	HL	ED 4A	ED 5A	ED 6A	ED 7A		
SUB with carry and set flags 'SBC'	HL	ED 42	ED 52	ED 62	ED 72		
Increment 'INC'		03	13	23	33	DD 23	FD 23
Decrement 'DEC'		08	1B	2B	3B	DD 2B	FD 2B

(Destination)

Figure 21.5 16-bit arithmetic instructions

WORKED EXAMPLE 21.2

Two numbers, 0AH and 04H, are held in memory locations 1900H and 1901H respectively. Write a program to subtract the contents of 1901H from 1900H and store the result in address 1902H. The program should be located at address 1800H in the target system.

Solution

(a) There are several ways to solve this problem. One way is to use a register pair (BC, DE or HL) as a **memory pointer**. Usually HL is used because it has more useful instructions than BC or DE for data transfers between the Z-80 registers and memory. For this case study both an **algorithm** and **flowchart** will be constructed. The algorithm could be:

(1) point HL to address 1900H
(2) load A with contents of address held in the pair
(3) incremenet HL
(4) subtract contents of address pointed to by HL pair from A
(5) increment HL
(6) store contents of A in the address pointed to by HL
(7) stop.

The flow chart for this algorithm is shown in Figure 21.6

(b) Study the program below carefully. The correct hex codes have been obtained from the charts in Figures 21.3 to 21.5. Make sure you can locate the correct hex code for each line of the program.

Address	Hex code	Mnemonic	Comment
		ORG 1800H	
1800	21 00 19	LD HL, 1900H	; point HL to 1900H
1803	7E	LD A, (HL)	; load A with contents of 1900H
1804	23	INC HL	; increment HL
1805	96	SUB (HL)	; subtract contents of HL from A
1806	23	INC HL	;
1807	77	LD (HL), A	; store contents of A at 1902H
1808	76	HALT	; stop

(c) Firstly, the data bytes 0AH and 04H need to be stored in address 1900H and 1901H. When the program is run and tested the result of the subtraction (06H) is stored in address 1902H.

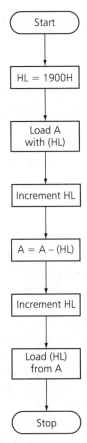

Figure 21.6 Flowchart for Worked Example 21.2

The flag register

All microprocessors will have a flag or **status register** which allows decisions to be made by the device. The flag register is a collection of independent bits which will be set or reset depending on the last operation that has taken place in the ALU. The Z-80 flag register is shown in Figure 21.7. Although eight bits were allocated to this register when the Z-80 was designed, only six of the individual

S = Sign flag   P = Parity flag

Z = Zero flag   N = Add/subtract flag

H = Half carry flag   C = Carry flag

Figure 21.7 Z-80 flag register

bits are used. Of these the zero flag and the carry flag are the ones that are most often used.

The zero flag

Whenever the ALU contains 00H, the zero flag is set to logic '1'. If the contents of the ALU do not equal 00H, the zero flag is reset to logic '0'. We will use the zero flag in the next case study.

The carry flag

The carry flag C is affected when a carry is produced from bit 7 of the ALU. For example, an addition or increment could produce a carry to the carry flag.

Only some of the Z-80 instructions will affect flags in the flag register. These are mainly arithmetic and logic instructions. Details of which flags (if any) are affected by a particular instruction are given in the full instruction set provided by the manufacturer.

Program flow instructions

Instructions are provided to allow the programmer to change the program counter contents and so jump from one part of a program to another. These are called jump or **branch instructions**. A jump instruction can be either unconditional or conditional. An **unconditional jump** is an instruction which, when encountered in a program, loads the program counter with the address given as part of the instruction. The program then continues to be executed from that address. For example, if the instruction JP 1950H is encountered in a program then the program counter is loaded by the CPU with the address 1950H and the program continues to be executed from there. The conditional jump instruction 'tests' a flag for certain conditions. If the condition holds, then the jump occurs. If the condition does not hold then the jump will not take place. For example JPZ, 1850H means that when this instruction is encountered in a program AND the zero flag Z is set to logic '1', the program will jump to address 1850H and continue to execute from there. Some jump instructions are shown in Figure 21.8.

Using the chart, some examples of coding are:

JP,   1950H   is   C3 50 19

JPZ,   1850H   is   CA 50 18

JPC,   180AH   is   DA 0A 18

          etc.

			Condition				
			Un-cond.	Carry	Non-carry	Zero	Non-zero
Jump 'JP'	Immed. Ext.	nn	C3 n n	DA n n	D2 n n	CA n n	C2 n n

Figure 21.8 Some jump instructions

Writing programs with loops

Very often it is necessary for a program to execute a particular task several times. One way of achieving this is to produce a **program loop** which instructs the CPU to perform a sequence of instructions a given number of times. A flowchart for a program loop is drawn in Figure 21.9(a).

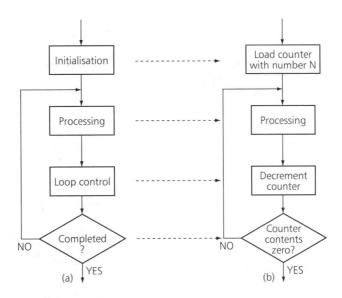

Figure 21.9 Creating a program loop

The loop has three parts: the **initialisation** part sets up starting values of whatever is going to control the number of 'passes' through the loop. This is called the **loop counter**. In many cases a register or register pair is loaded with a number. The **processing** part is the collection of instructions that performs the required operation. The **loop control** part updates the value of the loop counter after every pass through the loop. Figure 21.9(b) shows a typical technique where a loop counter is loaded with a starting number 'N'. After each pass through the loop, the counter is decremented. When the counter contains zero, the loop has been executed 'N' times and the program continues.

The following worked example uses a program loop.

WORKED EXAMPLE 21.3

Design a program which will add together the six numbers 01H, 02H, 03H, 04H, 05H, 06H stored in successive memory locations starting at address 1900H. The final result should be stored in address 1906H. The program should have its origin at 1800H.

Solution

(a) The best solution is to use a register as a loop counter. Initially, the counter is loaded with 06H and is decremented with each pass through the loop. When the register contains zero the processing is complete. A **flowchart** for the program is shown in Figure 21.10. The HL pair is used as a memory pointer and register B is used as a loop counter. Notice that since A is used to hold the result of each addition, it needs to be cleared at the beginning of the program. This is because A may contain data from previous operations.

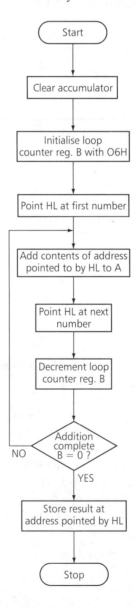

Figure 21.10 Flowchart for Worked Example 21.3.

(b) The final **program** with the correct **machine code instructions** is shown below. Make sure you understand the logic behind it and check each hex code with the instruction charts given. Notice how the 'JP NZ' instruction is used to pass through the loop each time until register B contains zero.

Address	Hex code	Mnemonic	Comment
		ORG 1800H	
1800	3E 00	LD A, 00H	; clear accumulator
1802	06 06	LD B, 06H	; initialise loop counter for a six count
1804	21 00 19	LD HL, 1900H	; point HL to first number
1807	86	FRED: ADD A, (HL)	; add contents of address held in HL to A
1808	23	INC HL	; point HL at next number
1809	05	DEC B	; decrease loop counter by 1
180A	C2 07 18	JP NZ, FRED	; if not zero jump back to label FRED at address 1807
180D	77	LD (HL), A	; store result at address held in HL
180E	76	HALT	; stop

(c) When **testing the program,** firstly, the bytes of data 01H to 06H need to be stored at addresses 1900H to 1905H before the program is run. The result of the addition is 21 decimal which is 15 hexadecimal. This result is stored at address 1906H when the program is run.

Input/output operations

Microprocessors really become useful when they can be used to take data in from external devices like switches and sensors etc., process that data and operate other external devices like lights and motors etc. Data enters and leaves a microprocessor-based system via **input and output ports**. All manufacturers provide a range of separate integrated circuits which can be easily connected to their CPU's for input/output interfacing. Although **interfacing techniques** will be the subject of the next element, a brief description of a Z-80 PIO is needed here so that more interesting worked examples can be attempted. We will use the Z-80 PIO to interface some switches and LEDs to a Z-80 CPU.

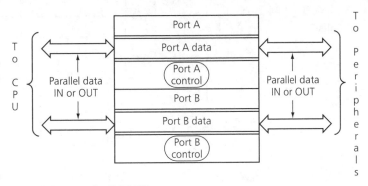

Figure 21.11 Simplified diagram of a Z-80 PIO

A simplified block diagram of a Z-80 PIO is shown in Figure 21.11. It has two ports called PORT A and PORT B which, in their simplest use, can be set up as input or output ports for 8 bits of parallel data. A port is set up as an input or output by writing '**command words**' to two **control ports** – one for port A and one for port B.

Each port has an address which, for our program examples, will be assumed to be:

PORT A   DATA   = 80H
PORT A   CONTROL = 82H

PORT B   DATA   = 81H
PORT B   CONTROL = 83H

For example, we would command port A to be an input port by sending a special command word to port A control at address 82H. Data can then enter the CPU via port A data port at address 80H. As another example, port B can be set up as an output port by sending commands to port B control at address 83H. Data can then leave the CPU via port B data port at address 81H.

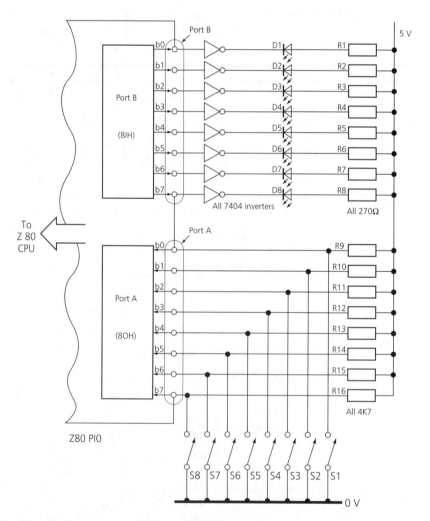

Figure 21.12 Interfacing switches and LEDs to a Z-80 PIO

The easiest input and output instructions to use are the immediate instructions. These use the accumulator to input and output data from the ports.

For example, assuming that port A has been set up as an input, then to take data in the instruction is:

IN A, (80H)   DB 80H

Assuming port B has already been set up as an output port, to send data out to port B the instruction would be:

OUT (81H), A   D3 81H

Figure 21.12 shows the hardware necessary to interface some switches for input to port A and some LEDs for outputs from port B.

The switches S_1 to S_8 are the input switches. When a switch is open the corresponding bit of port A is 'pulled up' to 5 V so that the input bit to the port has a '1' connected to it. When a switch is closed, the corresponding bit of the port has a '0' connected to it. If any of the bits of port B are at '1' then the corresponding LED is ON. This is because the '1' is inverted by the 7404 inverter to give a '0' on the cathode of the diode. Current flows through the LED and is 'sunk' by the output of the inverter gate.

Progress check 21.2

Why are the 270R resistors placed in series with each LED in Figure 21.12?

WORKED EXAMPLE 21.4

Write a program which takes data in from the switches on port A in Figure 21.12 and displays that data on the LEDs connected to port B. The program should start at address 1800H and run continuously.

Solution

(a) The program must input from port A and output to port B continuously. It must therefore be designed to run in a continuous loop. A suitable algorithm is as follows:

(1) Set up port A of the PIO as an input port.
(2) Set up port B of the PIO as an output port.
(3) Read in data from switches.
(4) Output data to the LEDs.
(5) Jump to (3) to run continuously.

(b) The process of setting up the ports of the PIO is called **'initialising the PIO'**. Bytes of data have to be sent to the respective control ports. For the moment you will need to accept that to set port A (80H) as an input port, the following instructions need to be executed:

LD A, 4FH
OUT (82H), A

To set up port B (81H) as an output port the instructions needed are:

LD A, 0FH
OUT (83H), A

The data bytes '4FH' and '0FH' are called **mode control bytes**. Why they are needed and have those particular values will be described in the next element. Notice that the mode control bytes are sent to the control ports (82H, 83H) of the PIO.

(c) The correctly coded program is shown below. Notice the unconditional jump that keeps the program running continuously. As before, check the coding for each instruction using the charts.

Address	Hex code	Mnemonic	Comment
		ORG 1800H	
1800	3E 4F	LD A, 4FH	;
1802	D3 82H	OUT (82H), A	; port A set as input
1804	3E 0FH	LD A, 0FH	;
1806	D3 83H	OUT (83H), A	; port B set as output
1808	DB 80H LOOP:	IN A, (80H)	; read switch input from port A data at 80H
180A	D3 81H	OUT (81H), A	; output data to LEDs at 81H
180C	C3 08 18	JP, LOOP	; repeat continuously

(d) Run and testing the program. When the program is correctly entered at address 1800H and run, it executes continuously. Whatever data is on the input switches appears on the LEDs.

Progress check 21.3

What change would you make so that the program in Worked Example 21.4 ran once?

Creating time delays

In many applications it is necessary to produce time delays of a fixed amount by using software. One way of easily achieving this is to 'force' a program to go through a loop a certain number of times. Since any instruction executed by a microprocessor takes a certain number of T-states, it is possible to predict exactly how long a particular program will take to run. The number of T-states required for each microprocessor instruction is given in the manufacturer's instruction set.

The basic principle of time delay production is to use the CPU as a counter. Consider the following program for the Z-80:

```
        LD C,    00H
LOOP:   DEC C
        JP NZ,   LOOP
```

The first decrement C causes its contents to change from 00H to FFH. A decrement instruction affects the flags and since C does not contain zero the 'JP NZ' instruction is obeyed and causes a decrement of C to FEH. The process continues until C finally decrements from 01H to 00H. Now the **zero flag** is set and the 'JP NZ' instruction is no longer valid. The CPU would then continue to execute any instructions that follow this one. In this example, the program has been 'forced' to go through a loop 256 times. By totalling the number of T-states, the time taken can be calculated. The instruction set gives the T-states per instruction as: .

```
        LD C,    00H   = 7T
LOOP:   DEC C          = 4T
        JP NZ,   LOOP  = 10T
```

The total number of T-states is:

$$7 + 256 \times (4 + 10) = 359 \text{ T-states}$$

Suppose that the CPU clock is 2 MHz. Each T-state lasts 500 ns, so the total time taken is:

$$3591 \times 500 \text{ ns} = 1.8 \text{ ms}$$

To create longer time delays a register pair is used rather than a single register. Unfortunately, the decrement of a register pair in the Z-80 *does not affect* the flags so checking that the pair contains zero needs two additional instructions. Consider the following program:

```
        LD BC,   0000H = 10T
LOOP:   DEC BC         = 6T
        LD A, B        = 4T
        OR C           = 4T
        JP NZ,   LOOP  = 10T
```

The two instructions LD A, B; OR C, are used to check when BC contains zero in the following way. Loading A from B and ORing with C produces a logical OR between registers B and C. Only when both B and C contain 00H will the result of the OR be zero, so setting the zero flag – cunning, isn't it! The total number of passes through the loop is 65,536. Counting the T-states as before gives a total time delay of approximately 0.8 s.

We have now succeeded in generating a time delay of close to 1 s. Suppose, however, a much longer delay is required. One simple way of achieving this is to surround the register pair decrement with a single register decrement. Consider the following program:

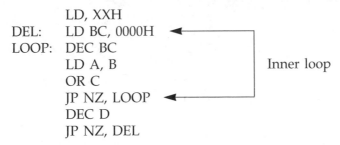

The 'XXH' in the first instruction means 'any number'. Suppose XX = 01H. Then we would pass through the inner loop just once giving a delay of about 0.8 s. Suppose, however, XX = 0AH. There would now be ten passes through the inner loop giving a total delay of about 8 s. If XX = 00H then there would be 256 passes through the loop giving a delay of over 200 s. So by varying the contents of register D we are able to produce a variable delay. Note that the time figures quoted have always been given approximately. If necessary, you can calculate the exact number of T-states required and so know exactly how long the delay actually is.

WORKED EXAMPLE 21.5

Write a program which causes the LEDs on port B in Figure 21.12 to flash ON and OFF continuously. The ON and OFF times should be about one second. The program should start at address 1800H.

Solution

The LEDs need to be turned ON, kept on for about a second, turned OFF and kept off for about a second. The sequence should then repeat continuously. The decrement of a register pair gives a delay of about a second so this technique can be used.

(a) A suitable algorithm would be:

(1) Set up port B (81H) of the PIO as an output port.
(2) Switch ON LEDs.
(3) Produce a time delay of about 1s.
(4) Switch OFF LEDs.
(5) Produce a 1s delay.
(6) Jump back to (2) to run continuously.

(b) The correctly coded program is shown below.

Address	Hex code		Mnemonic	Comment
			ORG 1800H	
1800	3E 0F		LD A, 0FH	;
1802	D3 83		OUT (83H), A	; port B (81H) is output
1804	3E FF	START:	LD A, FFH	;
1806	D3 81		OUT (81H), A	; switch on LEDs
1808	01 00 00		LD BC, 0000H	; this section produces the time delay
180B	0B	FRED:	DEC BC	;
180C	78		LD A, B	;
180D	B1		OR C	;
180E	C2 0B 18		JP NZ, FRED	; end of time delay section
1811	3E 00		LD A, 00H	;
1813	D3 81		OUT (81H), A	; switch OFF LEDs
1815	01 00 00		LD BC, 0000H	; time delay section
1818	0B	JIM:	DEC BC	;
1819	78		LD A, B	;
181A	B1		OR C	;
181B	C2 18 18		JP NZ, JIM	;
181E	C3 04 18		JP START	; run continuously

(c) When the program is entered and run for testing it flashes the LEDs ON and OFF continuously.

Can you spot why the program is not particulary efficient? Well, the code for the delay has had to be entered twice. This is not an efficient use of memory. In addition, the task of having to enter the same codes twice is tedious. The way out of the problem is to write the delay routine once and call that routine as necessary. Programs designed to be called and used over and over again in a program are called **subroutines**. The use of subroutines is a very important part of computer programming generally.

Subroutines and using the stack

A subroutine is a section of code, written only once, which can be used and reused whenever it is needed by the main program. The principle is illustrated in Figure 21.13.

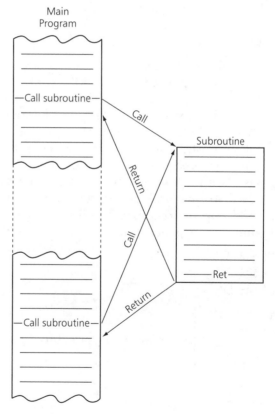

Figure 21.13 Principle of a subroutine

Whenever the main program needs to use the subroutine, a call instruction is included in it. When the call instruction is encountered, the microprocessor saves the address of *the next instruction automatically on the stack* and loads the program counter with the address of the subroutine. The program instructions in the subroutine are then executed one after another. At the end of the subroutine a return (RET) instruction must be included. When the microprocessor 'sees' this instruction, the address which was stored on the stack is placed in the program counter and the main program continues from where it left off. Thus the microprocessor automatically uses the stack to save the return address that it must go back to when the subroutine is complete.

Suppose, for example, a delay routine started at address 1900H. To CALL and RETURN from that routine, the instructions needed would be:

$$CD\ 00\ 19\ =\ CALL\ 1900H$$
$$and\ \ \ C9\ \ \ \ \ \ \ \ =\ RET$$

Since the stack is used automatically by a subroutine then the **stack pointer** (SP) 16-bit register must have been previously loaded with a valid address in RAM. This is usually chosen to be at the top of the RAM address range. In the Micro-professor training unit, the top of RAM is address 1FFFH. The stack pointer is usually set up right at the beginning of the main program just as, for example, the ports of the PIO are initialised at the start. To load the stack pointer the hex code is: LD SP, 1FFFH = 31 FF 1F. (Check this instruction yourself in 16-bit load group.)

Pushing and popping

Even though the stack is used automatically to save return addresses, it can also be used by a programmer to save the contents of CPU registers. The reason this may be required is that a subroutine might use some (or all) of the registers that are also used in the main program. Furthermore, it might be necessary to use the data that was in these registers after the subroutine is complete. To prevent the subroutine 'destroying' this data, it needs to be saved somewhere before the subroutine is used and then restored when the subroutine is complete. The obvious place to store the register data is on the stack. For the Z-80, the instructions used to save and restore the contents of registers on the stack are called **push and pop instructions**. The instructions are given in the 16-bit load group of Figure 21.3. A general subroutine may look something like this:

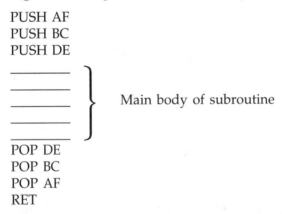

PUSH AF
PUSH BC
PUSH DE

} Main body of subroutine

POP DE
POP BC
POP AF
RET

Notice very carefully that the registers that need to be saved are 'pushed' in pairs before the main body of the subroutine starts. Since the stack operates on a last in/first out basis, then the registers have to be 'popped' in **reverse order** just before the return instruction. There are other ways of saving registers other than 'pushing' and 'popping'. The technique shown above is the best and is the method most often used.

Progress check 21.4

What are the instructions for: push AF, push BC, push DE, push HL, pop HL, pop DE, pop BC and pop AF?

WORKED EXAMPLE 21.6

Modify the program in Worked Example 21.5 so that the LEDs are ON for about 3 seconds and OFF for about a second. A variable delay subroutine should be used at address 1900H.

Solution

Since a delay of more than about a second is needed to keep the lights ON, a slightly more complicated delay routine is needed. The technique of surrounding a register pair decrement with a single register decrement will be used. The single register (in this case register D) can be loaded just before the call to the main delay section.

(a) The algorithm

    (1) Load stack pointer with 1FFFH.
    (2) Set up port B of PIO as an output port.
    (3) Switch on LEDs.
    (4) Load register for three passes through delay routine.
    (5) Call delay subroutine.
    (6) Switch off LEDs.
    (7) Load register for one pass through delay routine.
    (8) Call delay subroutine.
    (9) Jump back to (3) to run continuously.

(b) Program and program coding. Study the program below carefully. Notice how the value in register D controls the number of passes through the delay subroutine at 1900H. For this program it is not necessary to push any registers because their contents do not have to be preserved.

Address	Hex code		Mnemonic	Comment
			ORG 1800H	
1800	31 FF 1F		LD SP, 1FFFH	; set SP to top of RAM
1803	3E 0F		LD A, 0FH	;
1805	D3 83		OUT (83H), A	; port B is output
1807	3E FF	START:	LD A, FFH	;
1809	D3 81		OUT (81H), A	; switch on LEDs
180B	16 03		LD D, 03H	; set for three passes through delay
180D	CD 00 19		CALL 1900H	; call delay subroutine
1810	3E 00		LD A, 00H	;
1812	D3 81		OUT (81H), A	; switch off LEDs

1814	16 01		LD D, 01H	; set for one pass through delay
1816	CD 00 19		CALL 1900H	;
1819	C3 07 18		JP START	
		Delay subroutine		
			ORG 1900H	
1900	01 00 00	DEL:	LD BC, 0000H	; load pair for delay
1903	0B	LOOP:	DEC BC	;
1904	78		LD A, B	; checking that BC pair contains zero
1905	B1		OR C	;
1906	C2 03 19		JP NZ, LOOP	; dec again if not zero
1909	15		DEC D	; dec register controlling passes
190A	C2 00 19		JP NZ, DEL	; if D not zero, load pair again
190D	C9		RET	; return to main program

(c) Testing the program. The LEDs should continually flash ON and OFF. They are ON for about 3 seconds and OFF for about 1 second.

Activity 21.1

Modify Worked Example 21.6 so that the LEDs are ON for about 7 seconds and OFF for about 2 seconds.

Using an assembler

In each worked example so far the assembly language mnemonics have been translated into the correct hexadecimal code instructions by hand with aid of charts. In addition the correct addresses occupied by the codes have had to be worked out as well. The target system has been the Z-80 Micro-professor and the monitor program resident in the EPROMs allows the hex codes to be entered into the RAM chip which starts at address 1800H. The monitor program is then used to run and execute the program. Rather than using hand translation, a computer with an **assembler program** can be used instead. The use of an assembler offers several advantages over hand translation.

If a computer is going to be used to write an assembly language program, assemble that program and run it, then that computer must be equipped with three other programs called the **editor**, the **assembler** and the **monitor**. The editor allows the assembly language mnemonics to be entered into the computer's memory from a keyboard. When all the mnemonics have been entered correctly, the assembler is run to produce the correct machine code. Finally, the monitor is used to execute the program.

Cross-assembler

Small target systems like the Micro-professor only have a monitor program so an editor and assembler resident on another computer (a PC) are used. In this case the assembler program is called a **cross-assembler**. It is given this name because the PC's microprocessor is different to the target microprocessor. For example, a PC equipped with an Intel 80xx86 chip or Pentium may be used while the target machine many have a Z-80 microprocessor. The steps in using a cross-assembler are as follows:

- Enter assembly language mnemonics into computer using the editor program.
- Run the cross-assembler to produce the machine code for the target machine.
- Download the machine code into the target machine.
- Use the monitor in the target to execute the program.

If errors are detected and the program does not run correctly, then the editor must be used again and the whole sequence repeated.

A cross-assembler has been used to develop the programs for the last worked examples in this element.

Sequencing

Microprocessors and computers generally can be programmed to perform repetitive tasks continuously. These repetitive functions are usually called 'sequences' and the microprocessor is called a 'sequencer'. In the next worked example a Z-80 will be programmed to control a single set of traffic lights.

WORKED EXAMPLE 21.7

Imagine that the three LEDs connected to bit 0, bit 1 and bit 2 of port B of Figure 21.12 are the red, amber and green indicators of a single set of traffic lights. Write a program, organised at 1800H, so that the traffic lights operate continuously in the following approximately-timed sequence.

RED approx. 4 seconds
RED and AMBER approx. 3 seconds
GREEN approx. 8 seconds
AMBER approx. 2 seconds

Solution
We can use the same delay routine that was used in Worked Example 21.6. The number in the D register can be used as before to set the overall delay time. The codes that need to be outputted to port B for the light combinations are:

RED   →   0000 0001 = 01H
RED/AMBER →   0000 0011 = 03H
GREEN   →   0000 0100 = 04H
AMBER   →   0000 0010 = 02H

(a) A suitable flowchart is shown in Figure 21.14. Notice the symbol that is used to indicate a subroutine.

(b) This program was developed on a PC using a cross-assembler and downloaded to the Z-80 system for testing. A printout from the cross-assembler is reproduced below.

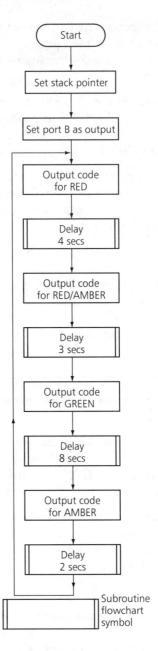

Figure 21.14 Flowchart for Worked Example 21.7

Line	Address	Hex code		Mnemonic		Comment
1			; TRAFFIC LIGHTS PROG 11/12/1996			
2			;			
3			;			
4	1800			ORG	1800H	
5	1800	31 FF 1F		LD	SP, 1FFFH	; SET STACK POINTER
6	1803	3E 0F		LD	A, 0FH	;
7	1805	D3 83		OUT	(83H), A	; PORT B OUTPUT
8			; START OF MAIN PROGRAM			
9	1807	3E 01	START:	LD	A, 01H	; RED CODE

10	1809	D3 81		OUT	(81H), A	;
11	180B	16 04		LD	D, 04H	;
12	180D	CD 00 19		CALL	DEL	; 4 SECS
13	1810	3E 03		LD	A, 03H	; RED/AMBER CODE
14	1812	D3 81		OUT	(81H), A	;
15	1814	16 03		LD	D, 03H	;
16	1816	CD 00 19		CALL	DEL	; 3 SECS
17	1819	3E 04		LD	A, 04H	; GREEN CODE
18	181B	D3 81		OUT	(81H), A	;
19	181D	16 08		LD	D, 08H	;
20	181F	CD 00 19		CALL	DEL	; 8 SECS
21	1822	3E 02		LD	A, 02H	; AMBER CODE
22	1824	D3 81		OUT	(81H), A	;
23	1826	16 02		LD	D, 02H	;
24	1828	CD 00 19		CALL DEL		; 2 SECS
25	182B	C3 07 18		JP	START	; REPEAT
26		; THE DELAY SUBRUTINE IS AT 1990H				
27	1900			ORG	1900H	;
28	1900	01 00 00	DEL:	LD	BC, 0000H	;
29	1903	0B	LOOP:	DEC	BC	;
30	1904	78		LD	A, B	;
31	1905	B1		OR	C	;
32	1906	C2 03 19		JP	NZ, LOOP	;
33	1909	15		DEC	D	;
34	190A	C2 00 19		JP	NZ, DEL	;
35	190D	C9		RET		;
36	190E					
37						
38						

Lines assembled: 38   Assembly errors: 0

Notice how the subroutine at 1900H has been given the label 'DEL' and DEL has been called from the main program.

Activity 21.2

Modify the program in Worked Example 21.7 so that the traffic light timings are:

RED	approx. 6 secs
RED and AMBER	approx. 2 secs
GREEN	approx. 10 secs
AMBER	approx. 3 secs.

Inputting data to a microcomputer

Some of the previous worked example have concentrated on programs which perform a certain task and then output data to the output port to drive LEDs. As well as outputting data, it is obviously very important that a microcomputer should be able to input data from various sources without any error. In practical situations, this data may come from digital inputs such as mechanical switches and digital transducers etc. or from analogue input devices such as temperature and pressure sensors etc. In the latter case the analogue input must first be converted into digital form using an analogue-to-digital converter. This will be examined in the next chapter.

Whatever the source of the digital data, its transfer into the microprocessor-based system must be correctly synchronised especially when 'streams' of data bytes need to be accepted. Several methods of achieving this are possible but one of the best is to set up a system such that the input device signals the microprocessor when data is available. Thus, as well as sending the data, the input device also sends a signal which informs the microprocessor that valid data needs to be accepted. This signal is usually called a **strobe** or **handshake signal**. The handshake signal essentially controls the orderly input of data to the microprocessor-based system. The technique is outlined in the next worked example.

WORKED EXAMPLE 21.8

Design a program which allows five numbers 01H, 02H, 03H, 04H and 05H, set up on the input switches of the circuit in Figure 21.12, to be input to a Z-80 system one after the other. Switch 8 (Port A bit 7) should be used as a strobe input and the running total of the inputs should be shown on the LEDs. When all five numbers have been accepted and the final total displayed, the program should halt.

Solution

(a) The numbers to be input are set up on switches S_1, S_2 and S_3 of port A (80H) in Figure 21.12. Switch S8 will be used as the strobe input to signal that data is ready to be input. One important point to take into account is that mechanical switches always have a certain amount of 'contact bounce'. What this means is that when they are opened or closed the contacts 'chatter' for a few milliseconds before they settle at their final logic levels. This is illustrated in Figure 21.15.

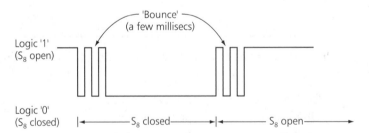

Figure 21.15 Contact bounce of switches

The microprocessor would interpret this 'bounce' as a series of '1's and '0's and so 'bounce' must be eliminated. Contact bounce can be eliminated by using some logic gates – a 'hardware solution' – or by software. In this worked example a software solution will be used. S8 is checked about every 20 ms to

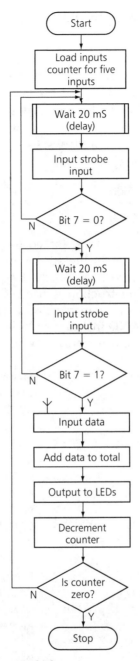

Figure 21.16 Flowchart for Worked Example 21.8

see if it is closed and then if it is opened. Thus the '1' to '0', '0' to '1' change tells the microprocessor to input data. A counter is used to count the five data inputs.

(b) The flowchart is drawn in Figure 21.16. Register B is used to count the number of inputs. A time delay of about 20 ms is used and this is written as a subroutine.

(c) Study the program and coding below carefully. The following points should be noted:

- The instructions LD A, 4F and OUT (82H), A, set up port A as an input.
- When data is input to A on the first two occasions it is ANDed with 80H. This technique is called **bit masking** and in this case the mask is 1000 0000. The AND instruction affects the flags and the result of the AND instruction is placed back into the accumulator. Suppose the input data is 1??? ????. The effect of AND 80H on the data is:

$$\begin{array}{r} 1???\ ???? \\ 1000\ 0000 \\ \hline 1000\ 0000 \end{array}$$

All the lower bits are 'forced to 0' and since bit 7 is 1, the zero flag will not be set. Only when the strobe, bit 7, is '0' will the contents of A be zero, so setting the zero flag. On the first occasion we are checking bit 7 for '0' and so a JP NZ instruction is used. On the second occasion we are checking bit 7 for '1' so a JP Z instruction is used.

- Notice the AND instruction (AND 7FH) is used in the main program to eliminate bit 7 from the addition calculation in the ADD part of the program.

Address	Hex code		Mnemonic	Comment
			ORG 1800H	
1800	31 FF 1F		LD SP, 1FFFH	; set SP to top of RAM
1803	3E 0F		LD A, 0FH	;
1805	D3 83		OUT (83H), A	; port B is output
1807	3E 4F		LD A, 4FH	;
1809	D3 82		OUT (82H), A	; port A is input
180B	06 05		LD B, 05H	; B counts number of inputs
180D	26 00		LD H, 00H	; clear reg H
180F	CD 00 19	JACK:	CALL DEL	; 20 mSecs wait
1812	DB 80	STROB0:	IN A, (80H)	;
1814	E6 80		AND 80H	; mask
1816	C2 12 18		JP NZ, STROB0	;
1819	CD 00 19		CALL DEL	; 20 mSecs wait
181C	DB 80	STROB1:	IN A, (80H)	;
181E	E6 80		AND 80H	; mask
1820	CA 1C 18		JP Z, STROB1	;
1823	DB 80	ADD:	IN A, (80H)	; valid input data
1825	E6 7F		AND A, H	; mask out strobe
1827	84		ADD A, H	;
1828	67		LD H, A	;
1829	D3 81		OUT (81H), A	; output running total to LEDs
182B	05		DEC B	; dec inputs counter
182C	C2 0F 18		JP NZ, JACK	; get another input
182D	76		HALT	; five numbers on input
			Delay subroutine	
			ORG 1900H	
1900	11 00 0A	DEL:	LD DE, 0A00	; load pair for about 20ms

1903	1B	LOOP:	DEC DE	;
1904	7A		LD A, D	;
1905	B3		OR E	; checking for zero
1906	C2 03 19		JP NZ, LOOP	;
1909	C9		RET	;

(d) In testing the program, when each input number (01H, 02H, etc.) is set up on port A and bit 7 is taken 'low' then 'high' again the number will be input to the system. The LEDs show the running total. When the five numbers have been accepted, the program halts and the LEDs indicate a result of FFH.

Assignment 21

This assignment provides evidence for:
Element 19.2: Write software to solve problems
and the following key skills:
Information Technology 3.1, 3.2, 3.3

Your Centre should provide you with a microprocessor-based system on which you can develop and test your own software.

The evidence indicators for this element require you to produce a portfolio of software written to solve three types of problem which cover the range of the element.

Report

This could include:

- a specification of the facilities provided by the software
- evidence that a structured approach has been adopted
- evidence that an appropriate language has been used
- notes with detail checks made to ensure that the software operates to specification.

Chapter 22: Interfacing microprocessor-based systems

This chapter covers:
Element 19.3: Investigate the interfacing of microprocessor-based systems.

... and is divided into the following sections:
- The general microprocessor interface
- The parallel interface
- Serial interfacing
- Interfacing protocols
- Analogue inputs and outputs.

The previous chapter (Element 19.2) served as an introduction to programming and the use of microprocessor instruction sets. In addition, some techniques to input data from external devices, such as switches, and output data to other external devices, such as LEDs, were examined. These external devices are generally called **peripherals**. Other examples of input devices are keyboards, disk drives and analogue sensors etc. Typical output devices are printers, visual display units, motors and disk drives. In fact, in many practical applications microprocessors may spend a considerable time inputting and outputting data to/from peripherals as well as processsing that data. Data to/from a microprocessor-based system enters and leaves via **I/O ports** (input/output ports) and these ports are provided in a range of integrated circuits called I/O chips. Each microprocessor manufacturer usually markets a range of **I/O chips** which allow easy connection to their microprocessors. You have already had some experience with using an I/O chip in Element 19.2. This was the Z-80 PIO which allowed data to be inputted and outputted from its two ports in parallel. This chapter begins by examining a general microprocessor interface.

When you complete this element you should be able to:

- describe the types of interface used in microprocessor-based systems
- describe interfacing protocols
- evaluate interface devices and software.

The general microprocessor interface

A general microprocessor interface to a peripheral device is illustrated in Figure 22.1.

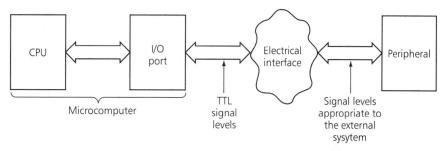

Figure 22.1 The general microprocessor interface

A microprocessor transfers data internally via its data bus and the width of the data bus depends on the microprocessor used, i.e. 8-bit, 16-bit, 32-bit etc. The Z-80, for example, transfers data bytes internally via its 8-bit data bus i.e. 8 data bits exist simultaneously on each of its data bus lines, D_0 to D_7. Data can enter or leave the microcomputing system via I/O ports and these ports are usually provided in specially designed integrated circuits. If the peripheral is such that data needs to leave or enter a port in parallel then the port is called a **parallel I/O port**. If data leaves or enters the port one bit at a time on a single wire then the port is called a **serial I/O port**. Both parallel and serial data transfer will be investigated a little later – you have already had some experience of parallel I/O, however, in Element 19.2 when switches and LEDs were interfaced to the Z-80 CPU.

Many peripherals cannot be directly connected to an I/O port and so some additional electronics may be required between a port and the peripheral. This is shown as the 'electrical interface' in Figure 22.1. Some reasons are as follows:

- *Voltage differences.* An I/O port normally operates at TTL levels (0 and 5 V). Some peripherals may need higher voltages for successful operation so extra electronics is needed to achieve this shift in voltage levels.
- *Current differences.* I/O ports are only capable of providing (sourcing) or absorbing (sinking) relatively small amounts of current (typically microamps/milliamps). If the peripheral operates at higher current levels then some sort of current amplifier will need to be employed.
- *Signal conversion.* Some peripherals may need analogue signals to operate them or conversely may produce analogue signals which the microcomputer has to process. Since the microcomputer is a purely digital device, the analogue-to-digital (ADC) and digital-to-analogue (DAC) converters will be needed. The use of ADCs and DACs is covered in detail in a later section of this element.

The interface requirement in general is very important as well as being very interesting. Since you have already had some experience of the Z-80 PIO in the previous chapter, we will investigate parallel ports first of all.

The parallel interface

Parallel ports have many uses and depending on whether they are input ports or output ports find many applications in microprocessor-based systems generally. Some typical uses of 8-bit I/O ports are as follows.

Input port

- Keyboard connection system.
- Connection of individual switches (as used in Chapter 21)
- Connection of an 8-bit ADC enabling an analogue signal to be read into the microprocessor

Output port

- Parallel connection to a printer.
- Connection of a DAC to produce an analogue signal.
- Driving 8 individual LEDs (as used in Chapter 21).
- Driving 7-segment displays (7-bits for the display and the eighth bit for the decimal point if needed).

Special interface circuits

One simple way of implementing an I/O port is to use one of the integrated circuits available in the TTL range of logic integrated circuits. The 74373 device, for example, is an octal (8) latch with tri-state outputs. The way it would be wired as an input port is shown in Figure 22.2.

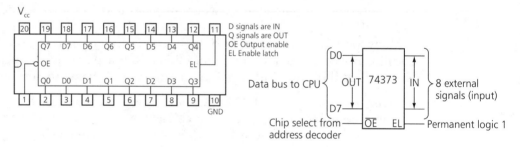

Figure 22.2 Using the 74373 as an input port

The output enable ($\overline{OE}$) would be fed from a decoder and taken to Logic '0' when the port is addressed. The enable latch (EL) would be permanently connected to Logic '1' (5 V) so that every time $\overline{OE}$ is taken to Logic '0' the 8 bits of data on its input pins would be latched to its output pins and hence to the CPU data bus.

Although simple, the device is not versatile in the sense that once it is wired in this way it is always an input port. It can, for example, be made into an output port but has to be physically rewired. To provide versatility and to ease the interfacing problem, microprocessor manufacturers produce specialist ICs that can be programmed to be input or output ports and provide additional facilities as well. These devices are so cheap and easily programmed that they are preferred to the non-programmable ICs like the 74373.

Zilog's parallel I/O chip is called the Z-80 PIO. Other manufacturers use different names. For example, Motorola's parallel devices are called PIAs or **peripheral interface adaptors**. The devices essentially all accomplish the same task of providing programmable parallel ports. Usually 2 or 3 ports will be provided. We will only investigate the Z-80 PIO in this element.

The Z-80 PIO

The Z-80 PIO was introduced and its basic programming was described in Chapter 21. The chip consists of two 8-bit parallel data ports that are very nearly identical. These two ports are called port A and port B. The simplified basic structure of port A is shown in Figure 22.3(a). The function of each of the blocks is as follows:

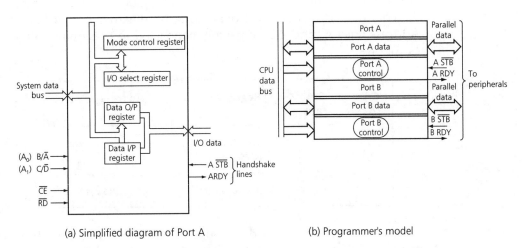

(a) Simplified diagram of Port A   (b) Programmer's model

Figure 22.3 The Z-80 PIO

Data input register

If the port is set up to be an input port then this register holds the data from the peripheral which is then transferred to the CPU. This register holds the data until the CPU reads it via an input instruction. The **strobe line** ($\overline{\text{ASTB}}$) can be used by the peripheral to control the transfer of data from itself to the data input register.

Data output register

When the CPU executes an OUT instruction this register holds the data until the peripheral wants to read it. The READY line (ARDY) can be used to signal to the peripheral that new data is available.

Mode control register

When the port is programmed the structure of the data byte in this register dictates in what mode the port is set. Four modes are possible – output, input, bi-directional and bit mode. These modes will be described a little later.

I/O select register

One very powerful mode is the **bit mode**. This allows any of the individual 8 bits to be set up as inputs or outputs. The I/O select register is used by the programmer to set which bits are outputs and which are inputs.

Notice the two inputs labelled 'B/$\overline{\text{A}}$' and 'C/$\overline{\text{D}}$'. Their function will become clearer later. It should be noted now though that these input pins are *usually* connected to address lines A_0 and A_1 from the CPU. $\overline{\text{CE}}$ is the 'chip enable' which must be at logic '0' to enable the PIO and $\overline{\text{RD}}$ (READ) which must be at logic '0' when reading from the PIO.

A programmer's model of the PIO

As outlined in Chapter 21, as far as the programmer is concerned the PIO appears as two **data ports** called A and B and two **control ports** called port A control and port B control. Each port has an address which depends on the system in which the PIO is used. The programmer's model is shown in Figure 21.3(b). The addresses of the ports are determined as follows. The pin $B/\overline{A}$ is used to select port B if it is at logic '1' or port A if it is at logic '0'. Similarly, the pin $C/\overline{D}$ selects the control port or data port depending on whether it is at logic '1' or logic '0'. Address lines A_0 and A_1 are usually connected to these pins so the addresses are as follows:

$C/\overline{D}$ (AI)	$B/\overline{A}$ (A0)	Register	Address
0	0	port A data	PIO base address
0	I	port B data	Base address + I
I	0	port A control	Base address + 2
I	I	port B control	Base address + 3

You will remember that the Z-80 uses only the lower eight bits of the address bus (A_0–A_7) for input/output addresses. Since A_0 and A_1 are used for the internal ports of the PIO, in real systems some (or all) of the higher order address lines A_2–A_7 are used as the inputs to a decoder. At the chosen base address of the PIO, one of the decoder outputs goes to logic '0' and this is used to chip enable ($\overline{CE}$) the PIO. In the Micro-professor target system that we have focused on, the single PIO is enabled at a base address of 80H. Thus for that system the port addresses are:

Port A data   80H
Port B data   81H
Port A control   82H
Port B control   83H

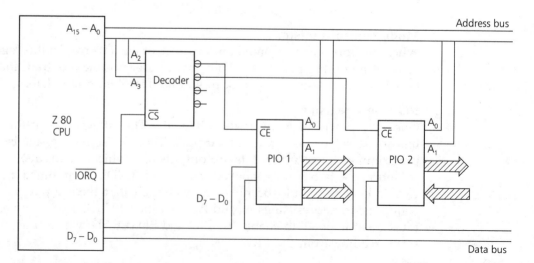

Figure 22.4 A system with two PIOs

These were the addresses used in our programming examples in Chapter 21. Therefore, because A_0 and A_1 are usually used to address the PIO ports, the overall addresses in a system will always be consecutive. The starting address depends on the decoding.

Figure 22.4 shows the 'bare bones' of a system which used two PIOs. Notice that each PIO has A_0 and A_1 connected to it to select the internal ports. Each device is chip enabled from the output of a decoder. Only two of the higher order address lines (A_2 and A_3 in this example) operate the single decoder. Every time the CPU executes an IN or OUT instruction, its control line $\overline{\text{IORQ}}$ goes to logic '0' to select the decoder and hence the required PIO. In this example PIO 1 has been set up to be two output ports. PIO 2 has been set up as an input port and an output port.

Progress check 22.1

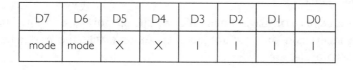

How many more PIOs could be connected in the system of Figure 22.4?

Progress check 22.2

In how many modes can a Z-80 PIO be set up?

Programming the Z-80 PIO

Each port of a PIO can be set up in three different ways or **modes**:

mode 0	output port
mode 1	input port
mode 3	bit port

Additionally, port A can also be set in a bidirectional mode which is called mode 2.

The operational mode is selected by writing a special 'mode control byte' to the control port. The mode control byte has the following form:

D7	D6	D5	D4	D3	D2	DI	D0
mode	mode	X	X	I	I	I	I

signifies a mode control byte

Bits D_4 and D_5 are 'don't cares' and can be chosen as '0' or '1' as desired. Bits D_6 and D_7 are used to identify the mode desired according to the following scheme:

447

D7	D6	MODE
0	0	0 (output mode)
0	I	I (input mode)
I	0	2 (bidirectional)
I	I	3 (bit-mode)

For example, to set up a port as an output port, the mode control byte (MCB) would be:

0 0 X X 1 1 1 1

Assuming '0' for the 'don't cares' then

MCB = 0 0 0 0 1 1 1 1 = OFH

This byte of data has to be sent to the control port as necessary.

Look back at the programs used in worked examples 21.5, 21.6, 21.7 and 21.8 of Chapter 21. At the beginning of each program, the desired port is set up as necessary by sending the correct MCB to the control port of port A or port B.

Progress check 22.3

What is the MCB if a port is to be set up as (a) an input port and (b) a bit-port?

The bit mode (mode 3)

The bit mode is designed for control applications in which each of the 8 bits can be chosen as an input or output. Once the bit mode has been selected for a port then another byte of data must be sent to the control port to 'tell' the PIO which bits are to be inputs and which are to be outputs. This byte of data is called the **I/O select byte**. The setting of mode 3 is illustrated in the next worked example.

WORKED EXAMPLE 22.1

Write the assembly language mnemonics to set up port A of a PIO such that bits 0, I and 2 are inputs and the rest are outputs. Assume that the port addresses are those for the Micro-professor.

Solution

(a) *Mode control byte*. The MCB for mode 3 is:

1 1 X X 1 1 1 1 .

Assuming the 'don't cares' are '1' then the MCB = FFH.

(b) *I/O select byte*. If a bit needs to be an output then a '0' is chosen. If a bit needs to be an input then a '1' is chosen.

I/O select = 0 0 0 0 0 1 1 1 = 07H

(c) *The program*. Port A data is 80H and port A control is 82H. A suitable program is:

```
LD A, FFH      ; MCB for mode 3
OUT (82H), A   ; send MCB to control port
LD A, 07H      ; I/O select byte
OUT (82H), A   ; send I/O select byte to port A control
```

The further use and programming of a PIO will be considered in further worked examples after serial interfacing has been described.

Serial interfacing

To transmit data in parallel from a parallel port to a peripheral would require one wire per bit, i.e. for a byte of data 8 wires would be needed. If the distance involved is not too great, say between a computer and a printer, then parallel transmission is usually used. Since all the data bits are received at exactly the same time, then the speed can be quite fast.

An alternative to this is **serial transmission** in which the data bits pass along a single conductor, one bit after another. Although slower than parallel data transfer, it is the preferred method of communication when distances are great or speed is not important. **Serial communication** is used to connect computers to visual display units (VDUs) and for data transmission over telephone lines. A basic system for serial transfer from one microcomputer to another is drawn in Figure 22.5. The system is called **asynchronous** because the clocks used at the sending and receiving ends are different.

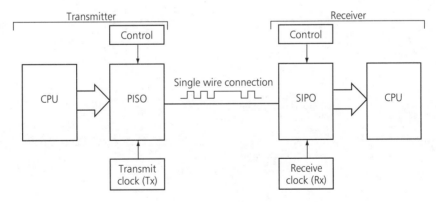

Figure 22.5 Serial transmission (asynechronous)

Essentially, two shift registers and some control circuits are required. The system works as follows. Data to be transmitted is loaded into a **parallel-in-serial-out** (PISO) register at the transmitting end. Under the control of the transmitter clock (TX clock), the data is shifted out one bit at a time and travels in series down the line. At the receive end, the data stream enters a **serial-in-parallel-out** (SIPO) register under the control of the RX clock and is passed in parallel to microcomputer 2 for processing.

If this arrangement is to work correctly then the following points have to be satisfied:

1 The frequency of the transmitter and receiver clocks must be the same.
2 The receiver must in some way 'know' when transmission has begun.
3 The receiver must somehow 'know' when transmission has stopped.

The consideration of the first point had led to the adoption of a set of **standard frequencies** for asynchronous serial data transmission. These frequencies are known as **baud rates** and indicate the number of bits transmitted per second when the line is carrying serial data. Thus:

$$\text{Baud rate} = \frac{1}{\text{bit time}}$$

At a data rate of, for example, 1800 Baud, the bit time is $1/1800 = 555.6\ \mu S$. To transmit a single byte of data at this Baud rate would take $8 \times 555.6\ \mu S = 4.44$ ms.

If the same byte of data was sent in parallel it would take only fractions of a microsecond to be transferred. Typical standard Baud rates are 75, 110, 150, 300, 600, 1200, 2400, 4800, 9600 and 19200.

Format for serial data

Consideration of points (2) and (3) above has led to a standard way of 'packaging' blocks of serial data. A typical asynchronous **serial data packet** is shown in Figure 22.6.

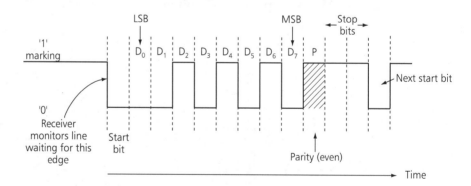

Figure 22.6 Asynchronous serial data packet

Notice that start and stop bits are added to the data and usually a **parity bit** that is used for error checking purposes. The system works as follows. When the line is not carrying data it is said to be in the **marking state** or logic '1' level. At the start of each transmission of data, the transmitter changes the signal on the line to logic '0' for one bit period, called the **start bit**. As soon as the receiver detects the change it 'knows' that transmission is about to begin. The data is sent, with the least significant bit (LSB) first. In the example, the data byte 54H has been chosen. Following the MSB, a parity bit is usually added by the transmitter. If even parity is chosen then the transmitter adds either a '1' or a '0' to keep the number of '1's in the data even. In Figure 22.6 the byte, 54 H, has an odd number of '1's' in it. The transmitter has added a '1' for the parity bit to keep the parity even. The receiver 'knows' that it should always receive an even number of '1's' – if it does not it knows an error has occurred. If **odd parity** is chosen, the number of '1's' is always odd.

After the parity bit, the line is held by the transmitter in the marking state for either 1, 1.5 or 2 bit times. Usually, one bit is used. During this time the receiver can **resynchronise** its circuits ready for the next detection of a start bit.

The electrical standard that is normally used is called the **RS-232 serial interface standard** which defines the voltage levels used as:

> logic '1' = –3 V to –25 V   typical value used = –12 V
>
> logic '0' = +3 V to +25 V   typical value used = +12 V

Because the voltage levels are different to the normal TTL levels of 0 V and 5 V, integrated circuits are produced to convert from TTL to RS-232 standard and vice-versa.

Progress check 22.4

For an even parity system, what will the value of the parity bit be if the data byte is 54H?

Serial I/O chips

The requirement for serial data transmission is so common that a range of integrated circuits is available which considerably eases the interface problem. The chip is called a Universal Asynchronous Receiver Transmitter or **UART** for short. A simplified block diagram is drawn in Figure 22.7(a). Figure 22.7(b) shows how two UARTs would be used for data communication over a telephone system.

The device is programmed in much the same way as a PIO by writing data words to a 'control register' within the UART. In general, the chip can be programmed for:

● data packages of 5, 6, 7 or 8 bits
● 1, 1.5 or 2 stop bits
● even, odd or no parity.

Various signals are provided to ease the interface problem between the UART and a modem for transmission over telephone lines. The manufacturer's data on the UART should always be consulted for precise details of programming and the various facilities offered. The UART commonly used with the Z-80 is the Intel 8251A.

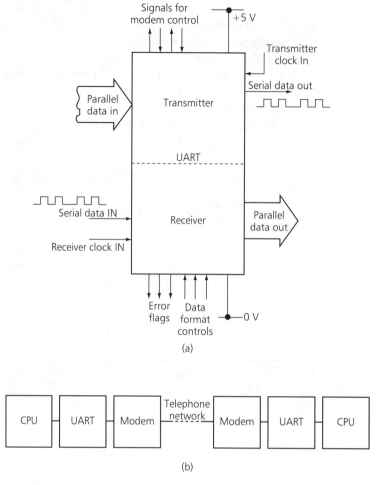

Figure 22.7 *Simplified block diagram of a UART*

Interfacing protocols

The vast majority of peripherals do not operate at the same speed as a micro-processor and, in fact, most operate at a much slower speed. Therefore, whenever data transfers have to be made, some sort of common timing arrangement between the devices is necessary if the data transfer is to be successful.

We have already described a possible solution for data inputs in Worked Example 21.8 of Chapter 21 and it is a good idea to review this example again. Here data was input from switches to the Z-80 and one of the switches was used as a 'strobe' to indicate to the microprocessor that data was available. This simple handshaking system ensured that data could be set up correctly on the remaining seven switches and then strobed into the microprocessor for servicing. The main disadvantage of this system was that in this example, the switch used as a strobe had to be reserved purely for this function and was not available as a data bit. Generally, there is a hardware overhead if this technique is adopted.

A slightly simpler technique, which is very useful when data is available on a very predictable and regular basis, is to dispense with a strobe bit and to include a fixed **time delay** between each read operation. A good example is when input data needs to be taken regularly from an analogue transducer – say for measuring temperature. The transducer would be interfaced to an ADC and the program would be designed so there is a fixed delay between every read operation from the ADC. This 'timed' protocol is fairly simple and is generally known as **data logging**.

Activity 22.1

Modify Worked Example 21.8 so that bit 7 is NOT used as a strobe. Incorporate a time delay of about 10–15 s between each input read to allow the data to be set on the switches. Write the delay as a subroutine.

The following case study outputs data using a simple fixed time delay to a set of LEDs connected to PORT B of a PIO.

WORKED EXAMPLE 22.2

Write a program which causes eight LEDs connected via port B of a PIO to count up slowly in binary with about a one second delay between each count. Assume that the address of port B is 81H and that the LEDs are interfaced to the Z-80 CPU as in Chapter 21.

Solution

The simplest solution is to increment the accumulator and output its contents every second. An approximate one second delay can be reproduced by decrementing a register pair. The delay is written as a sub-routine as usual.

(a) A suitable algorithm is:
  (1) Set stack pointer to the top of RAM because we will be using a sub-routine.
  (2) Set up port B as an output.
  (3) Clear the accumulator.
  (4) Output and call delay routine.
  (5) Increment the accumulator and jump back to step (4).

(b) The correctly coded program is shown below. Notice the use of the **exclusive or** (XOR) instruction which clears the accumulator. (An alternative would be, of course, to use the instruction LD A, 00H.)

Address	Hex code		Mnemonic	Comment
			ORG 1800H	
1800	31 FF 1F		LD SP, 1FFFH	; SP to top of RAM
1803	3E 0F		LD A, 0FH	;
1805	D3 83		OUT (83H), A	; port B is output
1807	AF		XOR A	; clear accumulator
1808	D3 81		OUT (81H), A	; out to port B
180A	CD 00 19	JACK:	CALL DELAY	; delay subroutine

180D	3C	INC A	; increment A
180E	D3 81	OUT (81H), A	;
1810	C3 0A 18	JP JACK	; repeat continuously

Delay subroutine

ORG 1900H

1900	F5	DELAY: PUSH AF	; save A and flags
1901	11 00 00	LD DE, 0000H	; load pair for approx. 1 second
1904	1B	LOOP: DEC DE	;
1905	7A	LD A, D	;
1906	B3	OR E	; checking for zero
1907	C2 04 19	JP NZ, LOOP	;
190A	F1	POP AF	; restore A and flags
190B	C9	RET	; return to main program

(c) For testing when the program is entered into the Microprofessor and run, the LEDs connected to port B continually count up in binary.

Polling

Very often a microprocessor will need to accept data from, or provide data to, more than one peripheral device to which it is connected. One way of achieving this is to use a system known as **software polling**. With this technique the microprocessor checks each device in turn to see if it is ready to send or receive data. Generally, a peripheral will indicate that it is 'ready for servicing' by setting a status 'flag' at the I/O port to which it is connected. The CPU will then continually check (or poll) the flags and 'service' the peripheral as necessary. Since the CPU itself initiates the polling of the peripherals, there is no problem in interrupting the flow of the main program that the CPU may be executing. However,

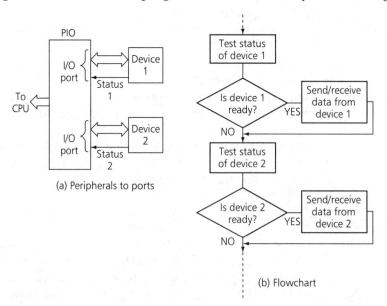

Figure 22.8 The concept of polling

the method does considerably slow down the operation of the main program since it is necessary to suspend the main program execution while polling takes place at regular intervals, even though the peripherals may not actually require servicing at that instant. The concept of polling is illustrated in Figure 22.8. Here only two peripherals connected to a PIO have been shown.

Notice that each device has to wait for its turn before it can be serviced. The general flowchart in Figure 22.8(b) can be extended for other peripherals connected to further ports in the system as necessary.

In the following worked example, a simple polling routine is used to interrogate a single digital sensor.

WORKED EXAMPLE 22.3

A digital sensor connected to bit 7 of port A counts objects as they pass on a conveyor belt. As an object passes the sensor it produces a logic '0'. Write a program which polls the sensor and produces a binary count on the LEDs connected to port B, i.e. a running total is kept.

Solution
The sensor can be simulated with the switch connected to bit 7 of port A (80H). Since the switch is mechanical we have to eliminate the effect of 'bounce' in a similar way to that used in Worked Example 21.8 of Chapter 21.

(a) *The algorithm/flowchart.* You are required to construct this in the next activity – Activity 22.2.

(b) *Program and program coding.* The correctly coded program is shown below. As usual, the necessary delay has been written as a subroutine and called when required.

Address	Hex code		Mnemonic	Comment
			ORG 1800H	
1800	31 FF 1F		LD SP, 1FFFH	; SP to top of RAM
8033	E 4F		LD A, 4FH	;
1805	D3 82		OUT (82H), A	; port A is input
1807	3E 0F		LD A, 0FH	;
1809	D3 83		OUT (83H), A	; port B is output
180B	06 00		LD B, 00H	; clear 'pulse' counter
180D	78	DIS:	LD A, B	;
180E	D3 81		OUT (81H), A	;
1810	CD 00 19		CALL DEL	; 20 mSec wait
1813	DB 80	POLO:	IN A, (80H)	;
1815	E6 80		AND 80H	; checking bit 7
1817	C2 13 18		JP NZ, POLO	;
181A	CD 00 19		CALL DEL	;
181D	DB 80	POLO1:	IN A, (80H)	;
181F	E6 80		AND 80H	;
1821	CA 1D 18		JP Z, POLO1	;
1824	04		INC B	; increment pulse counter

1825	C3 0D 18		JP DIS	; repeat
			Delay subroutine	
			ORG 1900H	
1900	F5	DEL:	PUSH AF	; save accumulator
1901	11 00 0A		LD DE, 0A00	;
1904	1B	LOOP:	DEC DE	;
1905	7A		LD A, D	;
1906	B3		OR E	; check for zero
1907	C2 04 19		JP NZ, LOOP	;
190A	F1		POP AF	; restore accumulator
190B	C9		RET	; return to main program

(c) *Testing the program.* When run, the LEDs on port B will increment every time bit 7 is taken to logic '0' and returned to logic '1'.

Activity 22.2

Produce an algorithm and flowchart for the program in Worked Example 22.3.

Interrupts

Polling is initiated and controlled by the CPU and as such can be wasteful of CPU time since the microprocessor has to leave its main program and perform the polling routine frequently. An alternative is to use a hardware technique called an interrupt which is initiated by the peripheral. Here, each peripheral has an

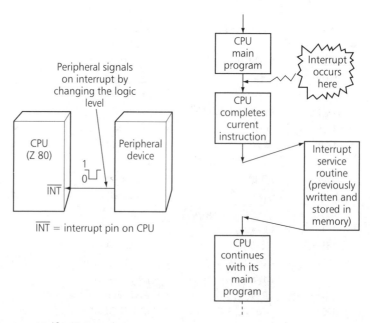

Figure 22.9 The concept of an interrupt

electrical connection to a special pin on the CPU called the **interrupt pin**. When a peripheral device needs servicing, it changes the logic state on this input pin. The CPU then suspends the execution of its normal program and executes a routine called an **interrupt service routine** (ISR). When the ISR is complete, the CPU then continues with its main program. The concept of an interrupt is illustrated in Figure 22.9.

The CPU must, of course, be set up so that it can accept the interrupt and the ISR must have been previously written and stored in memory.

The use of interrupts has several advantages when compared to software polling. The main one is that the CPU uses its time more efficiently since it only has to leave the execution of its main program when the peripheral signals to it that it needs to communicate with it. The following worked example illustrates how one type of interrupt can be set up for a Z-80 CPU and PIO.

WORKED EXAMPLE 22.4

(a) Briefly describe the difference between **non-maskable** and **maskable** interrupts. Explain how a Z-80 CPU can be set up and how it responds to MODE 2 interrupts.

(b) Write a program so that:

   (1) Data inputted to switches on bits 3 to 7 of port A (80H) of PIO is displayed directly on the corresponding LEDs connected to port B (81H).
   (2) If any of the bits 0, 1 or 2 of port A are pulsed from logic '0' to logic '1' an interrupt is generated.
   (3) The interrupt causes all the LEDs on port B to flash ON and OFF five times (each LED should be ON or OFF for a period of about one second). The CPU should then return to its main program — (1).

Solution

(a) Types of interrupts

Interrupts can be conveniently divided into two types:

Non-maskable interrupts: These are always recognised by the CPU as soon as possible. *They cannot be prevented (masked out) by software*. This interrupt is generally reserved for very important functions that must be serviced whenever they occur.

Maskable interrupts: Sometimes it is convenient to be able to 'switch-off' an interrupt capability, i.e. 'mask it out'. This is possible with a maskable interrupt by including a special instruction at an appropriate point in the program.

The Z-80 interrupt pins

On the Z-80 CPU the interrupt pins are labelled: $\overline{\text{NMI}}$ (non-maskable for pin 17) and $\overline{\text{INT}}$ (maskable, pin 16).

Also important in connection with interrupts are two flip-flops inside the CPU called IFF_1 and IFF_2. The action of these flip-flops depends on whether or not maskable or non-maskable interrupts are operating. Figure 22.10 illustrates the

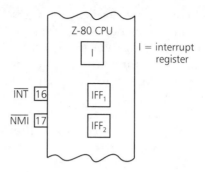

Figure 22.10 Important parts of the Z-80 concerned with interrupts

interrupt pins and flip-flops symbolically. Also important for mode 2 interrupts is a special register inside the Z-80 called the **interrupt register** I.

Maskable interrupts (INT) and IFF_1 and IFF_2

At power-up (or when reset) the Z-80 resets IFF_1 and IFF_2 to logic '0' so that any interrupt signals on the INT pin are disabled. To enable interrupts, the instruction EI (opcode FB) must be executed. When this occurs, both IFF_1 and IFF_2 are enabled and set to logic '1'. When an interrupt is accepted by the Z-80 on the INT pin, both IFF_1 and IFF_2 are reset. This action prevents any further interrupts from being accepted *until you want them to*. To do this you must give another EI instruction.

Z-80 response to maskable interrupts (INT)

The Z-80 can be programmed to respond to a maskable interrupt (INT) in one of three ways, or modes, as in the table.

Mode	Instruction	Opcode
mode 0	IM0	ED 46
mode 1	IM1	ED 56
mode 2	IM2	ED 5E

Of these three modes, mode 2 is the most powerful. It is especially designed to operate with the Z-80 PIO and CTC etc. *Before mode 2 interrupts can be used, both the Z-80 CPU and the interrupting peripheral device (e.g. a PIO port) must be correctly initialised.*

We will first describe the general operation of MODE 2 and then look at the CPU and PIO initialisation in detail.

The mode 2 interrupt

For the CPU to enter this mode, the following instructions must have been executed:

IM 2   (opcode ED 5E)
EI   (opcode FB)

The Z-80 checks the status of its interrupt system at the end of each instruction. If the interrupt system is enabled and an interrupt is received, the following sequence of events occurs:

1   The Z-80 acknowledges the interrupt by making [$\overline{\text{MI}}$] and [$\overline{\text{IORQ}}$] active simultaneously. (This allows the peripheral to respond to the interrupt.)
2   IFF_1 and IFF_2 are automatically reset, so that if another interrupt occurs, it cannot affect the system.
3   The interrupting device (e.g. PIO) places a byte of data called a vector (*which has already been loaded into it – see below*) on to the data bus.
4   The contents of the PC are pushed to the stack.
5   The contents of the Z-80 I register (*which has already been loaded with a data byte* – see below) is combined with the vector to form a **two-byte memory pointer**. This pointer is used to access a table of **interrupt service routines** starting addresses *which have been located in memory*.
6   The PC is loaded with the address pointed to from this table. This is the address of the ISR.
7   The CPU 'jumps' to the start of the ISR and executes it.

At the end of the ISR, the EI instruction is included again so that more interrupts can be accepted.

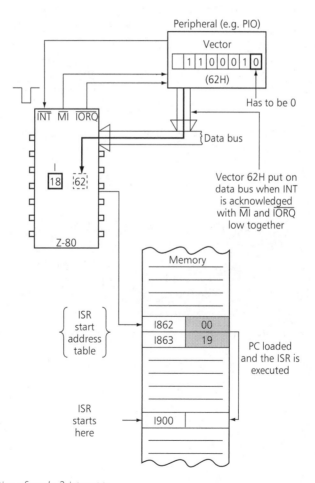

Figure 22.11 Operation of mode 2 interrupts

The last instruction in the ISR should be the **return from interrupt** (RETI). This causes the stack to be 'popped' so that the CPU continues from where it left off in the main program.

As an example, consider the diagram in Figure 22.11.

1 The peripheral has been loaded with its vector; in this case 62H.
2 The I register of the Z-80 has been loaded with a byte – 18H in this case.
3 When interrupt occurs ($\overline{\text{INT}}$ active), Z-80 acknowledges by taking $\overline{\text{MI}}$ and $\overline{\text{IORQ}}$ low and places 62H on data bus.
4 Z-80 combines this data byte with its I register to form address 1862H.
5 Z-80 takes data stored at 1862H (in this case 00H) and treats this as the lower byte of a new address.
6 Z-80 next reads data at 1862H + 1 (in this case 19H) and treats this as the upper byte of the address. This memory address 1900H is the location of the ISR.
7 Z-80 jumps to 1900H and executes the ISR.

To **initialise** the Z-80, you would carry out the following instructions in this case:

IM2	; select mode 2 interrupts
LDA, 18	; ISR START ADDRESS TABLE – High byte
LDI, A	; INTERRUPT REGISTER loaded with 18H
LDHL, 1900H	; Load HL pair with ISR start address
LD (1862), HL	; Load ISR start address table
EI	; Enable Z-80 to accept interrupt

Z-80 PIO in mode 3 with interrupts

In mode 3 each bit of a port can be chosen as an input or an output. We have to send data bytes to the control port associated with each data port.

(1) Send **PIO mode control byte** to set operating mode of the port:

(2) When mode 3 selected, next send **PIO I/O select byte** to the port.

With interrupts we now have to send more control bytes.

(3) *PIO interrupt vector* – actually this can be sent before (1) if you want:

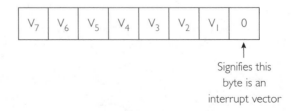

Signifies this byte is an interrupt vector

The vector represents the low part of an address at which the address of the interrupt service routine will be located. (Note: Because LSB = 0, an even address will always result.

(4) PIO interrupt control byte

ENABLE INTERUPT	AND/ OR	HIGH/ LOW	MASK FOLLOWS	0	I	I	I
d_7	d_6	d_5	d_4	d_3	d_2	d_1	d_0

Indicates interrupt
control byte

d_7 = 1   Enables PIO interrupts
d_7 = 0   Disables PIO interrupts
d_6 = 1   AND – means that *all* monitored I/O lines need to be active to cause an interrupt.
d_6 = 0   OR – means that *any* of the monitored I/O lines will generate an interrupt when they become active.
d_5 = 1 or 0   PIO can be programmed for active HIGH or active low signals to produce an interrupt, '1' – high, '0' – low.
d_4 = 1   Means that the next control word that follows is an interrupt mask. This next control byte is such that only lines with a mask bit of 0 will be monitored.
d_4 = 0   Means a mask does not follow and so all inputs may generate an interrupt.

(5) *PIO interrupt mask byte*
If d_4 = 1 then the next byte is the interrupt mask:

M_7	M_6	M_5	M_4	M_3	M_2	M_1	M_0

Only those port lines whose mask bit = '0' will be monitored for generating an interrupt.

M = '1' – ignore
M = '0' – monitor for an interrupt

(b) The program

The program shown below is quite complicated and you should study it carefully. The initialisation sequence is as follows:

Set the stack pointer
Port A initialisation
Port B initialisation
Z-80 initialisation
Main program

Even though Port A is configured as I/P's and B as O/P's mode 3 is chosen for the PIO ports. No handshaking is then involved.

Address	Hex code		Mnemonic	Comment
			ORG 1800H	
1800	31 FF 1F		LD SP, 1FFFH	; SP to top of RAM
	Port A initialisation			
1803	3E FF	PTA INIT:	LD A, FFH	; choose MODE 3
1805	D3 82		OUT (82H), A	; Port A in MODE 3
1807	D3 82		OUT (82H), A	; all bits as inputs
1809	3E 62		LD A, 62H	; low byte of start address table
180B	D3 82		OUT (82H), A	; Port A vector loaded
180D	3E B7		LD A, B7H	; 1011 0111 interrupt control byte
180F	D3 82		OUT (82H), A	; output control byte
1811	3E F8		LD A, F8	; 1111 1000 mask control
1813	D3 82		OUT (82H), A	; only bits 0, 1, 2 will be monitored
	Port B initialisation			
1815	3E FF	PTB INIT:	LD A, FFH	; choose MODE 3
1817	D3 83		OUT (83H), A	; Port B in MODE 3
1819	AF		XOR A	; clear accumulator
181A	D3 83		OUT (83H), A	; Port B is now set as all outputs
	Z-80 CPU initialisation			
181C	3E 18	INIT CPU:	LD A, 18H	;
181E	ED 47		LD I, A	; high byte of start address table
1820	21 00 19		LD HL, 1900H	;
1823	22 62 18		LD (1862), HL	; address of ISR loaded in 1862 & 1863
1826	ED 5E		IM 2	; MODE 2 selected for CPU
1828	FB		EI	; CPU will now accept interrupts
	Main program			
1829	DB 80	PROG:	IN A, (80H)	; input from switches
182B	E6 F8		AND F8	; mask out bits 0 — 2
182D	D3 81		OUT (81H), A	; output to LEDs
182F	C3 29 18		JP PROG	; repeat continuously
	Interrupt service routine			
1900	F5	ISR:	PUSH AF	; save A and flags
1901	16 05		LD D, 05H	; D holds number of flashes
1903	3E FF	LOOP:	LD A, FFH	;
1905	D3 81		OUT (81H), A	; switch all LEDs ON
1907	CD 50 19		CALL DELAY	; 1 second delay at 1950H
190A	D3 81		OUT (81H), A	; switch all LEDs OFF
190C	CD 50 19		CALL DELAY	; 1 second delay
190F	15		DEC D	; decrement counter for a number of flashes
1910	C2 03 19		JP NZ, LOOP	;
1913	F1		POP AF	; restore A and flags
1914	F3		EI	; enable interrupt
1915	ED 4D		RETI	; return from ISR
	Delay subroutine			
1950	01 00 00	DELAY:	LD BC, 0000H	; load pair for about 1 second
1953	0B	DEL:	DEC BC	;
1954	78		LD A, B	;
1955	B1		OR C	; check for zero
1956	C2 53 19		JP NZ, DEL	;
1959	C9		RET	; return from subroutine

(c) Enter program

This program should be entered into a Microprofessor. Check that 'pulsing' any of bits 0, 1 or 2 will cause the interrupt routine to be executed.

Progress check 22.4

What instruction needs to be changed in the ISR so that the LEDs flash 10 times?

Activity 22.3

Locate a data sheet for a Z-80 PIO. Produce a report describing how the Z-80 CPU and Z-80 PIO are set up for mode 2 interrupts.

Analogue inputs and outputs

Up to now all of the signals processed by the microprocessor-based system have been digital (ON/OFF) signals. For example, in some of the worked examples data has been inputted from switches which are either open or closed and outputted to LEDs which can be either on or off.

In many applications, it may be necessary to process analogue voltages or currents produced by **sensors** for temperature, pressure and humidity etc. and generate analogue voltages or currents to drive various **actuators** such as motors

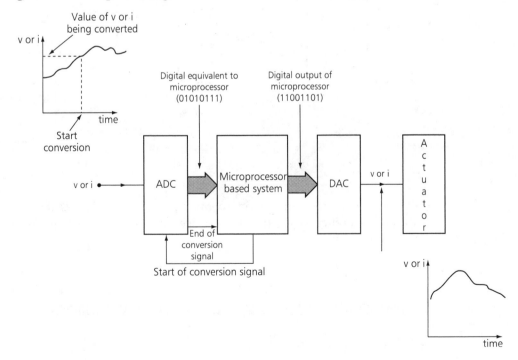

Figure 22.12 Using an ADC and a DAC

etc. Since the microprocessor is a purely digital device, additional integrated circuits are needed to convert the analogue input into a digital equivalent which the microprocessor can process and then take the digital output and convert it to an analogue signal for the actuator. Such devices are called **analogue-to-digital converters** (ADCs) and **digital-to-analogue converters** (DACs). A block diagram of a typical system is drawn in Figure 22.12.

The analogue voltage or current signal, perhaps generated by a temperature sensor of some sort, is connected to the input of the ADC. When the microprocessor needs the digital equivalent of this signal it generates a **start conversion pulse** (usually via a PIO) to instruct the ADC to convert. After a period of time known as the **conversion time**, the ADC signals to the microprocessor that it has the digital equivalent available – in this diagram an 8-bit word is shown. The conversion time is an important parameter of an ADC and varies from nanoseconds to milliseconds. Generally, the faster the device, the more expensive the ADC is. The digital equivalent is then processed by the microcomputer system according to the user's program. If an analogue output is required to drive an actuator, a digital word is output to the DAC from the microprocessor usually via a PIO. The output from the DAC is then used to drive the actuator. Often additional electronics is needed between the DAC and the actuator to produce the necessary power levels etc. The basic operating principles of DACs and ADCs will now be briefly described.

Digital-to-analogue conversion

A block diagram of an 'n-bit' DAC is drawn in Figure 22.13(a) with Figure 22.13(b) showing a simple '3-bit' DAC.

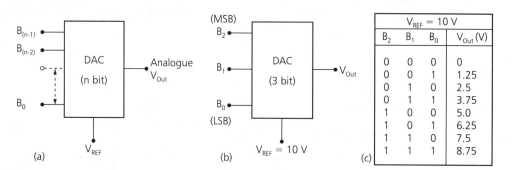

Figure 22.13 Basic DAC operation

For any DAC, the binary code connected to it determines the proportion of the applied reference (V_{REF}) that is produced at the output (V_{out}). For an 'n-bit' DAC, V_{out} is given by:

$$V_{out} = \left(\frac{B_{(n-1)}}{2} + \frac{B_{(n-2)}}{4} + \frac{B_{(n-3)}}{8} + \ldots + \frac{B_0}{2^n} \right) \times V_{REF}$$

We will use a '3-bit' DAC as an example so V_{out} will be:

$$V_{out} = \left(\frac{B_2}{2} + \frac{B_1}{4} + \frac{B_0}{8} \right) \times V_{REF}$$

The 3 bits B_0, B_1 and B_2 can have $2^3 = 8$ combinations. If V_{REF} is chosen to be, say, 10 V then the possible values of V_{out} will be as given in Figure 22.13(c). Notice that each discrete step of V_{out} changes by 1.25 V. Obviously, the greater the number of bits used for a given V_{REF}, the smaller the steps will be. An 8-bit device will give a step value of $10/256 = 0.0391$ V, for example. The step size is called the **voltage resolution** of the DAC.

Analogue-to-digital conversion

An ADC will take an analogue voltage or current at its input and produce at output a digital equivalent of that signal. In effect, the process is opposite to that performed by a DAC.

A block diagram of an ADC is shown in Figure 22.14(a).

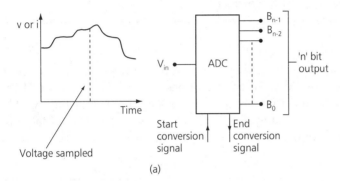

Figure 22.14 Basic ADC operation

When the start conversion signal is received, the ADC will sample the analogue input and start the conversion process. When conversion is complete, the ADC signals this fact to the microprocessor via an end of conversion signal. The digital equivalent of the analogue input will then be available at the ADC outputs.

For illustration purposes, suppose we have a 3-bit ADC with each change in the digital output corresponding to a change of 0.5 V in the analogue input. A complete table of values can be drawn as in Figure 22.14(b). Thus, when the analogue input is 2.5 V, the code 101 will appear at the ADC output. If V_{in} changes to 3 V, the code becomes 110 etc.

You should now begin to see how a fundamental problem arises. What happens if, say, V_{in} lies between 2.5 V and 3.0 V? What code will the ADC produce? At 2.75 V (half-way) it could produce either 101 or 110. This is known as the **quantisation error** and we have already investigated this in our work on communication systems. Obviously increasing the number of bits will bring the allowed voltage steps closer together, so reducing the error.

Activity 22.4

Use manufacturers' data sheets (e.g. RS Components) to select an ADC and DAC chip which *could* be used with a Z-80 microprocessor.

In the final worked example of this element we will briefly examine a Z-80 system in which several I/O devices are used. Activities are integrated within this worked example.

WORKED EXAMPLE 22.5

Determine the addresses of the four devices used in the system of Figure 22.15 and briefly describe how each is initialised (set up) to perform its function.

Solution

The system contains two devices already familiar to you (the Z-80 PIO and 8251A UART) and two new devices – the 8255 PPI and the Z-80 CTC. Firstly consider how each device is 'chip selected'.

General decoding

Each device is 'chip selected' by one of the outputs of a 2-to-4 decoder. The inputs of the decoder are fed from address lines A_6 and A_7. Each chip has address lines A_0 and A_1 connected to it except the UART which only has A_0. The first step in deducing the address ranges of each device is to construct a table for the various address combinations. Since A_2 to A_5 are not used, they are known as 'don't care' (X).

A7	A6	A5	A4	A3	A2	A1	A0	Device selected
0	0	X	X	X	X	0	0	
		X	X	X	X	0	1	8255 PPI
		X	X	X	X	1	0	
		X	X	X	X	1	1	
0	1	X	X	X	X	0	0	
		X	X	X	X	0	1	CTC
		X	X	X	X	1	0	
		X	X	X	X	1	1	
1	0	X	X	X	X	0	0	
		X	X	X	X	0	1	Z-80 PIO
		X	X	X	X	1	0	
		X	X	X	X	1	1	
1	1	X	X	X	X	X	0	8251 UART
		X	X	X	X	X	1	

The Z-80 PIO

Taking 'X' to be '0' gives the PIO address as

$$
\begin{array}{rcll}
1\,0\,0\,0\,0\,0\,0\,0 & = & 80\text{H} & = & \text{Port A data} \\
1\,0\,0\,0\,0\,0\,0\,1 & = & 81\text{H} & = & \text{Port B data} \\
1\,0\,0\,0\,0\,0\,1\,0 & = & 82\text{H} & = & \text{Port A control} \\
1\,0\,0\,0\,0\,0\,1\,1 & = & 83\text{H} & = & \text{Port B control}
\end{array}
$$

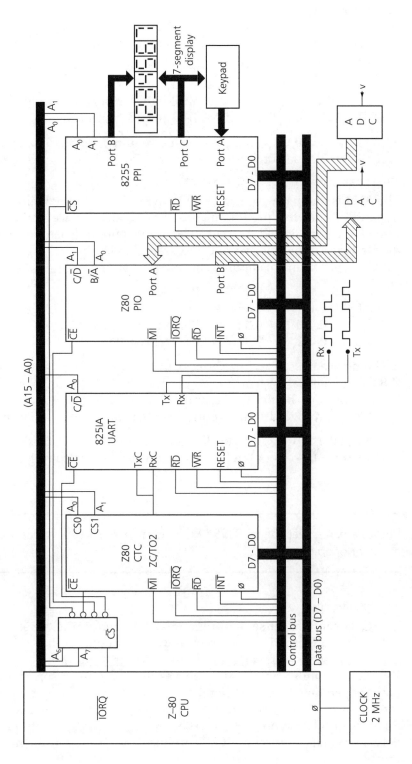

Figure 22.15 Z-80 system for Worked Example 22.5

These addresses are by now very familiar to you because they are the same as those used on the Micro-professor learning system which we have focused on throughout the section on microprocessors. In this application port A is set up as an input port to take data from an ADC and port B is an output port to provide data for a DAC. A suitable basic initialisation routine would be:

```
LD A, 4FH        ; mode control byte for port A as input
OUT (82H), A     ; send MCB to port A control
LD A, 0FH        ; mode control byte for port B as output
OUT (83H), A     ; send MCB to port B control
```

Suitable software, of course, would be required for the analogue-to-digital and digital-to-analogue conversion processes. This will not be considered here.

The 8255 PPI

The 8255 is another very useful I/O device that was originally manufactured by Intel. PPI stands for '**programmable peripheral interface**'. Briefly, the device has three 8-bit ports called port A, port B and port C. There are several operating modes that can be selected and these are defined by writing control words to a control register. In this application the port addresses are:

```
0 0 0 0 0 0 0 0  =  00H  =  Port A
0 0 0 0 0 0 0 1  =  01H  =  Port B
0 0 0 0 0 0 1 0  =  02H  =  Port C
0 0 0 0 0 0 1 1  =  03H  =  Control register
```

The device is used to interface seven, 7-segment displays and a keypad to the Z-80. (An identical system is used on the Micro-professor.) Port A is set up as an input with ports B and C as outputs. The initialisation is performed by sending a single control word to the control register. A suitable initialisation routine is:

```
LD A, 90H        ; set up the control word
OUT (O3H), A     ; send control word to the control register
```

Activity 22.5

Locate a data sheet for an 8255 PPI. Produce a short report describing the mode 0 mode of operation.

The Z-80 CTC

The CPU can, of course, be programmed to perform a variety of counting and timing operations. Manufacturers, however, provide other chips which can be used for these tasks, freeing the microprocessor to carry out other operations. The Z-80 CTC (**counter timer chip**) relieves the microprocessor of routine counting/ timing operations and is a useful addition to any Z-80 based system. In effect, it can count external events and when a chosen number has been counted, signal this fact to the microprocessor. Alternatively, it can generate precise time intervals. In this application the CTC produces pulses at a fixed rate to provide the transmit and receive clocks for a UART – more about this later.

A block diagram of a CTC is drawn in Figure 22.16.

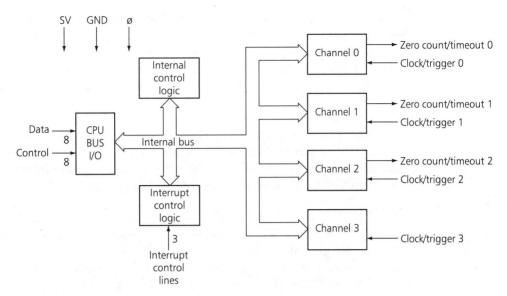

(a) Block diagram of a Z-80 CTC

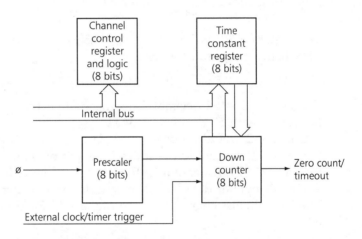

(b) Block diagram of a single channel

Figure 22.16 The Z-80 CTC

The device has four (almost identical) channels called ch0, ch1, ch2 and ch3. The channels are selected by the CS0 and CS1 pins on the chip. In our application these are connected to address lines A_0 and A_1 so the CTC addresses are:

```
0 1 0 0 0 0 0 0  =  40H  =  ch0
0 1 0 0 0 0 0 1  =  41H  =  ch1
0 1 0 0 0 0 1 0  =  42H  =  ch2
0 1 0 0 0 0 1 1  =  43H  =  ch3
```

As shown in Figure 22.16(b), each channel consists of two registers, two counters and some additional logic. Each channel can be programmed to be a counter of external events with the external clock input or as a timer with the system clock as the input. We will only consider the timer mode.

The system clock (2 MHz in our application) first passes through the prescaler counter which is programmed to divide the system clock by either 16 or 256 depending on how it is programmed. The output pulses from the prescaler decrement a **down counter** which is initially loaded from the **time constant register**. When the down counter output reaches zero the ZC/TO output pulses to logic '1'. (This could be used to produce an interrupt if required.) The contents of the time constant register are then automatically loaded into the down counter and the whole process repeats itself. Thus, a stream of pulses can be produced at the ZC/TO output pin of a channel whose frequency is the system clock divided by 16 or 256 and further divided by the value loaded into the time constant register. The period of the pulses from the ZC/TO pin is given by:

$$\text{pulse period} = T \times p \times TC$$

where T = system clock periodic time, p = prescaler value chosen (16 or 256), TC = time constant register value.

Suppose that in our application we need to provide a pulse train of 2400 Hz to the TX and RX clock inputs of the UART from channel 2 of the CTC. The pulse period is:

$$\text{pulse period} = \frac{1}{2400} = 416.7\ \mu s \quad (417\ \mu s \text{ say})$$

The system clock is 2 MHz so T = 0.5 µs. Choosing a prescaler value of 16 (set up when we program) gives:

$$417\ \mu s = 0.5\ \mu s \times 16 \times TC$$

Therefore TC = 52 decimal. This means that the time constant register needs to be loaded with 52 decimal or 34 H.

Channel 2 would be programmed as follows. The format of the control word that needs to be sent to the control register of channel 2 is given in the data sheet for the CTC. In this case it is 07 H. (You'll have to investigate why in a moment.) After this byte of data has been sent, the time constant value must then be sent. Thus, to initialise channel 2 to produce a 2400 Hz pulse train the program required is:

```
LD A, 07H       ; load A with control word
OUT (42H), A    ; output control word to channel 2
LD A, 34H       ; load A with the time constant value
OUT (42H), A    ; send time constant and start the timer
```

Thus in summary, channel 2 is dividing the 2 MHz system clock by 16 × 52 and producing a 2400 Hz pulse train for the UART.

Activity 22.6

Locate a data sheet for a Z-80 CTC. Produce a short report describing how it may be used in the timer mode.

The 8251A UART

The UART only has one address line (A_0) connected to the $C/\overline{D}$ (control/data) input line. When the $C/\overline{D}$ is at logic '1', data is read from or written to a control register in the chip. If $C/\overline{D}$ is at logic '0', data can be read from or written to a data register in the chip.

The chip is enabled by the output of the decoder which is active when $A_6 = A_7 = 1$. Taking the 'don't cares' as '0' gives the addresses of the control and data registers as:

$$1\ 1\ 0\ 0\ 0\ 0\ 0\ 0\ =\ \text{C0H}\ =\ \text{Data register}$$
$$1\ 1\ 0\ 0\ 0\ 0\ 0\ 1\ =\ \text{C1H}\ =\ \text{Control register}$$

The control register is used to initialise the device to set up the baud rate, number of data bits, parity and number of stop bits. In this system the baud rate has been chosen to be 2400 Hz which is provided by the CTC. To set up the device for this baud rate, with seven data bits, even parity and two stop bits, the program required is:

```
LD A, F9H        ; mode control word
OUT (C1H), A     ; send to the control register
```

This sets up the UART. To transmit and receive data another command word is written to the control register and further software is needed to accomplish this. Full details are given in the 8251 data sheet (see next activity).

Activity 22.7

Locate a data sheet for an 8251 UART. Produce a short report describing how the device is programmed to transmit and receive data.

Assignment 22

This assignment provides evidence for:
Element 19.3: Investigate the interfacing of microprocessor-based systems
and the following key skills:
Communication 3.2
Information Technology 3.1, 3.2, 3.3, 3.4

The evidence required for this element requires you to produce a report evaluating the devices and software used in three types of interface used with a microprocessor-based system. Your Centre will advise you of the various possibilities available there.

Report

Your report should include:
- a description of the type of interface used
- a description of the interfacing protocols
- an evaluation of the software used
- an evaluation of the interfacing devices used

PART EIGHT: PROGRAMMABLE LOGIC CONTROLLERS

Chapter 23:   Main features of programmable logic controllers
Chapter 24:   PLC information flow and communication
Chapter 25:   PLC programming techniques

- Programmable logic controllers or PLCs are now used in great numbers all over the world to control industrial processes. They have developed considerably over the last 15 years and now offer a wide range of facilities which allow them to be used to control anything from a lift system to a nuclear power station. An understanding of their basic principles of operation, programming methods and applications are therefore essential.
- Part eight investigates the main features, applications and more importantly methods of programming PLCs. Other aspects are also covered including the developing area of inter-PLC communications. Basic principles are developed and applied with numerous practical examples and activities.

Chapter 23: Main features of programmable logic controllers

This chapter covers:
Element 32.1: Investigate the main features of programmable logic controllers.

... and is divided into the following sections:
- Design characteristics
- Main components
- Forms of signals used in PLC systems
- Input and output devices.

The basic concept of the **programmable logic controller** (PLC) was first devised way back in the 1970's. It's main purpose then was to simply replace the many **relays** used to control a production line such as that found in a car assembly plant. The main advantage of the PLC over the previous relay-based system was that it was programmable. In other words, if changes to the production sequence were required, it would mean a major rewire of the relay-based system compared with a simple change to the PLC programme. It is this fact as well as it's simple programming language that has made the PLC so popular with industry.

The modern day PLC is a far cry from the very simple device first used in the 1970s. It has many facilities including the ability to **communicate** with other PLCs and computers. You will learn more about these facilities as you progress through this section of the book.

In this element you will be solely concerned with the physical characteristics of the PLC.

When you complete this element you should be able to:

- identify the main uses of programmable logic controllers
- identify the design characteristics of programmable logic controllers
- identify the main components of the internal architecture of programmable logic controllers
- describe the forms of signals used in programmable logic controllers
- investigate input and output devices.

Design characteristics

There are many PLC manufactures, and each has their own particular design methodology. However, each manufacturer broadly speaking produces three basic physical PLC designs called unitary, modular and rack-mounted.

We will now consider each of these designs in turn.

Unitary type

This type of PLC is usually at the bottom end of the manufactures range and is normally small and relatively cheap. It is usually a self-contained unit and is non-expandable. An example of this type of PLC would be the Mitsubishi FX0 series, shown in Figure 23.1(a).

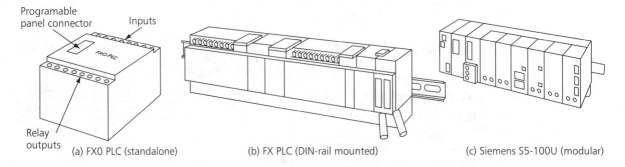

(a) FX0 PLC (standalone)   (b) FX PLC (DIN-rail mounted)   (c) Siemens S5-100U (modular)

Figure 23.1 PLC systems (not to scale)

Modular type

In this case the complete system is constructed from a number of different modules depending on the application. The individual modules are then mounted on a **DIN rail**. Upgrading or system expansion is therefore easy. The required units are simply added to the DIN rail as and when required. Examples of this type of PLC system would include the Mitsubishi FX system and the Seimens S5–100U Simatic range, examples of which are shown in Figure 23.1(b) and (c).

Rack-mounted type

This system is similar to the modular system with the exception that the individual modules are mounted on a rigid back plane which itself can then be

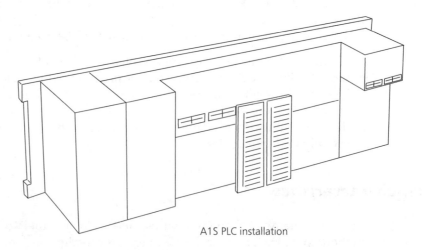

A1S PLC installation

Figure 23.2 A1S PLC installation

rack-mounted. Typical systems include the Mitsubishi A1S range shown in Figure 23.2.

Activity 23.1

Find a manufacturing company in your local area that uses PLCs to control a particular process. Visit the company and investigate a process controlled by a PLC. Note the specifications of the PLC system used and produce a short report describing the type of process controlled.

Main components

A simplified block diagram of a PLC is illustrated in Figure 23.3.

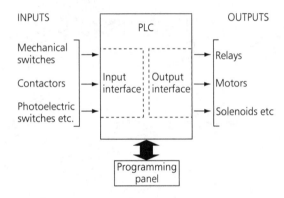

Figure 23.3 Block diagram of a PLC system

The PLC contains a microprocessor or CPU, memory devices for storing the programme, input/output interface circuits and a power supply unit or PSU. Input devices, such as mechanical switches, sensors and output devices such as relays and solenoids from the system that is being controlled are directly connected to the PLC unit via the input/output interface circuitry.

Previously, a sequence of instructions, called the program, would have been stored in the memory device via an external programming panel or some other means. The program dictates how the outputs will be controlled depending on the state of the inputs.

When a PLC is running a program, it will continually monitor the state of the inputs connected to it and switch the outputs according to the instructions contained in the program.

Because of the simplicity of modifying the program, the sequence for the process being controlled can be easily changed without having to rewire any of the input and output devices. Thus the PLC can be used easily and efficiently to control repetitive tasks of varying complexity.

Figure 23.4 shows an exploded view of a typical PLC indicating the various component parts.

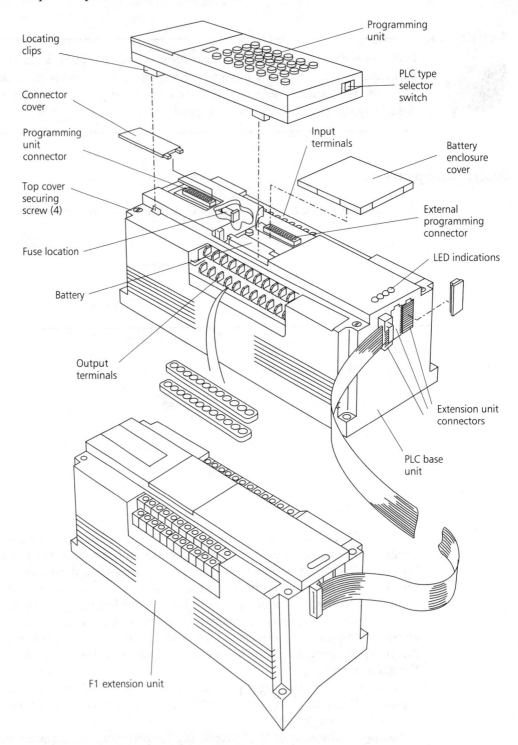

Locating clips

Connector cover

Programming unit connector

Top cover securing screw (4)

Fuse location

Battery

Output terminals

Programming unit

PLC type selector switch

Input terminals

Battery enclosure cover

External programming connector

LED indications

Extension unit connectors

PLC base unit

F1 extension unit

Figure 23.4 F1 series Mitsubishi PLC, exploded view

Activity 23.2

In Figure 23.4, which shows an F1 series PLC, at the top of the diagram there is a device called a programming unit or panel. Using an RS Components catalogue, find out what this device is used for, a part number and the specification for the panel.

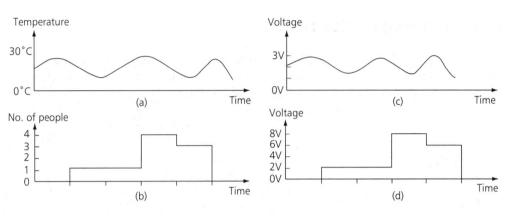

Figure 23.5 Continuos and discrete quantities

Forms of signals used in PLC systems

Changes that take place in nature either take place smoothly and continuously or in discrete steps. As an example, the temperature of a room over a period of time might vary as in Figure 23.5 (a)

In Figure 23.5(a), the temperature lies between about 0°C and 30°C and as time goes on changes in temperature occur smoothly. In contrast, the number of people that occupy the room is a discrete quantity and may vary as indicated in Figure 23.5(b). Here the number of people varies between 0 and 4. Any change is discrete, there is no such thing as 1/10th of a person!

To process temperature or people electrically, each quantity has to somehow be changed into an equivalent voltage or current signal. A device that converts signals from one energy form into another is called a **transducer**. The output of suitable transducers for variables in Figure 23.5(a) and (b) are shown in Figure 23.5(c) and (d).

Each voltage signal is an **analogue** of (an analogy of) the quantity that it represents. There is a one-to-one correspondence between the value of the voltage in each signal waveform and the quantity it is representing. In Figure 23.5(c), the transducer converts the 0°C to 30°C range into a 'span' of 0 V to 3 V. The 'people transducer' in Figure 23.5(d) generates, four discrete voltages of 0 V, 2 V, 4 V and 8 V to represent the number of people.

The most important type of discrete signal is the one which can only have one of two possible values at any instant. Such a voltage (or current) is called a **digital voltage** (or current) which has been described in Chapter 16.

A PLC is a digital device that typically operates internally with voltages of 0 V and 5 V. It is important to realise that any signal with two distinct levels is a digital signal. Transducers that generate digital signals can only be in any one of two distinguishable states at any instant.

Progress check 23.1

Recall the differences between an analogue signal and a digital signal

Input and output devices

A PLC has suitable digital input devices connected to it and is programmed to control output devices wired to its output terminals, as dictated by the user's program and the input signal conditions.

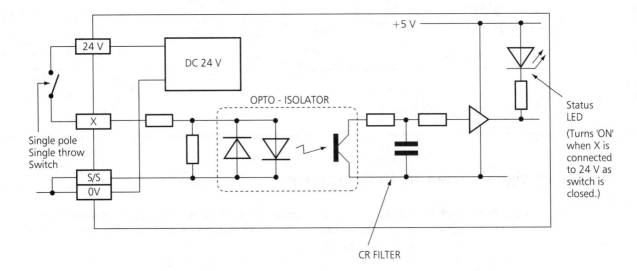

Figure 23.6 Opto-isolation used within PLC interface circuit

An understanding of the basic electronics of the input and output interface circuitry within the PLC is important and will be considered before investigating actual input and output devices.

It is standard practice in PLCs for all inputs and outputs to be electrically isolated from the rest of the interface circuitry with **opto-isolator devices**. This is clearly shown in Figure 23.6.

An opto-isolator consists essentially of a **light emitting diode** (LED) and a **photo transistor** in a single package. When the input switch in Figure 23.6 is closed, the LED will turn ON and emit light. This light falls on the base of the photo transistor causing it to turn ON as well. Thus, the state of the input switch (on or off) is passed to the PLC, but there is NO electrical connection between them. This provides a high degree of protection against voltage surges and transients which would otherwise damage the PLC.

Also note the **capacitor resistor (CR) filter** which reduces the effects of noise and 'contact bounce' and the status LED which shows whether the input is ON or OFF.

Activity 23.3

In figure 23.6 what type of filter is created with the resistor and capacitor?

Input devices

Proximity detectors

In many manufacturing situations there is a need to detect the presence of an object at a particular location. An example would be the detection of a product passing by a particular point on a conveyor belt. A proximity detector is an ideal device for this type of application.

Proximity detectors are non-contact devices that use inductive, capacitive or optical effects to switch their outputs when an object moves into the sensing range. When compared to mechanical switches, the switching of the output is controlled electronically so that there is no mechanical wear.

Inductive proximity detector

Inductive proximity detectors are designed to detect the presence of metallic objects by detecting changes in the magnetic field as the object moves into the sensing range. The basic principle is shown in Figure 23.7.

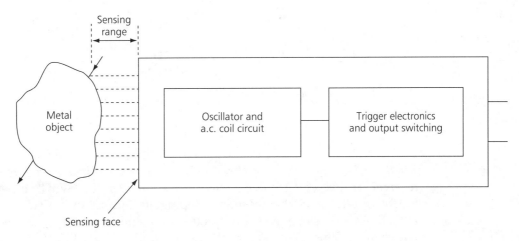

Figure 23.7 Inductive proximity detector

The oscillator circuit produces a high frequency magnetic field which radiates from the sensing face. As a metallic object passes the face, the magnetic field varies and these changes are detected by the a.c. coil. The trigger/output circuitry is activated so switching the output. The sensing range depends on the type of device, but distances between about 1 mm to 20 mm are common.

Capacitive proximity detector

Capacitive proximity detector circuits are similar in design to inductive detectors except that the coil circuit is replaced by capacitance. The device senses a change in the dielectric constant as an object moves into the sensing area and the output is switched accordingly. Because all material possess a dielectric constant, capacitive detectors can be used to detect metallic and non-metallic objects with sensing distances similar to those of inductive detectors.

Activity 23.4

Using an RS Components catalogue find a sensor suitable for detecting the presence of a metal peg at a distance of 1 mm. The sensor must operate from 24 V d.c. and produce a digital output.

Photoelectric sensors

Photoelectric sensors can also be used for object detection. They operate on a principle of the 'breaking' or reflection of a light beam when an object moves into the 'sensing area'. The three main types of photoelectric sensor are illustrated in Figure 23.8.

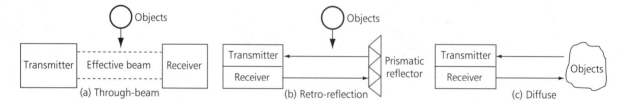

Figure 23.8 Types of photoelectric sensor

The through-beam sensor. The through-beam sensor uses a separate transmitter and receiver, and objects to be detected move between the two. The beam from the transmitter falls on the receiver, keeping its output in a fixed state (LOW or HIGH). As an object moves through the beam and interrupts it, the receiver output will switch. Thus a digital signal can be provided to a PLC.

The retro-reflective sensor. The retro-reflective sensor has the transmitter and receiver in the same location, with the beam from the transmitter directed at a prismatic reflector. The reflector reflects the light back to the receiver in a beam that is parallel to the one it receives. An object which breaks the beam causes the output of the receiver to switch as required, again providing a digital output to a PLC.

The diffuse detector. The diffuse detector relies on the reflection of light from the object itself to cause the output of the receiver to switch as required. Obviously, the sensing distance depends on the reflectivity of that object. Ambient lighting can also cause problems with this type of device.

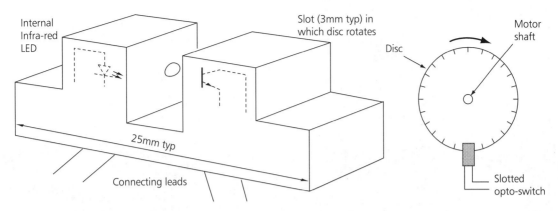

Figure 23.9 Example of the use of a through-beam sensor: the slotted opto-switch

An example of the use of a through-beam sensor is illustrated in Figure 23.9. Here the device is used as part of a system to measure the speed of rotation and position of the shaft of a motor. A disc, which is marked with evenly spaced dark bands, is rigidly fixed to the shaft of the motor and the sensor is positioned so that the disc will rotate in the sensor slot. As the motor turns, the phototransistor will turn on and off and produce pulses. These pulses can be input to the PLC and by correctly programming the PLC the speed of shaft rotation can be deduced.

Progress check 23.2

What is the major difference between an inductive proximity sensor and a capacitive sensor?
Which type of photo-electric sensor would you NOT use for an application where the ambient light level was very high?

Output devices

Relays

It was mention at the beginning of this chapter that the PLC was originally conceived as a replacement for 'hardwired' relay systems. The relay however, is one of the most important devices associated with the PLC. Most PLC's actually use relays as outputs to which other output devices can be connected and controlled via the PLC program.

The relay is an electrical component that has been used in electrical engineering for many years. The basic principle of operation is illustrated in Figure 23.10(a)

When switch SW1 is closed, current flows through the coil Y and causes its core to be magnetised. The soft iron armature is attracted towards the core so that the normally closed contact OPENS, and the normally open contact CLOSES. The lamp shown in the diagram will now light.

Another type of relay called a **reed relay** is shown in Figure 23.10(b). It consists of two overlapping, but non-contacting, ferromagnetic 'reeds' sealed within a glass tube. A magnetic field produced by a magnet or current through a coil surrounding the tube causes the reed to 'flex' and make contact.

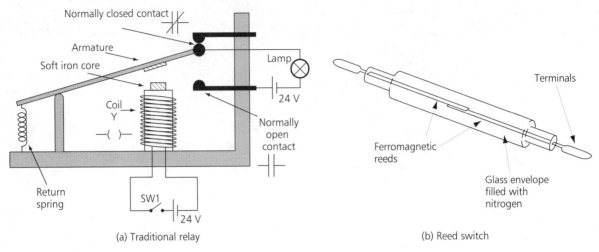

Figure 23.10 Relays

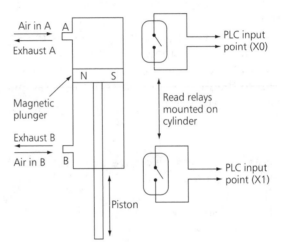

Figure 23.11 Application of reed relays

The operation time is much faster than a conventional relay and a millisecond or so is typical. Reed relays are often used with pneumatic devices to detect the position of a piston as illustrated in Figure 23.11. When pressurised air is input to the cylinder at points A or B, a **magnetic plunger** attached to the piston is moved. The two extreme positions of the piston are indicated by either relay A or relay B being activated. These relays could be connected to the inputs of a PLC as indicated.

Progress check 23.3

What are the major differences between a reed relay and a normal electromechanical relay?

The solenoid

Another magnetically operated device that is frequently used by a PLC output is a solenoid. The basic structure is shown in Figure 23.12.

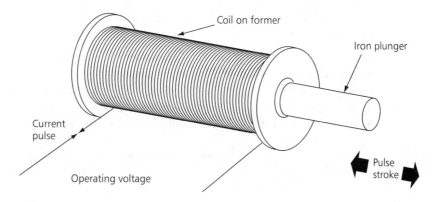

Figure 23.12 A basic solenoid

A solenoid consists of a coil within which an iron armature can move freely. When current is passed through the coil, the magnetic field produced causes the plunger to move through a distance called its stroke. Typically, solenoids produce strokes up to about 25 mm. Usually, a return spring or some other method is used to return the plunger to its resting position when current through the coil is removed.

Solenoids are often used to reject faulty articles moving along conveyor belt systems.

The solenoid described converts electrical energy into linear motion and is the most common type. It is possible, however, to use the solenoid principle to produce rotation and lever type action if desired.

Other output devices

The solenoid principle is also commonly used in an output device called a **contactor** which is used for switching large currents. This device was described in chapter 7 where it was used to control electric motors.

Another common output device is the **stepper motor** which was describe in chapter 19 and is basically a d.c. motor controlled with a digital signal from a PLC and is widely used in positional control systems.

Assignment 23

This assignment provides evidence for:
Element 32.1: Investigate the main features of programmable logic controllers
and the following key skills:
Communication 3.2, 3.3
Information Technology 3.1, 3.2, 3.3

Fine Plastic Products Ltd requires a process to be controlled by a PLC system. The company however is not sure which type of PLC to use. They have asked you to produce a report investigating the main features of two different types of PLC system and to produce a list with prices of the various system parts.

The company has drawn up a specification for the basic requirements of the PLC system as follows:

- a means of entering up to 1000 program steps
- digital inputs of 15 × 24 V opto-isolated
- digital outputs of 7 × (24 V relay with 1 A rated contacts)
- analogue inputs of 3 × (maximum i/p +10 V)
- analogue output of 1 × (maximum o/p +10 V)

Report

The evidence indicators for this element suggest that a report is produced including:

- an identification and explanation of the uses of PLCs
- identification and explanation of their design characteristics
- identification of the main components of the internal architecture
- a description of the forms of signals used
- a description of input/output devices used for a given application.

Chapter 24:   PLC information flow and communications

This chapter covers:
Element 32.2:   Investigate programmable logic controller information and communication techniques.

. . . and is divided into the following sections:
- Operational functions of PLCs
- Program execution
- Protocols.

For a PLC to effectively control an industrial process it must be programmed correctly. To make the PLC work efficiently a thorough understanding of how it processes information is required as well as an ability to use the correct instructions to form programs which are well structured.

Modern processes can become very complex, requiring more than one PLC to control it. Where multiple PLC systems are used there is a requirement for the PLCs to communicate with each other. An understanding of the techniques associated with PLC communications is therefore very important.

In this chapter you will be concerned with the basic **operational characteristics** of a PLC, simple **programming techniques** using logic instructions and PLC **communication techniques**.

When you complete this element you should be able to:

- investigate the operational functions of the PLC's central processing unit and the flow of instructional and data information
- derive simple programmes using logic functions based on relay ladder logic
- describe typical protocols used in signal communications
- describe types of communication links.

Operational functions of PLCs

The PLC is actually a small computer system which has at its core a device called the **central processing unit** (CPU) which is usually a **microprocessor** and is responsible for all calculations, logical operations, execution of the user's program and overall system control.

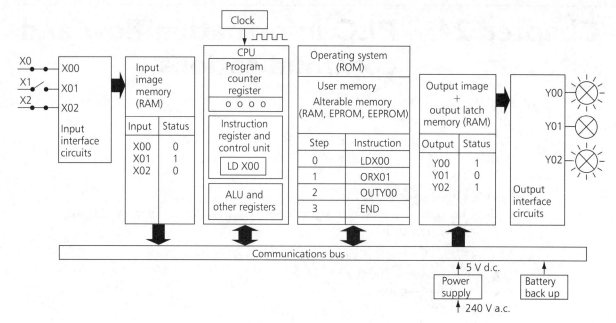

Figure 24.1 Simplified block diagram of a PLC

Notice from Figure 24.1 how the CPU communicates with the rest of the system via the internal **communications bus**. (A bus is a collection of conductors and are usually copper tracks on a printed circuit board.)

The CPU is supplied by a **clock** which determines the overall speed at which the system operates. The clock is a crystal controlled oscillator which generates a very stable square wave that is used by the CPU to control its own operation and the rest of the system. Typical clock frequencies are in the range 1 to 12 MHz. The clock must be stable since it provides all the timing and synchronisation signals that the system needs.

A microprocessor is a complex integrated circuit, but it is possible to obtain a good understanding of the overall system operation by using a simple model of a CPU. If you study Figure 24.1 it can be seen that the CPU consists of three main parts.

The arithmetic logic unit (ALU)

The ALU performs the arithmetic and logic operations that are required by the program. Typical operations include logical ORing and ANDing, arithmetic addition and subtraction to name but a few.

The instruction register/control unit

These determine how the CPU will operate. When, for example an instruction is received from the PLCs memory, the control unit passes this instruction into the **instruction register**. Here the instruction would be decoded and then the control unit would set up all of the necessary signals so that the instruction could be executed.

The program counter register

This is one of the most important registers in the CPU since it contains the current address of the memory location from which the CPU has received its instruction. When the CPU has executed the current instruction the program counter will automatically increment and point to the next address. Thus the CPU will systematically step through the program which has been stored in the memory.

Program execution

When a PLC is set to RUN, it will execute each program instruction in turn. If the instruction requires it to check the condition of an input, it will store the status of that input in what is called **input image memory**. When an output instruction is finally executed the required status is then output to a physical relay. When an END instruction is encountered, the program counter is cleared to 0000 and the program starts again from the beginning. The time to complete one scan of the program is called the **scan time**. The PLC will continue scanning the inputs, executing the program and updating the outputs until the PLC is switched to STOP.

The operation of the PLC cycle is summarised as:

- scan input
- execute instructions
- update outputs.

The type of processing described is called **batch I/O or mass I/O processing** and is the method favoured by most PLC manufacturers.

An important point here is to realise that the status of an input cannot be changed within the same scan cycle. Thus if an input is changed after an input copy, it will not be detected until the next scan cycle. The length of a scan cycle depends on the type of PLC and the program length and is typically in the range 1 ms to 10 ms.

Progress check 24.1

What are the three most important components of the CPU within a PLC system?
Which of these components performs the arithmetic and logic functions?

Simple programs

To help you to understand how to write simple programs the basic rules of logic will be introduced using relays. This work is important and should be fully understood before progressing.

Consider the basic relay in Figure 24.2(a) which has one normally open contact (NO) and one normally closed contact (NC) associated with its coil Y. The circuit which energises the coil Y has two series connected switches, S1 and S2.

Whether Y is active or not will depend on the condition of the switches S1 and S2. Using the convention:

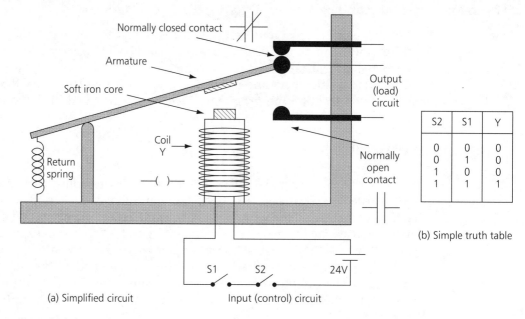

S2	S1	Y
0	0	0
0	1	0
1	0	0
1	1	1

(b) Simple truth table

(a) Simplified circuit

Input (control) circuit

Figure 24.2 Basic relay operation

'0' means switch open
'1' means switch closed
'0' means relay off
'1' means relay on

the operation of the circuit can be fully described by the table in Figure 24.2(b).

The table is called a **truth table**. From the table it can be seen that the coil Y will only be active when both switches are closed at the same time.

Ladder diagrams

When writing PLC programs it is normal practice to use a ladder diagram to describe circuit operation. A ladder diagram representing the above circuit is shown in Figure 24.3.

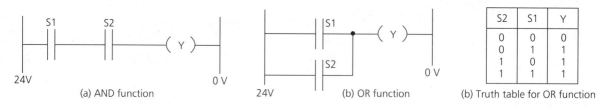

S2	S1	Y
0	0	0
0	1	1
1	0	1
1	1	1

(a) AND function   (b) OR function   (b) Truth table for OR function

Figure 24.3 Simple ladder diagrams

Effectively current only flows through the relay coil Y when both switches S1 and S2 are closed.

This arrangement performs a logic function called the **AND function**. To describe its effect concisely, the form of algebra called **Boolean algebra** is used.

The equation used to describe the AND function:

$$Y = S1 \cdot S2$$

Note that '·' means AND and the equation means:

$$Y = 1 \text{ when } S1 \text{ AND } S2 = 1$$

If the switches in Figure 24.2 are rearranged so that they are in parallel, a logic function called the **OR function** can be implemented as shown in Figure 24.3(b).

Coil Y is energised if S1 OR S2 or both are closed, the Boolean equation for the OR function is:

$$Y = S1 + S2$$

Note that the '+' sign is used for the OR function.

Another very useful logic function is the **NOT or invert function**. In Boolean algebra this is represented by placing a bar '$\overline{}$' above the variable as required, e.g. $\overline{S1}$ means NOT S1.

In general, the NOT function will always invert the logic state of the input. To generate a NOT function therefore, we need a circuit whose output will be ON when the input is OFF and OFF when the input is ON. This can be achieved using relays as shown by the ladder in Figure 24.4(a). The truth table and Boolean equation for the NOT function are shown below the circuit diagram.

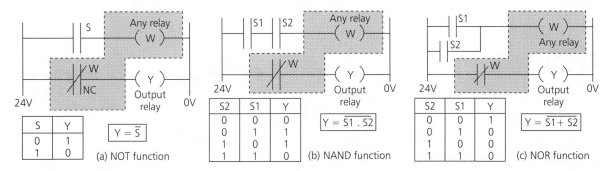

Figure 24.4 Ladder for NOT , NAND and NOR functions

The ladder in Figure 24.4(a) operates by using an additional relay W. This will energise when S is closed. A normally closed contact on relay W will open de-energising the output relay Y. When S1 is opened, relay W de-energises causing the contact connected to relay Y to close which in turn will energise relay Y.

Having now investigated three of the basic logic functions, we can use them to create other useful logic functions such as the NAND and NOR function. Figure 24.4(b) shows the truth table for the NAND function. From this it can be seen that the NAND function is in fact the inverse of the AND function. The NAND function can therefore be generated by using AND and NOT functions as shown by the ladder in Figure 24.4(b).

Notice how, in the ladder, S1 and S2 form an AND function and Relay W and the normally closed contact W form the NOT function. Relay Y will therefore remain energised until both S1 and S2 are turned ON. The Boolean equation for the NAND function is written as:

$$Y = \overline{S1 \cdot S2}$$

The bar signifies 'inverse of' i.e. Y is active (or logic) '1' when S1 AND S2 are NOT active (that is, when S1 and S2 are logic '0').

Using an OR function and a NOT function the logical NOR function can be created. This is illustrated by the ladder in Figure 24.4(c). The truth table and Boolean equation for the NOR function is also shown and from this it can be clearly seen that the NOR function is in fact the inverse of the OR function.

Finally the Boolean equation for the NOR function is written as:

$$Y = \overline{S1 + S2}$$

One other very useful logic function is the XOR or exclusive OR function. The truth table and Boolean equation for this function are shown below.

$$Y = (\overline{S1} \cdot S2) + (S1 \cdot \overline{S2})$$

SI	S2	Y
0	0	0
I	0	I
0	I	I
I	I	0

Progress check 24.2

Using the information in both the truth table and Boolean equation above create a ladder diagram which will implement the XOR function.

Protocols

Communications is a rapidly expanding area of application of PLC systems. It is therefore important to have an understanding of the basic techniques used to allow systems to communicate with each other. The method used by one system to communicate with another is called a protocol. In this section we will consider four such protocols, RS-232, IEEE 488, RS-422 and 20ma.

Communications interfaces are broadly divided into two categories, **serial** and **parallel**. Of the four protocols or standards mentioned above, RS-232, RS-422 and 20ma apply to serial interfaces and IEEE 488 is parallel.

Serial interfaces

We need to start by considering, in general terms, how data is communicated over a serial interface link. The first problem to consider is that the voltage level of the data signals within the PLC system is only 5 V. This is too low for a serial communication system to operate over a sensible distance. The RS-232 standard defines that the voltage level should be increased to ± 12 V and converted to a suitable current level in the case of RS-422 and 20ma. All three of these standards require that the data to be sent is arranged in what is called a **data frame** which

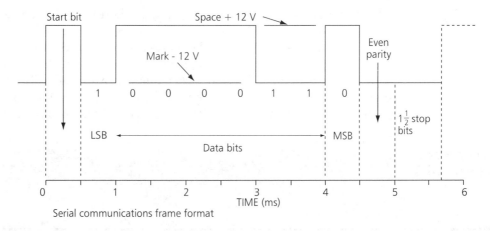

Serial communications frame format

Figure 24.5 Serial communication frame format

has a **start bit**, an optional **parity bit** and 1 to 2 **stop bits**. This frame format is clearly shown in Figure 24.5.

This particular method of communication is known as **asynchronous communication** and operates as follows.

Prior to data being transmitted, the signal is at the **mark level**, which represents the idle state. The start of transmitted data is indicated by the signal going to the **space level** for a single time interval. This is called the start bit. This is followed by the actual data, using a mark to indicate a binary '1' and a space to represent a binary '0', the least significant bit of the data being sent first. The next time interval is occupied by the **parity bit** – this is used for error checking. To complete the frame, the signal remains in the mark state for a minimum period

	BINARY			D7	0	0	0	0	0	0	0	0
				D6	0	0	0	0	1	1	1	1
				D5	0	0	1	1	0	0	1	1
				D4	0	1	0	1	0	1	0	1
D3	D2	D1	D0	HEX	0	1	2	3	4	5	6	7
0	0	0	0	0	NUL	DLE	SP	0	@	P	`	p
0	0	0	1	1	SOH	DC1	!	1	A	Q	a	q
0	0	1	0	2	STX	DC2	"	2	B	R	b	r
0	0	1	1	3	ETX	DC3	#	3	C	S	c	s
0	1	0	0	4	EOT	DC4	$	4	D	T	d	t
0	1	0	1	5	ENQ	NAK	%	5	E	U	e	u
0	1	1	0	6	ACK	SYN	&	6	F	V	f	v
0	1	1	1	7	BEL	ETB	'	7	G	W	g	w
1	0	0	0	8	BS	CAN	(	8	H	X	h	x
1	0	0	1	9	HT	EM	)	9	I	Y	i	y
1	0	1	0	A	LF	SUB	*	:	J	Z	j	z
1	0	1	1	B	VT	ESC	+	;	K	[	k	{
1	1	0	0	C	FF	FS	,	<	L	/	l	}
1	1	0	1	D	CR	GS	-	=	M	]	m	)
1	1	1	0	E	SO	RS	.	>	N		n	
1	1	1	1	F	SI	US	/	?	O	_	o	DEL

Figure 24.6 ASCII code table

of time before the start of the next frame. The bits at the end of the frame are known as stop bits: 1½ stop bits are shown in Figure 24.5.

The data signals within the PLC system consist of binary 0's and 1's; for this data to make any sense it has to be coded in some way. One of the most common codes used is the **American Standard Code for Information Interchange (ASCII)**. This code allocates a binary number to each character in the alphabet and to other symbols and numbers. A complete ASCII code table is shown in Figure 24.6.

Progress check 24.3

Why is it important to have a START bit and a STOP bit in a serial data frame?

WORKED EXAMPLE 24.1

Represent the ASCII character 'A' in hex and binary.

Solution
With reference to the ASCII table the character 'A' would be represented by 41 hexadecimal or 01000001 binary.

Transmission speed
When talking about serial communications one other factor has to be considered and that is speed of transmission. The **baud rate** is a measure of this and has the units of Bits/sec. High baud rates are preferable but are more susceptible to noise and this worsens with distance. Typical baud rates are: 75, 110, 150, 300, 600, 1200, 2400, 4800, 9600, 19600 and 28800 bits/sec.

Although the **RS-232 standard** is probably one of the most common, it has serious limitations when used in an Industrial environment. This is due to the fact that the RS-232 standard specifies an **unbalanced system** where by one of the cables in the communications link is earthed. This causes electrical interference (noise) to be induced as shown in Figure 24.7(a). The noise appears as a differential voltage at the input of the receiver and eventually corrupts the data.

This problem can be overcome by using a balanced system where neither cable is earthed. In this way the noise appears on both lines and effectively cancels

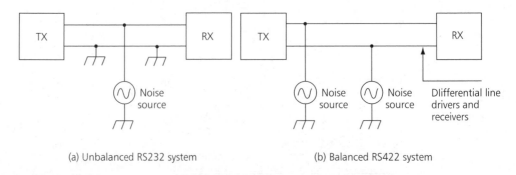

(a) Unbalanced RS232 system   (b) Balanced RS422 system

Figure 24.7 Unbalanced and balanced communication systems

itself out. This method in specified in the RS 422 standard and is shown in Figure 24.7(b).

This particular method is favoured by PLC manufacturers and therefore most PLCs have a communications interface which supports the RS-422 standard.

Another interface standard is the **20ma current loop**. This interface has its origin in the days of the old teleprinters. As its name suggests, it is possible to configure a loop with a transmitter and a number of receivers. Unfortunately, due to the lack of standardisation of this type of interface, resistance values and voltages required vary considerably. In terms of wiring, this type of interface requires two wires for transmitting and receiving. Hence, a simple bidirectional interface would require four wires. It has to be said that this standard is not often used for PLC communications.

Progress check 24.4

What is the main advantage of using a balanced serial communication system as opposed to an unbalanced system?

The final standard to consider is the **IEEE 488 interface** which specifies a parallel interface and was originally developed by Hewlett Packard in the late 1970's to use with programmable instruments. The general form of the interface is shown in Figure 24.8.

The basic idea behind this standard was that a microcomputer would remotely control a number of instruments to undertake such tasks as automatically testing equipment.

The interface has two groups of **data lines**, eight for actual data and eight for control. Up to 15 devices may be connected to the interface with a maximum distance between devices of 20 metres. The devices connected to the interface are able to perform one or more of the following communication tasks:

- talking
- listening
- control.

A **talker** sends data over the interface, a **listener** only accepts data and the **controller** addresses other devices and grants permission for talkers to use the interface. Only one controller and one talker may be active at any one time.

Although a popular interface standard with desktop computers the IEEE 488 it is not widely used in PLC systems for communications.

Communication links

PLCs are now being used in greater and greater numbers to control industrial processes. At one time a single PLC was used to control a single process. This is now changing rapidly to a point where multiple PLCs are used to control very complex processes. This has lead to a requirement for PLC systems to communicate with each other in a variety of ways. It is important therefore to gain an understanding of the concepts involved in interconnecting and networking PLC systems. In this section we will be investigating the various methods of interconnecting PLCs using twisted pair cable, co-axial cable and fibre optics as well as networking techniques.

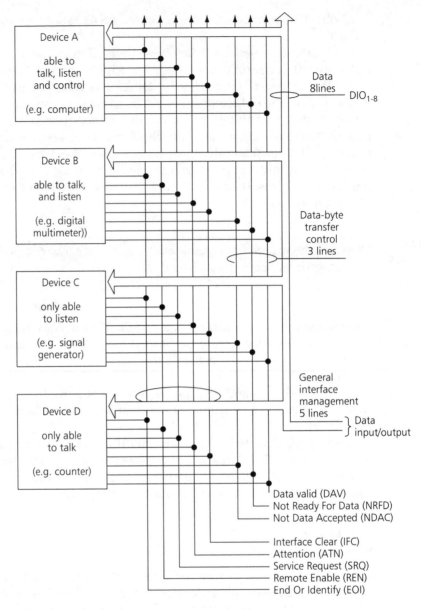

Figure 24.8 IEEE 488 interface

In standalone systems, each PLC works independently and has no way of transmitting information about one process to the PLC controlling another process. A very simple way of passing information between PLC's is to use the output from one PLC to drive the input of the next PLC. This is quite adequate where the result of a single contact closure is passed on to another PLC but is clearly not suitable for passing large amounts of information. It may be that the temperature of process 1 is important in the setting up of process 2. In this case data representing temperature would need to be passed from one PLC to the next. This data would normally be stored in what is called a **data register** within the PLC, holding between 16 and 32 'bits' of information. PLCs exist which have the ability of transmitting and receiving information via data registers using a **serial communications link**, Figure 24.9.

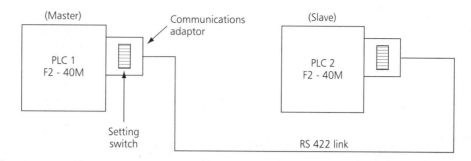

Figure 24.9 PLC serial communications link

This type of communications system is available in a PLC such as the Mitsubishi F1-40. With this type of PLC a special communications adaptor module is required.

The communications adaptor uses the RS-422 interface standard and simple **twisted pair cable** between the two adaptors. This allows for communications over a distance of some 300 m at a Baud rate of 9600 Bits/s.

Each adaptor can be configured via the setting switch to be either **master** or **slave** and to communicate using either 16 bits or 32 bits of data.

Communications take place by transferring data stored in the **data register** into what are called **link relays**. The flow of data can be seen in Figure 24.10.

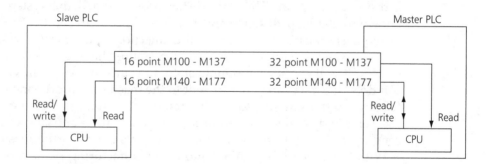

Figure 24.10 PLC link relays

As can be seen the link relay numbers are common to both systems. Hence, the master station will drive relays M140 – M177 and the slave station will receive on link relays M140 – M177. The slave station will drive link relays M100 – M137 and the master station will receive on M100 – M137. In this way complete two-way communication can be maintained. When two systems can communicate in either direction at the same time this is known as **full duplex communication**.

The obvious limitation of this simple **master/slave** communication system is that the PLCs can only be linked in pairs. To enable more PLCs to communicate with each other a much more complex communication system is required whereby each PLC has its own unique address within the system. This added complexity requires the PLC to have a much more powerful 16-bit microprocessor.

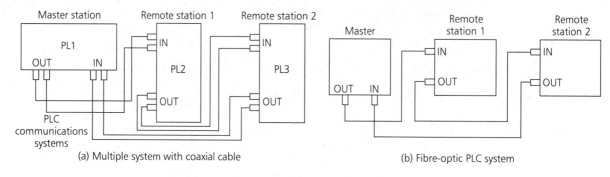

(a) Multiple system with coaxial cable   (b) Fibre-optic PLC system

Figure 24.11 PLC communication systems

This type of system typically uses coaxial cables to interconnect the PLCs in a complete loop as shown in Figure 24.11(a) which shows how a complete communication loop is formed between the master station and the last remote station in the system. Typically this type of system can support one master station and up to 64 remote stations. One of the drawbacks of this type of system is that four coaxial connections have to be made on each PLC station in the system. However, by using **fibre optic cable**, the number of connections can be reduced as shown in Figure 24.11(b).

The address of the master station is set to '0' and the addresses of the remote stations are set to 1–64 using miniature rotary switches inside each PLC system. The baud rate of such a system is typically 1.25 Mbits/s and it will operate over a distance of 10 km. The total distance over which the system will operate is known as the loop distance and is the distance between the output connections of the master station and the input connection to the master station.

The system operates by the master station having complete control of the system. It will check each remote station in turn to see if it has any data to send. If it does, the data can either be sent to the master for processing or via the master for onward transmission to another remote station for processing. Thus the master can send and receive data from any remote station and any remote station can send or receive data from any other remote station via the master.

With this type of system a complete manufacturing plant could be controlled, with any PLC in the system being able to talk to any other PLC in the system.

Progress check 24.5

Why is it important for PLC systems to communicate?
What is the main advantage of using a fibre optic PLC communication system as opposed to a coaxial cable system?

PLC communication networks

This type of communication system can now be extended to incorporate a standard computer network. Most companies now have their own computer networks which allow computers within the company to talk to each other. However, in many manufacturing companies they have a **PLC communication network** which also connects into the computer network allowing data to be sent to any PLC in the network from any computer within the network. This brings many benefits

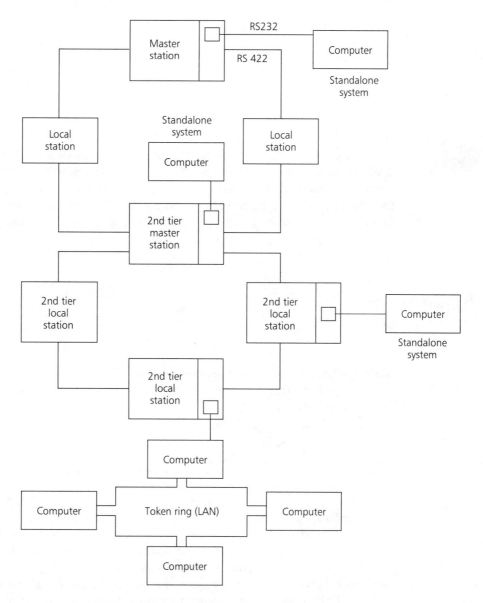

Figure 24.12 Networked PLC system

to the company permitting PLC systems to be controlled from desktop computers anywhere within the company, allowing data to be gathered from any PLC system for analysis and for overall monitoring or supervision of the manufacturing process. Such a system as this is often called a **SCADA system**, the initials stand for Supervisory, Control and Data Acquisition. Figure 24.12 shows an example of how such a system would be interconnected.

In this type of system there can be a total of three tiers or networks containing a maximum of 64 PLCs in each tier or network. Each tier has a master station which controls its own tier and allows communication between tiers and resolves the differences between the addresses of each station within the tiers. Notice that any station within any tier can be connected to a standard computer network such as the **token ring computer network** shown in Figure 24.12. Hence any computer can communicate with any PLC or computer anywhere within the network. This type of complex PLC/computer network would normally be used to control a large chemical plant or nuclear power station.

Activity 24.1

ACE Components Ltd has six different manufacturing processes; each one is to be controlled by a separate PLC system. Each process depends on the outcome of the previous process, therefore each PLC must be able to pass information on to the next. The company would also like to be able to monitor each process via a computer connected to its standard computer network.

Decide on the best type of PLC communication system to use for this application and produce a sketch of the system showing the various interconnections.

Assignment 24

This assignment provides evidence for:
Element 32.2: Investigate programmable logic controller information and communication techniques
and the following key skills:
Communication 3.2, 3.4
Information Technology 3.1, 3.2, 3.3

Using manufactures information select two different types of PLC system and investigate and contrast the different facilities available in each type. In particular consider the differences in operational functions of their CPUs and the way in which simple programs are constructed. Also, investigate the communiation facilities of each type and consider the protocols used and the types of communication links possible and write up your findings in the form of a report.

Report

The evidence indicators for this element suggest that a report is produced, investigating programmable logic controller information flow and communcation techniques in one PLC system including:

- an identification of the operational functions of the central processor unit and flow of instructional and data information
- a derivation of simple programs using logic functions based on relay logic
- a description of typical protocols used in signal communication
- a description of types of communication link.

Chapter 25: PLC programming techniques

This chapter covers:
Element 32.3: Investigate programmable logic controller programming techniques.

... and is divided into the following sections:
- Methods of programming PLCs
- PLC elements and ladder diagrams
- Other programming methods
- Text and documentation.

For a PLC to control an industrial process effectively it must first be programmed. The code which makes up this program has to be stored in the memory of the PLC. Various methods are available which allow program code to be entered and stored; these will be investigated in this section. It is also important to understand the characteristics of all the elements available within a typical PLC and to be able to use them in a logical way to create working programs.

When you complete this element you should be able to:

- Describe methods of programming PLC's
- Describe the advantages of off-line programming
- Identify and apply methods of producing and storing text and documentation
- Identify and apply methods of testing and debugging hardware and software
- Identify elements associated with the preparation of a PLC program
- Produce and demonstrate a program for a PLC.

Methods of programming PLCs

There are two basic methods used to program PLCs: one is the **programming panel** and the other is what is referred to as the off-line method. We shall consider each of these methods in turn starting with the Programming Panel.

Programming panels

The programming panel was the very first method devised to program PLCs and is the simplest and cheapest. A simple programming panel is shown in Figure 25.1.

The programming panel is a simple device which plugs into the front of the PLC and has a number of buttons on it. The buttons are arranged to be either

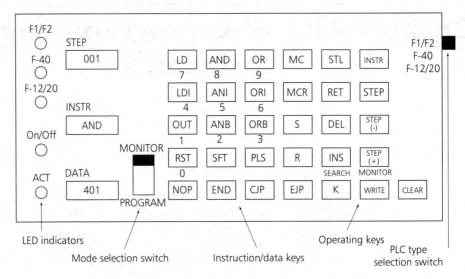

Figure 25.1 Simple programming panel

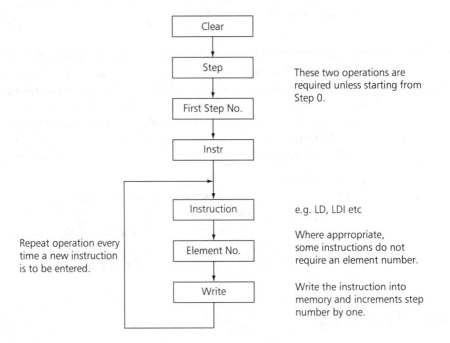

Figure 25.2 Button press flow chart

instruction input buttons or control buttons. A typical sequence of button presses required to enter a single instruction is described by the flow chart shown in Figure 25.2

This sequence would be repeated a number of times until a complete program had been entered. The program/monitor switch would then be switched from Program to Monitor and the PLC would be set to run the program. This type of programming panel also has simple editing facilities such as INSERT, DELETE and SEARCH but no storage facilities. Thus any changes made to a program could not be easily recorded and therefore de-bugging a program becomes difficult. This type of programming panel is therefore of limited use and would normally only be used to make very simple changes to a program.

The graphical programming panel

This type of panel is more complex than the previous one and allows a programme to be constructed using ladder logic symbols, the rungs of the ladder being displayed on the LCD screen of the programming panel as shown in Figure 25.3. The program is actually stored in the programming panel and when complete is transferred to PLC memory.

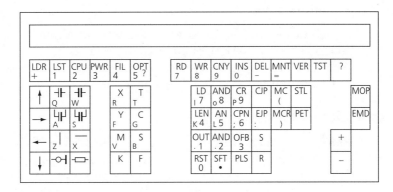

Figure 25.3 Graphical programming panel

This unit shown in Figure 25.3 is a multi-function graphic programmer providing a number of facilities in one portable unit. It incorporates a dot matrix LCD screen which displays ladder diagrams or programme lists. The screen display makes programming, editing, machine diagnostics and trouble-shooting very easy.

In addition, by using the interface functions available it is possible to store programmes onto EPROMS, audio cassette tapes or floppy disks. A printer may also be used to document programmes. Functions include:

- off-line programming and editing
- programme transfer to or from the PC
- on-line monitoring, forcing, and changing timer/counter constants while in the RUN mode
- audio cassette recorder interface
- EPROM writer interface
- printer interface
- floppy disk drive unit interface
- comment and device name notation.

Although programming panels are useful is has become standard practice these days to use a technique called **off-line programming** which requires the use of a computer and associated software. This particular method of programming is better understood after some experience has been gained preparing PLC programs. The off-line method will therefore be described in more detail towards the end of this chapter.

PLC elements and ladder diagrams

It is important to understand that a PLC contains a number of different elements which also includes the inputs and outputs of the PLC. Each element is identified by a letter and a number. For instance the input element may be assigned as X01 and an output may be assigned as Y01 etc. In addition to inputs and outputs a typical PLC may have other elements including:

- auxiliary relay contacts
- timers
- counters
- shift registers.

Each of these is assigned with a letter and a number, such as T50 for a timer, C60 for a counter, M100 for an auxiliary relay contact and M300 for a shift register. Programs are normally created using different combinations of these elements and each will be investigated in turn.

Input/output elements

Ladder diagrams were considered in the previous chapter when investigating basic logic functions. As an example, consider a situation where two inputs are ORd together to produce an output. Using input X01 and X02 for the inputs and Y01 for the output, the ladder diagram representing this combination is as shown in Figure 25.4(a). (Figure 24.3 refers.)

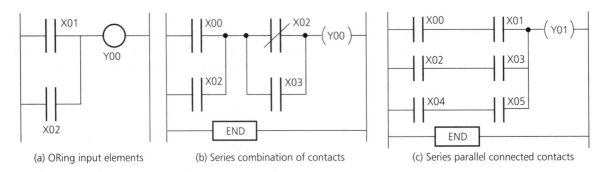

(a) ORing input elements   (b) Series combination of contacts   (c) Series parallel connected contacts

Figure 25.4 Input/output elements

It is clear that if X01 contact is made or X02 contact is made then the output element Y01 will be energised.

Many industrial process control problems can be solved using input/output elements and combinations of basic logic functions. However, problems can arise when complex arrangements of logic functions are required making use of series and parallel combinations of contacts. As an example consider the case shown in Figure 25.4(b), which shows two sets of parallel contacts connected in series. The example shown is in fact not directly programmable using simple logic instructions. For the PLC to interpret this ladder an additional instruction is used called the **AND block instruction** (ANB). Converting the ladder shown in Figure 25.4(b) into PLC instructions gives:

```
LD X00
OR X01
LDI X02
OR X03
ANB
OUT Y00
```

Notice that the ANB instruction will AND two blocks together, effectively connecting the blocks in series.

A similar problem exists when connecting series connected contacts in parallel as shown in Figure 25.4(c). In this example the problem is solved by using an **OR block instruction** (ORB) as follows:

```
LD X00
AND X01
LD X02
AND X03
ORB
LD X04
AND X05
ORB
OUT Y01
```

The ORB instruction connects the two blocks of series connected contacts in parallel. Using both the ANB and ORB instructions very complex arrangements of contacts can be programmed.

Activity 25.1

Convert each of the ladders shown in Figure 25.5 into PLC instructions using the ANB or ORB instructions as required.

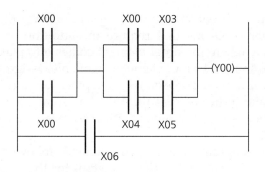 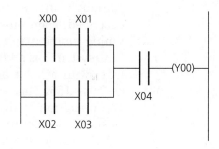

Figure 25.5 Example ladders

Auxiliary relay contact elements

In addition to the physical output relays available within the PLC most PLCs have additional software relays called auxiliary relays. These relays only exist in the PLC memory and are assigned to a range such as M0 – M495. These can be used instead of the actual output relays where the number of output relays are limited. In addition to these auxiliary relays, software auxiliary contacts are available associated with each physical output relay. This facility allows a single output relay to have as many contacts associated with it as may be required by the applications. To understand how these work in practice consider the ladders shown in Figure 25.6.

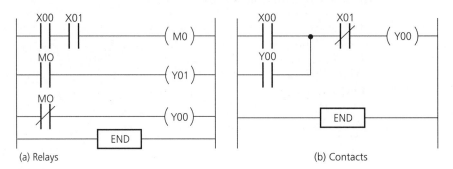

(a) Relays   (b) Contacts

Figure 25.6 Use of auxiliary relays

In Figure 25.6(a) the auxiliary relay is used to transfer the result of a logical combination of contacts to the next rung of the ladder.

Figure 25.6(b) illustrates the use of an auxiliary contact controlled by an output relay Y00. The effect of the circuit is for the output relay to self-hold or **latch** when contact X00 closes. The relay can only be de-energised by opening contact X01. This particular application is a very useful building block circuit which can be used in many different situations.

The timer element

Although many industrial problems can be solved using the instructions described, there are other problems that require the addition of some type of timing function and this is where the timer element comes into play. The average PLC system has a large number of timer elements available, assigned in the usual way, with a letter and a number. For example a typical range of timers would be T0 – T55. This example indicates that the PLC has some 56 timer elements. The operation of the timer is best explained with reference to a ladder diagram as shown in Figure 25.7.

Notice from the ladder diagram that there is a letter K followed by a number 100 below the timer relay T0. This is the **time constant** for the timer and defines the **delay time** for the circuit. The ladder operates in the following way: when contact X00 closes the timer T0 starts counting in 0.1 s steps; the number of steps counted is defined by constant K. This means that as K = 100 it will take $100 \times 0.1 = 10$ s for the timer relay to activate. The effect of this is that output Y00 will activate 10 s after X00 closes. Y00 will remain active until X00 is opened; this will cause the timer to reset. If X00 is opened before the timer has timed out,

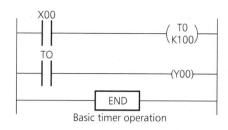

Basic timer operation

Figure 25.7 Basic timer operation

T0 resets itself. The timer configuration just described is known technically as an **ON delay timer**.

There are many different types of PLCs available but all of them have timer elements. The operation of the element internally may be different but the overall effect is the same.

Timers are crucial to the operation of manufacturing processes and are used for time delays, timing the duration of a particular sequence in a manufacturing process and so on. However, to make best use of the timer elements they have to be configured in different ways; each configuration has a different name and a different purpose.

OFF delay timer

The OFF delay timer operates an output relay for a time defined by the constant K as soon as a contact closes. This is illustrated in the ladder diagram shown in Figure 25.8a.

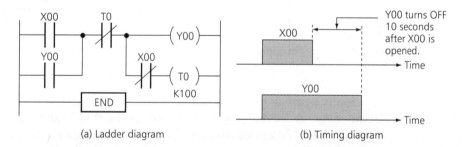

(a) Ladder diagram   (b) Timing diagram

Figure 25.8 OFF delay timer

When contact X00 closes, output relay Y00 activates which in turn closes an auxiliary contact Y00. This latches Y00 and the timer T0 starts timing out; after a delay of 10 s, T0 times out causing the auxiliary contact T0 to open which in turn de-activates the output relay Y00. This type of timer configuration effectively generates a pulse, the width of which is defined by the timer constant K. This effect is shown in Figure 25.8(b).

Activity 25.2

Produce a ladder diagram which will operate an output relay Y00 for 2 s every 2 s. In effect the output relay will continuously turn ON then OFF every 2 s. This can be achieved by using two timers which are interconnected. Have a go and see what you can do!

Another industrial application of timers is to control a sequence. This is achieved by turning outputs ON and OFF one after the other. This type of operation is illustrated in Figure 25.9(a).

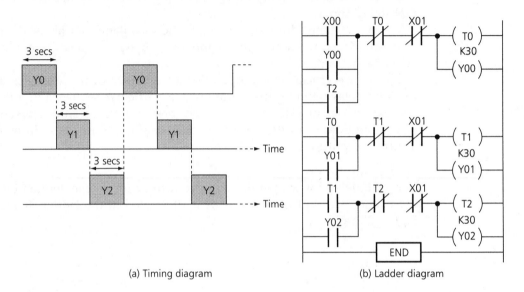

(a) Timing diagram   (b) Ladder diagram

Figure 25.9 Sequencing outputs

To actually achieve this effect three timers will be required, each having a time constant of 3 s. A ladder diagram to implement this is shown in Figure 25.9(b).

When contact X00 closes Y00 will latch ON and timer T0 will start timing out; after 3 s T0 will time out unlatching Y00 which will de-activate. The auxiliary contact T0 used in the second rung will then latch Y01 ON and start T1 timing out. T1 will time out after 3 s which will unlatch Y01 and auxiliary contact T1 will activate Y02 and timer T2 in the final rung. T2 will time out after 3 s which will cause Y02 to de-activate, completing the sequence. However, because auxiliary contact T2 is also connected across contact X0 in the first rung it will start the sequence all over again. The sequence will therefore run continuously until contact X01 is opened; this will de-activate all output relays and reset all of the timers.

This type of sequencer configuration is known as a **three-stage sequencer** and could form the basis for a controller to control a manufacturing sequence which would cycle continuously until a STOP button was pressed.

In the sequencer just described, each output turns ON and then OFF in turn. There are however situations where the sequenced outputs need to overlap as shown in Figure 25.10.

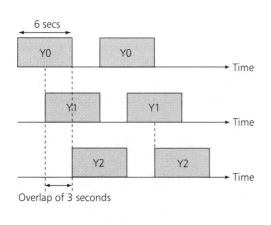

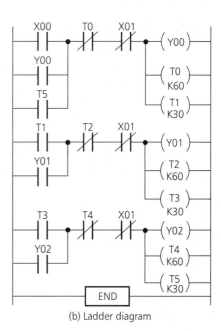

(a) Timing diagram

(b) Ladder diagram

Figure 25.10 Sequence with overlaps

To achieve overlaps additional timers are required as illustrated in Figure 25.10(b).

When contact X00 closes Y00 is latched ON and both Timers T0 and T1 start timing out. T0 will time out after 6 s after which time Y00 will turn OFF, T1 times out after 3 s latching Y01 ON. Hence Y01 turns ON 3 s after Y00 turns ON generating a 3 s overlap. This process is repeated down the ladder until finally timer T5 times out causing the whole sequence to be repeated.

Activity 25.3

You are required to design an overlapping sequencer to control a single set of traffic lights. The sequence is to run continuously with each light remaining on for a length of time defined as in the table below:

RED	ON for 4 s
RED and AMBER	ON together for 4 s
GREEN	ON for 12 s
AMBER	ON for 4 s

Sequence then repeats

The RED light will be controlled by output Y00; the AMBER light will be controlled by output Y01; the GREEN light will be controlled by output Y02

No input contacts are required to be used and your PLC has four timers assigned as T0 – T3.

Your task is to produce a ladder diagram and list of PLC instructions which will operate the traffic lights in the correct sequence.

The counter element

There is a requirement in many industrial applications to perform **batch counting**. This type of application requires a facility which will detect the passing of an object and record how many have passed. This requirement is met by the counter element. All PLCs have counter elements and typically they are assigned the letter C followed by a number. The Mitsubishi FXO PLC for instance has 16 counters in the range C0 – C15, each counter being capable of counting from 1 to 32,767. The operation of the counter is best illustrated using a ladder diagram as shown in Figure 25.11.

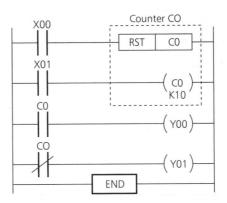

Figure 25.11 Basic counter element

Unlike the timer element, the counter has two coils associated with it. One is to reset the counter the other is to count incoming pulses from a sensor. The counter is first reset by closing and then opening X00. A suitable sensor is connected to input contact X01 which detects a passing object. Every time an object passes the sensor X01 will close causing the counter to register a count of one. When the count is equal to the value of the constant K the relay contact C0 associated with the counter activates, which in turn activates output relay Y0.

Most counter elements operate in a down counting mode. In this mode the counter is loaded with the value of constant K when reset and counts down to zero. Some PLCs also have counters which operate in the up counting mode where the counter counts from zero up to K.

Activity 25.4

An example of a batch counting application is shown in Figure 25.12, the system has two conveyor belts which run continuously. A through beam sensor detects the passing of objects on the conveyor belt and a solenoid is used to direct objects onto conveyor belt 2 and into bin 2.

The requirement is that 8 objects are to be batched into Bin 1 and 10 components into Bin 2. A reset switch is required to enable the counters to be reset after a batch run. A START and STOP switch are also required for the conveyor belts. The various devices will be allocated as follows:

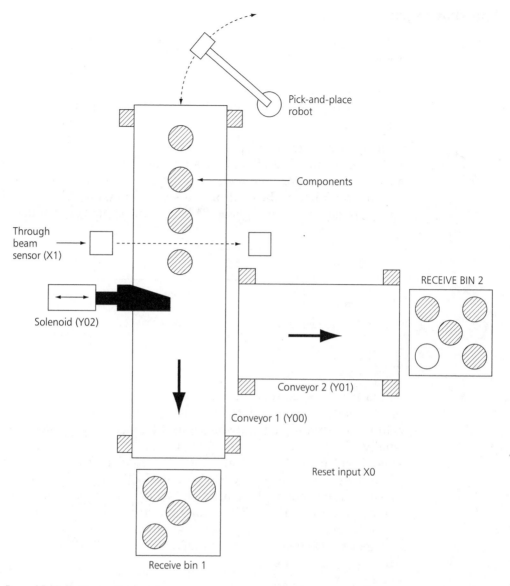

Figure 25.12 Batch counting system

X0	Reset switch
X1	Through beam sensor
X2	Start switch
X3	Stop switch
Y00	Conveyor 1
Y01	Conveyor 2
Y02	Solenoid

You are required to devise a ladder diagram and list of PLC instructions to meet the specification for the batch counting system.

The shift register element

This is probably the most complex and versatile element available within a PLC system. It is used in programs which are used to control complex process sequences and is more often used for this application compared with the timer element which has already been investigated. To understand how this element operates within a PLC system an understanding of the basic concept of a shift register is required.

Basic shift register concepts

A register is simply a storage location for a group of bits and one of the most important operations on a register is to be able to move these data bits to the left or to the right under the control of a **shift command**. When used in this way the register is called a shift register. The basic principle is illustrated in Figure 25.13.

Figure 25.13 Shift register

Data input to the register enters on an input line, and is moved through the register under the control of the shift command input. In many PLCs there are facilities for shifting left and shifting right. Since shift registers are of a finite size, usually 16 bits, after a certain number of shift commands the original data in the register will be shifted out and lost. The particular advantage of shift registers lies in the fact that the contacts associated with the elements comprising the shift register can be activated/de-activated as data is moved through the register.

A shift register can easily be configured to produce a sequence of outputs which turn ON and then OFF one after the other. Considering an example using the shift register element in a basic PLC, such as the Mitsubishi F1–20, the solution is as shown in Figure 25.14.

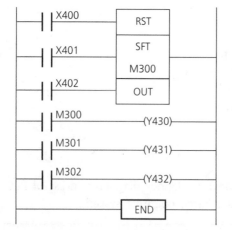

Figure 25.14 Simple sequencer using a shift register

When input X400 is closed and then opened all the relays associated with the shift register are de-energised or reset. Closing contact X402 sets the data input to ON; this causes the first relay M300 to energise. If contact X401 is now closed and then opened a right shift is executed and relay M301 energises, which in turn energises output Y430. The data input contact X402 is now opened which causes relay M300 to de-energise. If the shift input contact X401 is now opened and then closed another right shift is executed. This right shift causes relay M301 to de-energise turning output Y430 OFF and relay M302 to energise turning output Y431 ON. This process can be continued until each of the relays associated with the shift register have been energised in turn. This example generates the required type of sequenced output but is achieved using manual control of the three input contacts. In an industrial situation this type of sequence would have to be generated automatically. To convert this manual sequencer into an automatic sequencer requires the introduction of additional instructions namely SET and RESET.

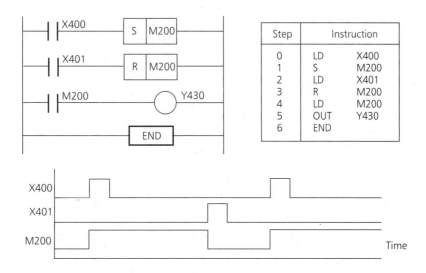

Figure 25.15 Application of SET and RESET instructions

SET and RESET instructions

These two instructions complement each other and are used to latch and unlatch auxiliary or output relays. A simple ladder illustrating their use is shown in Figure 25.15.

When input contact X400 is momentarily closed and then opened auxiliary relay M200 is latched ON. When input contact X401 is momentarily closed and then opened auxiliary relay M200 is unlatched or turned OFF. These instructions can therefore be used to latch relays ON for a momentary closure of a contact. This can be put to good use in creating an automatic sequencer using a shift register. A functional diagram is helpful to illustrate the connections required between the various relays associated with the shift register to create an automatic sequencer. Figure 25.16 portrays the sequencer.

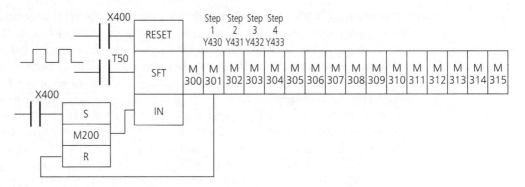

Figure 25.16 Functional diagram of sequencer

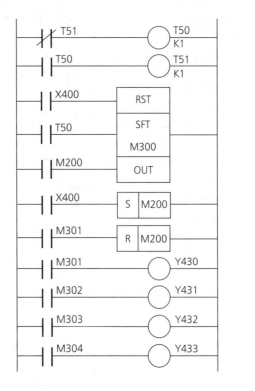

Figure 25.17 Ladder diagram for automatic sequencer

The shift register is reset manually by contact X400 which also now sets relay M200. Notice that relay M200 is reset by the shift register relay M301 and M200 supplies the data input. The shift input is now controlled by a contact from a timer; this contact will in fact close and open every second. The program to drive this sequencer is shown in the ladder diagram in Figure 25.17.

Following the ladder diagram shown in Figure 25.17 it can be seen that two timers T50 and T51 are cross-connected so as to cause the contact associated with T50 to turn ON and then OFF at one second intervals. This same contact is then used to control the shift input to the shift register executing a right shift every two seconds. When relay M301 is energised after a shift right this also resets M200 which in turn de-energises M300 so that on the next shift right M301 will turn OFF and M302 will turn ON. Hence, every time a shift right is executed from now onwards only one relay in the sequence will be activated at one time.

Activity 25.5

Can you see a major problem with the automatic sequence generated in the shift register configuration shown in Figure 25.17?

Try to modify the ladder diagram to overcome this problem.

Having seen how the shift register element can be used to create a sequencer it would now be instructive to see how this type of sequencer could control a set of traffic lights using a Mitsubishi F1–20 PLC.

Traffic Light Sequencer

Starting with the basic traffic light sequence as defined in the table below:

RED
RED and AMBER
GREEN
AMBER
REPEAT

The output relays will be allocated as follows:

RED LIGHT   Y430
AMBER LIGHT   Y431
GREEN LIGHT   Y432

A suitable ladder diagram is shown in Figures 25.18(a) and 25.18(b).

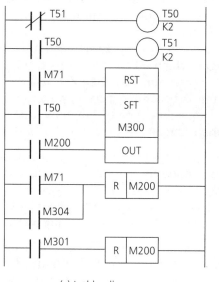

(a) Ladder diagram

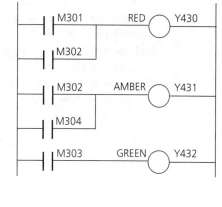

(b) Traffic light outputs

Figure 25.18 Shift register traffic light sequencer

Notice that the shift register is configured as an automatic sequencer with the addition of feedback from M304 which causes the sequence to automatically repeat. The traffic light outputs are controlled by logically combining the outputs from M301, 302, 303 and 304 as shown in Figure 25.18(b). This ensures that only the required lights operate at each point in the sequence.

Activity 25.6

Now try and modify the ladders shown in Figure 25.18 to control two sets of traffic lights which control traffic in two directions. Output relays Y433, Y434 and Y435 can be used for the other set of lights. Remember that as one set of lights goes through the normal sequence the other set must remain on red.

Having now investigated the different elements found in a typical PLC system it is important to understand more of the techniques associated with actually programming a PLC system.

Other programming methods

Two methods of programming have already been introduced: the ladder diagram and the statement list. The statement list is in fact simply the list of instructions used to program the PLC. Examples of **statement lists** have been given with each example of a ladder diagram. However, there are a number of other methods which can be used to program a PLC although not all of them are supported by every PLC. The type of method used is often limited by the type of programming system used; for instance a basic programming panel may only allow a statement list to be used where as a graphical programming panel may allow both ladder diagram and/or statement list to be used.

As PLCs become more sophisticated so the type of programming language also becomes more sophisticated. The use of more sophisticated programming techniques also requires the use of **off-line programming** normally using specialised software running on a personal computer. This type of programming system allows PLC programs to be created using **function diagrams/charts** and **graphical programming languages** which create programs from a flow chart.

Examples of this type of programming are shown in Figure 25.19. This graphical method of programming is becoming very popular.

Some PLCs use a Boolean algebra type of language; the Siemens PLC system is a typical example. The instructions are formulated as a Boolean expression. An example is shown below:

AN	I2.0	This representations a normally closed contact in series
A	I2.1	with a normally open contact ie. two contacts ANDed
=	Q 4.0	together controlling an output relay Q 4.0, notice the = sign is used to represent the output instruction.
A	I2.3	This represents two normally open constants in series, the
A	I2.4	'O' represents the OR function, the third normally closed
O		contact is connected in parallel and the result output to
AN	I2.6	relay Q 4.1.
=	Q4.1	

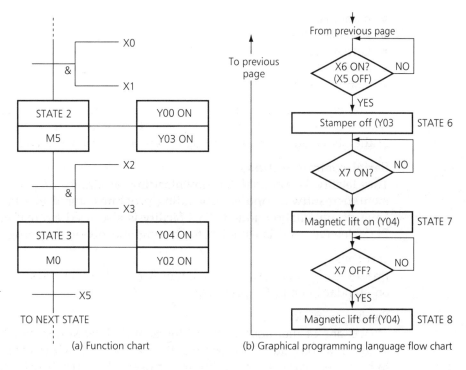

Figure 25.19 Function diagram and graphical programming language

Finally, some of the most sophisticated and expensive PLC systems allow programs to be created using high-level programming languages such as 'C' and BASIC or a low-level language called assembler. The following instructions will instruct the CPU in a microprocessor to add 1 to the number stored in the accumulator:

High-level language   LET A= A + 1
Assembler language   INC A

Unlike ladder diagram or statement list programming, using high-level languages or assembler requires a great deal of training and expertise. This is why ladder and statement lists are still the most widely used methods of programming although graphical programming methods are rapidly taking over.

Text and documentation

Most PLC manufacturers also sell documentation software for their particular range of equipment. Typically this software allows programs to be prepared off-line, as well as documentation to be added and a certain amount of testing and debugging to take place. One such software package is MEDOC which supports a wide range of Mitsubishi PLC systems and is fairly representative in terms of its facilities. Typically these include:

● serial communications
● production of hard copy
● saving programs to computer disk

- name list
- statement list
- ladder diagram
- comments
- testing/debugging

This type of software is now in widespread use and has become an accepted way of developing and documenting PLC programs. It is useful to consider each of the above facilities in some detail starting with serial communications.

Serial communications
This facility is essential for downloading programs prepared using the documentation software and for uploading programs stored in a PLC which can then have documentation added. This facility is also used for real time monitoring of PLC elements and is essential for testing and debugging programs.

Hard copy
Printouts of contact names, statement lists and ladder diagrams can be produced on standard computer printers.

Saving to disk
This is one of the most useful facilities and allows programs to be stored safely away on a computer disk. Some software packages allow for parts of a program to be stored away so that a library of useful building blocks can be built up for use at a later date. This ultimately speeds up development of new programs as ready proven blocks can be used from the library.

Name list
A table of contacts, relays and elements to be used in a program is created and then meaningful names are given to them, such as START, STOP, RED, GREEN etc. These names can then be used in the program as well as the normal letters and numbers. An example of a typical name list is shown below.

I/O	Name	Comment
X0	START	START BUTTON
X1	STOP	STOP BUTTON
Y0	ON/OFF	
Y1	FORWARD	FORWARD CONTROL
Y2	REVERSE	REVERSE CONTROL
Y3	SPEED	TWO-SPEED CONTROL
Y4	BUZZER	BUZZER CONTROL
Y5	START NOW	START NOW LIGHT
M0	1ST PRESS	
M1	BUZZER ON	
M2	STRT NOW	
M3	SEQ RESET	
M4	MOTOR ON	
M5	RESET	RESETS SYSTEM WHEN MOTOR STOPS
T0	0.5 SEC	TEST FOR PRESS
T1	2.0 SEC	BUZZER TIMER
T2	4.0 SEC	QUIET PERIOD TIMER
T3	10 SEC	SEQ RESET TIMER
T4	10 SECFAST	MOTOR FAST FOR 10 SECS
T5	10 SEC SLOW	MOTOR SLOW FOR 10 SECS

Statement list

This is a list of instructions which actually make up the required PLC program. As the various instructions are typed into the computer contact, relay and element names are automatically added to the statement list. To make the program more readable and to assist with de-bugging, comments can also be added to the statement list. Comments can also help to divide the program up into defined blocks and help to give some structure to the program. An example of a printout of a statement list from a typical documentation package is shown below.

Step	Instr	I/O	Name	Comment
		Block 1 Start button control		
0	LD	X0	START	Start button
1	AND	X1	STOP	Stop button
2	ANI	M4	MOTOR ON	
3	ANI	M3	SEQ RESET	
4	ANI	M5	RESET	
5	OUT	T0	05.SEC	Test for press
		K5		
8	LD	X0	START	Start button
9	OR	M0	1ST PRESS	
10	ANI	T0	0.5 SEC	Test for press
11	AND	X1	STOP	Stop button
12	ANI	M4	MOTOR ON	
13	ANI	M5	RESET	Resets system when motor stops
14	OUT	M0	1ST PRESS	
		Block 2 Buzzer control		
15	LD	M0	1ST PRESS	
16	AND	X1	STOP	Stop button
17	ANI	M3	SEQ RESET	
18	ANI	M5	RESET	Resets system when motor stops
19	OUT	T1	2.0 SEC	Buzzer timer
		K20		
22	ANI	T1	2.0 SEC	Buzzer timer
23	OUT	M1	BUZZER ON	
		Block 3 2 sec quiet period		
24	LD	M0	1ST PRESS	
25	AND	X1	STOP	Stop button
26	ANI	M3	SEQ RESET	
27	ANI	M5	RESET	Resets system when motor stops
28	OUT	T2	4.0 SEC	Quiet period timer
		K40		
31	LD	T2	4.0 SEC	Quiet period timer
32	AND	T1	2.0 SEC	Buzzer timer
33	OUT	M2	STRT NOW	
		Block 4 10 second timer control		
34	LD	M0	1ST PRESS	
35	AND	X1	STOP	Stop button

BLOCK 1 start button control

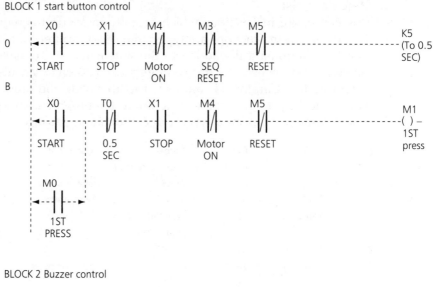

BLOCK 2 Buzzer control

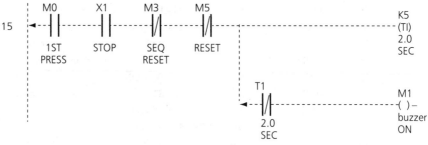

Figure 25.20 Fully documented ladder diagram

Ladder diagram

In most documentation packages the ladder diagram is automatically created as the statement list is developed. Contact names and comments are also automatically transferred to the ladder diagram. A hard copy of a fully documented ladder diagram is shown in Figure 25.20.

Testing/de-bugging

Having developed a program and produced all of the documentation, the software package can then be used to check the operation of the program in a number of different ways.

Firstly, the program is usually tested for correct use of instructions and this is done automatically by the software. Any errors are usually either highlighted using special **error codes** or the line number in the statements list which has the error is displayed.

The program is then transferred to the PLC by downloading via the computer's RS-232 interface. After downloading the program stored in the PLC is compared with the file stored on the computer. If they are both identical a message such as 'Programs Equal' is often displayed.

Once the program has been downloaded the software is typically put into monitor mode. This mode allows the condition of contacts, relays and elements to be actively displayed on the screen of the computer. The operation of the program can then be carefully checked to see that contacts are operating correctly and that the output relays are operating in the correct sequence. Elements such as counters and timers can also be checked to ensure that the correct constants are being loaded and that timing periods etc. are correct. If not, data such as constants can be changed and the program downloaded again.

Forcing inputs/outputs

Another facility which is available, depending on the type of PLC system used, is that of forcing inputs/outputs. This is useful when the PLC is actually physically connected to the devices it is required to control. To ensure that the PLC output relays are connected to the correct device and that they operate the device in the correct way, Output relays can be forced ON using the software directly rather than actually running the PLC program.

This is very useful for eliminating wiring problems and avoids running the actual program which may cause damage to the external devices if they where wired incorrectly. Inputs can also be forced ON or OFF, but again this depends on the PLC system used. This facility is not supported by every PLC. If available this allows the operation of the program to be fully tested without having any external devices connected to the inputs. This is useful if there is no access to the actual input devices. However, where this facility is not available, it is common practice to wire some temporary switches and/or push buttons directly to the PLC inputs.

Assignment 25

This assignment provides evidence for:
Element 32.3: Investigate programmable logic controller programming techniques
and the following key skills:
Communication 3.2
Information Technology 3.1, 3.2, 3.3

For this assignment you will need access to a PLC system and supporting documentation software.

Documentation. You are required to use any one of the programs developed in previous activities and using a suitable software package fully document the program, store it on disk and produce a hardcopy printout of a name list, statement list and ladder diagram also remember to include suitable comments in your statement list.

Testing/debugging. Having produced a fully documented program you are now required to download the program into a PLC system and fully test the program to ensure correct operation using any of the previously described techniques which are applicable to your particular PLC system. Make a list of any errors found and identify solutions for correcting the errors.

Report

The evidence indicators for this element suggest that a report is produced investigating PLC programming techniques including:

- a description of method of programming PLCs
- a description of the advantages of off-line programming
- an identification of methods of producing and storing text and documentation and the application of one such method to document a PLC program
- an identification of methods of testing and debugging harware and software and an application of one such method to a PLC program
- an identification of elements associated with the preparation of a PLC program.

Answers to Progress Checks

These answers include *calculated answers only*.

1.1 By Ohm's Law $V = I \times R$, therefore:

$$I = \frac{V}{R} = \frac{1.5}{100} = 0.015 \text{ A} \quad (15 \text{ mA})$$

1.2 Total resistance of circuit:

$R = 1 + 3 = 4 \, \Omega$

By Ohm's Law, circuit current:

$I = 12/4 = 3 \text{ A}$

Voltage drop across $1 \, \Omega = 3 \times 1 = 3 \text{ V}$
Voltage drop across $3 \, \Omega = 3 \times 3 = 9 \text{ V}$
(Check: $12 = 3 + 9$)

1.3 For the two resistances:

$$\frac{1}{R} = \frac{1}{3} + \frac{1}{3} = \frac{2}{3}$$

$$R = \frac{3}{2} \, \Omega$$

For the three resistances:

$$\frac{1}{R} = \frac{1}{3} + \frac{1}{3} + \frac{1}{3} = \frac{3}{3}$$

$$R = \frac{3}{3} = 1 \, \Omega$$

The general rule is that if 'n' equal resistances are in parallel, then the equivalent resistance is $1/n$ times the common resistance value.

1.4 $P_1 = I_1^2 \times R_1$

$P_1 = (1.5)^2 \times 2 = 4.5 \text{ W}$

$P_2 = I_2^2 \times R_2$

$P_2 = (0.5)^2 \times 6 = 1.5 \text{ W}$

Alternatively,

$P_1 = Vp^2/R_1 = 3^2/2 = 9/2 = 4.5 \text{ W}$

$P_2 = Vp^2/R_2 = 3^2/6 = 9/6 = 1.5 \text{ W}$

1.5 Using the same choice of currents as in Worked Example 1.5, the loop equations are:

Loop ABEF

$6 - 2I_1 - 4 + 3I_2 = 0$
$2 = 2I_1 - 3I_2$ (1)

Outer loop ACDF

$6 - 2I_1 - 10(I_1 + I_2) = 0$
$6 = 12I_1 + 10I_2$ (2)

Multiplying (1) by 6 gives:

$12 + 18I_2 = 12I_1$

From (2)

$6 - 10I_2 = 12I_1$

Equating

$12 + 18I_2 = 6 - 10I_2$

$I_2 = -6/28 = -0.2143 \text{ A}$

Substitution in (1) gives:

$$I_1 = \frac{2 - 3I_2}{2} = 0.6786 \text{ A}$$

$I_3 = I_1 + I_2 = 0.6786 - 0.2143$

$= +0.4643 \text{ A}$

Power dissipated $= I_3^2 \times 10 = 2.156 \text{ W}$

1.6 The $1 \, \Omega$ resistance is first removed and the open circuit voltage V_{OC} between X and Y is determined. V_{OC} is the voltage across the $2 \, \Omega$ resistance. The current I flowing in the loop is:

$$I = \frac{9}{3 + 4 + 2} = \frac{9}{9} = 1 \text{ A}$$

Therefore,

$V_{OC} = 2 \times I = 2 \text{ V}$

r is $2 \, \Omega$ in parallel with $3 \, \Omega$ and $4 \, \Omega$ in series:

$$r = \frac{2 \times 7}{2 + 7} = \frac{14}{9} = 1.556 \ \Omega$$

The current through the 1 Ω resistance is:

$$\frac{V_{OC}}{r + 1} = \frac{2}{1.556 + 1} = 0.782 \ A$$

Power dissipated $= (0.782)^2 \times 1 = 0.612 \ W$

2.1 (a) $T = \dfrac{1}{f} = \dfrac{1}{50} = 0.02 \ s = 20 \ ms.$

(b) A half cycle takes $20 \ ms/2 = 10 \ ms.$

(c) From the waveform diagram the voltage starts from zero and rises to its peak value V_M after a quarter of a cycle, i.e. after $20/4 = 5 \ ms.$

2.2 No matter how many sections the waveform is divided into the mid-ordinate value will always be 1V. Suppose we have six mid-ordinates then:

$$V_{AV} = \frac{1 + 1 + 1 + 1 + 1 + 1}{6} = 1 \ V$$

2.3 $V_{RMS} = 0.707 \ V_M$ so $V_M = 28.28/0.707 = 40 \ V$

$$\omega = \frac{2\pi}{T} = \frac{2\pi}{10^{-3}} = 2000\pi$$

Since v is negative at $t = 0$ then the phase angle ϕ is negative.

At $t = 0$, $-20 = 40 \sin(\phi)$

$\phi = \text{arc sin}(-0.5)$

$\phi = 0.5236 = -\pi/6 = -30°$

Therefore, $v = 40 \sin\left(200\pi - \dfrac{\pi}{6}\right)$

2.4 (a) There is no phase angle between $V_1 = 40 \sin \omega t$ and $V_2 = 50 \sin \omega t$. We would say that 'v_1 and v_2 are in phase with one another'. The resultant is just the numerical addition of the two voltages.

So, $v_r = 90 \sin \omega t \ V$

(b) Here v_2 lags v_1 by 90°. v_2 has no horizontal component and v_1 has no vertical component.

Therefore,

Hor $= 40 \ V$

Vert $= -50 \ V$

$$V_R = \sqrt{[(40)^2 + (-50)^2]} = 64.03 \ V$$

$$\tan \phi = \frac{Vert}{Hor} = -\frac{50}{40}$$

$\phi = -51.34° = -0.896 \ rads$

So $v_r = 64.03 \sin(\omega t - 0.896) \ V$

2.5 The voltage and current are in phase with each other. This means that they are perfectly in step reaching their peaks at exactly the same instant.

2.6 (a) $f_o = \dfrac{1}{2\pi\sqrt{(LC)}}$

$$= \frac{2\pi}{\sqrt{(25 \times 10^{-3} \times 0.1 \times 10^{-6})}}$$

$$= 3.18 \ kHz$$

(b) Current $I = \dfrac{V}{R} = \dfrac{10}{15} = 667 \ mA$

(c) Since $V_C = V_L$ at resonance we only need to calculate, say, V_L.

$V_L = I \times X_L$

$= 2\pi \times 3.18 \times 10^3 \times 25 \times 10^{-3}$

$= 333.18 \ V$

Notice that V_C and V_L are over 33 times larger than V.

2.7 $Q = \dfrac{I}{R}\sqrt{\left(\dfrac{L}{C}\right)} = \dfrac{1}{15}\sqrt{\left(\dfrac{25 \times 10^{-3}}{0.1 \times 10^{-6}}\right)} = 33.33$

The other equations for Q could be used as well.

2.8 At resonant the circuit is a pure resistance of 10 Ω and $\phi = 0°$

$P = (0.707 \times 10)^2 \times 10 = 499.85 \ W$

or $P = (0.707 \times 100) \times (0.707 \times 10) \times \cos 0°$

$= 499.85 \ W$

2.9 (a) $f_o = \dfrac{1}{2\pi}\sqrt{\left(\dfrac{1}{LC}\right)} = 3.18 \ kHz$

(b) $R_D = \dfrac{L}{CR} = 16.7 \ k\Omega$

(c) $I = \dfrac{V}{R_D} = \dfrac{100}{16.7 \times 10^3} = 6\text{ mA}$

(d) $Q = \dfrac{2\pi \times 3.18 \times 10^3 \times 15 \times 10^{-3}}{15} = 33.3$

(e) Circulating current is the capacitor current I_C. Using $Q = Ic/I$ gives $I_C = 200$ mA.

3.1 (a) $i(t{=}0) = \dfrac{V}{R} = \dfrac{100}{40 \times 10^3} = 2.5\text{ mA}$

(b) Initial rate of rise of
$$v_C = \dfrac{V}{CR} = \dfrac{100}{0.4} = 250\text{ V/s}$$

(c) Initial rate of fall of
$i = V/CR^2 = 6.25\text{ mA/s}$

(d) $0.4 \times 100 = 100\,(1 - e^{-t_1/CR})$

so $e^{+t_1/CR} = \dfrac{1}{0.6}$

$t_1 = 0.511 \times CR = 0.204\text{ s}$

3.2 (a) Initial discharge current =
$$\dfrac{V}{R} = \dfrac{100}{50 \times 10^3} = 2\text{ mA}$$

(b) Time constant $= C \times R = 10 \times 10^{-6} \times 50 \times 10^3 = 0.5$ s. We need to calculate v_C at this time:
$$v_C = 100\,e^{-0.5/0.5} = 100\,e^{-1} = \dfrac{100}{e} = 36.8\text{ V}$$

This gives an alternative definition for the time constant. It is the time taken for the capacitor voltage to fall to 36.8% of its initial value when discharging.

(c) $i = \dfrac{V}{R}e^{-t/CR} = \dfrac{100}{50 \times 10^3}e^{-2/0.5}$

$\quad = 2 \times e^{-4} \times 10^{-3} = 36.6\ \mu A$

3.3 (a) Time constant $= L/R = 0.5$ s

(b) Final current $= V/R = 20/4 = 5$ A

(c) $U = \dfrac{1}{2}LI^2 = \dfrac{1}{2} \times 25^2 = 25$ joules

(d) $i = 5(1 - e^{-1/0.5}) = 5(1 - e^{-2}) = 4.323$ A

4.1 Phase voltage = 240 V
Phase current = 8 A
Line current = 8 A

4.2 Phase voltage = 415 V
Phase current = 6.92 A
Line current = 11.98 A

4.3 Total power = 18.56 KW
Total power = 500 W
Power factor = 0.95

5.1 Turns ratio = 20:1

5.2 Secondary current = 22.5A

5.4 Voltage regulation = 5%
Efficiency = 95%

6.3 Regulation = 8.3%

7.2 Speed = 8.33 rad/sec or 500 revs/min

7.3 Slip = 30%

8.1 The fundamental frequency
$$f_F = \dfrac{1}{T} = \dfrac{1}{1 \times 10^{-3}} = 1\text{ kHz}$$

The three other sinusoidal components are 3 kHz, 5 kHz and 7 kHz, i.e. the third, fifth and seventh harmonic.

8.2 Maximum fundamental frequency is one half the bit rate
$$= \dfrac{9600}{2} = 4.8\text{ kHz}$$

Minimum fundamental frequency is 0 Hz, i.e. d.c.

8.3 Power gain $= 10\log_{10}(50/5) = +10$ dB

When P_{out} is doubled,

power gain $= 10\log_{10}(100/5) = +13$ dB

Notice that doubling the ratio of P_{out} to P_{in} *increases* the gain by 3 dB. In the same way if the ratio is halved the gain would *decrease* by 3 dB.

8.4 $10 \log_{10} (P_{out}/P_{in}) = -10 + 20 - 20 - 30$

$$= -40 \text{ dB}$$

$\log_{10} (P_{out}/P_{in}) = -40/10 = -4.$

Therefore, $P_{out}/P_{in} = 1/10,000$

In this example, the overall system has a *loss* of 40 dB.

For $P_{in} = 1W$, $P_{out} = 1 \times 10^{-4} \text{ W}$

8.5 Power gain or loss $= 10 \log_{10} \left(\dfrac{i^2_{out} \times R_{out}}{i^2_{in} \times R_{in}} \right)$

$$= 10 \log_{10} \left(\dfrac{i^2_{out}}{i^2_{in}} \right) + 10 \log_{10} \left(\dfrac{R_{out}}{R_{in}} \right)$$

Power gain or loss =
$$20 \log_{10} \left(\dfrac{i_{out}}{i_{in}} \right) + 10 \log_{10} \left(\dfrac{R_{out}}{R_{in}} \right)$$

8.6 (a) $\dfrac{S}{N} = 10 \log_{10} \left(\dfrac{10}{1 \times 10^{-3}} \right)$

$$= 10 \log_{10} (10 \times 10^3) = 40 \text{ dB}$$

(b) $\dfrac{S}{N} = 10 \log_{10} \left(\dfrac{10}{0.5 \times 10^{-3}} \right)$

$$= 10 \log_{10} (20 \times 10^3) = 43 \text{ dB}$$

Since the noise power is halved, the signal-to-noise ratio improves by 3 dB.

8.7 The audio frequency range is defined as lying between 20 Hz to 20 kHz. It must be sampled at a minimum frequency of 2×20 kHz = 40 kHz i.e. 40,000 samples per second.

8.8 Number of levels is $2^{16} = 65,536$ levels.

9.1 (a) $V = f \times \lambda$ so $\lambda = \dfrac{v}{f}$,

$$= \dfrac{0.3 \times 10^8}{100 \times 10^6} = 0.3 \text{ m}$$

(b) Since velocity = distance/time then time taken 't' is

$$t = \dfrac{\text{distance}}{\text{velocity}} = \dfrac{30 \times 10^3}{0.3 \times 10^8} = 1 \text{ ms}$$

9.2 $Z_0 = \dfrac{60}{\sqrt{3}} \log_e \left(\dfrac{9}{1.5} \right) = 62.07 \ \Omega$

9.3 Each telephone channel has a bandwidth of 4 kHz. The channel capacity is 60 MHz divided by 4 kHz = 15000 channels. (The technique used to assemble the channels together will be investigated in the following element.)

9.4 $c = f \times \lambda$

so $\lambda = \dfrac{c}{f} = \dfrac{3 \times 10^8}{10 \times 10^9} = 0.03 \text{ m}$

9.5 $c = f \times \lambda$

so $f = \dfrac{c}{\lambda} = \dfrac{3 \times 10^8}{1.6 \times 10^{-6}} = 1.9 \times 10^{14} \text{ Hz}$

10.1 The waveform contains:

$$f_c = 60 \text{ kHz}, f_c + f_m = 65 \text{ kHz},$$
$$f_c - f_m = 55 \text{ kHz}.$$

The bandwidth $= 2 \times 5$ kHz = 10 kHz.

10.2 The commercial speech bandwidth is 0 to 4 kHz but the voice only occupies the frequency range 300 Hz to 3.4 kHz. Your sketch should be the same as that in Fig. 10.2(e) with $f_c = 108$ kHz, $f_c + f_1 = 108.3$ kHz, $f_c + f_2 = 111.4$ kHz, $f_c - f_1 = 107.7$ kHz, $f_c - f_2 = 104.6$ kHz.

A voice frequency of 1 kHz appears as 109 kHz in the upper sideband and 107 kHz in the lower sideband.

10.3 Channel 2 modulates a 104 kHz carrier and produces a USB of 104 – 108 kHz and an LSB of 100 – 104 kHz. Only the LSB is passed by the filter so the 1 kHz signal appears as 103 kHz.

For channel 12, it appears as 63 kHz.

10.4 The net carrier frequency is the average of the two frequencies,

i.e. ½ (2100 + 1300) = 1700 Hz.

14.1 (a) No, because the current depends on the number of electron–hole pairs and not on the reverse voltage.

(b) As the diode gets hotter more electron–hole pairs will be produced so the current will increase. If heated too much the current will be so large that the diode will 'burn out'.

14.2 (a) For $I_F = 10$ mA, V_F is about 0.74 V. Therefore,

$$R_{dc} = 0.74/(10 \times 10^{-3}) = 74 \; \Omega$$

and

$$P_{dc} = 0.74 \times 10 \times 10^{-3} = 7.4 \; mW$$

(b) For $I_F^* = 50$ mA, V_F^* is about 0.84 V. Therefore,

$$R_{dc} = 0.84/(50 \times 10^{-3}) = 16.8 \; \Omega$$

and

$$P_{dc} = 0.84 \times 50 \times 10^{-3} = 42 \; mW$$

14.3 For diodes supplied by ITT Semiconductors, *all* diodes have $I_{FAV} = 1$ A. $V_{RRM} = 100$ V for the 1N4001, and the V_{RRM} rating progressively increases to 1300 V for the 1N4007.

14.4 Every diode in the range has an I_{FAV} rating of 1 A up to 75°C. As far as this rating is concerned *any* could be used. The diode will be repeatedly reverse biased to a peak value of $1.414 \times 170 = 240.4$ V. The chosen diode must have a V_{RRM} rating of at lease this value. The 1N4001 and 1N4002 could *not* be used but any other could. We would probably choose a 1N4003 which has a V_{RRM} of 400 V.

14.5 (a) Twice the mains frequency – 100 Hz.

(b) Electrolytics are polarised and have to be connected correctly in the circuit, i.e. the terminal marked '–' must be connected to the most negative voltage point. Also, its d.c. working voltage must be greater than the peak value of the transformer secondary voltage.

15.1 The terminal of each battery will be *reversed*.

15.2 I_B is 1% of 40 mA (i.e. 1/100th of 40 mA)

$$I_B = \frac{40 \times 10^{-3}}{100}$$

$$I_B = 400 \; \mu A$$

Since $I_E = I_B + I_C$, then

$$I_C = I_E - I_B$$

$$= 40 \times 10^{-3} - 400 \times 10^{-6} = 39.6 \; mA$$

15.3 $I_E = I_B + I_C$

$$= 20 \times 10^{-3} + 60 \times 10^{-6} = 20.06 \; mA$$

$$h_{FE} = \frac{I_C}{I_B} = \frac{20 \times 10^{-3}}{60 \times 10^{-6}} = 333.3$$

15.4 The polarity of each power supply would have to be reversed.

15.5 Using the typical values for the BC108A: $h_{ie} = 2700$, $h_{fe} = 220$ and $R_c = 2.2K$ gives

$$A_v = \frac{220 \times 2200}{2700}$$

$$A_v = 179.2$$

Note: Your values of h_{ie} and h_{fe} could actually be anywhere within the minimum to maximum range.

16.1 For 3 inputs there are 8 possible combinations. The output $Y = 1$ only when $A = B = C = 1$. The truth table is:

A	B	C	Y
0	0	0	0
0	0	1	0
0	1	0	0
0	1	1	0
1	0	0	0
1	0	1	0
1	1	0	0
1	1	1	1

The Boolean expression is $Y = A.B.C.$ (or $Y = ABC$)

16.2 You just need to tie the 2 inputs together.

A	Y
0	1
1	0

No matter how many inputs the gate has, provided they are all tied together the NAND will function as an inverter. A NOR gate with all its inputs tied together will also act as a NOT gate. (Check this yourself.)

527

16.3 (a) logic '1', (b) logic '1', (c) logic '0', (d) logic '1'.

16.4 Logic '1' about 3.3 V up to 5 V.
Logic '0' 0 V up to about 1.67 V.

16.5 Inverters: 1×7404 chip.
3-input ANDs: 2×7411 chips.
The 4-input OR can be constructed by using three of the four 2-input ORs in a 7432 chip as shown.

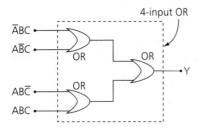

16.6 A floating TTL input assumes a logic '1' level.

16.7 The complete waveform for Q is:

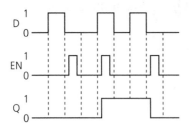

When the first enable pulse occurs D is '0' so Q remains at 0. D is '1' at the next enable pulse so Q follows D to '1' and SETS. At the third enable pulse D is '0' and so Q follows D to '0' and RESETS.

16.8 On the negative edge of clock pulse 1, Q sets to '1'. It resets to '0' on the negative edge of pulse 2. The same sequence happens at edges 3 and 4 and 5 and 6. One complete output pulse is produced for every two input pulses. The frequency of the input pulses is halved. It is a 'divide-by-two' counter.

17.2 68°F; -173°C; 263°K

17.3 Pressure = 5 Pa

17.4 Angular velocity = 314 rad/sec

17.9 Smallest voltage = 2.5 mV

20.1 Each bit in the accumlator would be inverted.

20.2 From the pin-out diagram we find that the clock should be connected to pin 6.
For the power supply, +5 V and GND should be connected to pins 11 and 29 respectively.

20.3 The lowest number is when all 8 bits (D_0 to D_7) of the data bus are '0'. This is 00 Hex. The highest is when each bit is '1'. This is FF Hex. For 8 bits there are $2^8 = 256$ combinations so all of the Hex numbers from 00 to FF are possible.

20.4 The capacity is $2^{10} = 1024 = 1K$.

20.5 For a 4 MHz clock, each T-state last 250 ns. The instruction takes 7 T-states = 1750 ns = 1.75 μs.

21.1 (a) The contents of address 1850H would be loaded into register A.

(b) Load register H with the contents of register L.

(c) Load memory address 190AH with the contents of A.

21.2 To limit the current through each LED to a safe value.

21.3 The JP, LOOP instruction needs to be replaced with a halt. Only the switch data into port A when the program executes once will appear on the LEDs.

21.4 Push AF = F5, push BC = C5, push DE = D5, push HL = E5; pop HL = E1, pop DE = D1, pop BC = C1, pop AF = F1.

22.1 The decoder has two spare outputs so two other PIOs could be 'chip enabled' from these outputs if necessary.

22.2 Four modes

22.3 For an input port the MCB = 0 1 X X 1 1 1. If the 'X's' are chosen as '1' then MCB = 7FH.

For a bit port, MCB = 1 1 X X 1 1 1 1. Letting the 'X's' be '1' gives an MCB = FFH.

In each case, the MCB has to be sent to the control port as necessary.

22.4 Register D needs to be loaded with 0AH.

Index